T0140280

Green Energy and Technology

More information about this series at http://www.springer.com/series/8059

Eduardo Jacob-Lopes · Leila Queiroz Zepka
Maria Isabel Queiroz
Editors

Energy from Microalgae

Editors
Eduardo Jacob-Lopes
Federal University of Santa Maria
Santa Maria
Brazil

Maria Isabel Queiroz
Federal University of Rio Grande
Rio Grande
Brazil

Leila Queiroz Zepka
Federal University of Santa Maria
Santa Maria
Brazil

ISSN 1865-3529 ISSN 1865-3537 (electronic)
Green Energy and Technology
ISBN 978-3-030-09869-8 ISBN 978-3-319-69093-3 (eBook)
https://doi.org/10.1007/978-3-319-69093-3

Printed on acid-free paper

This Springer imprint is published by Springer Nature
The registered company is Springer International Publishing AG
The registered company address is: Gewerbestrasse 11, 6330 Cham, Switzerland

Preface

As the world takes steps to improve our energy security, biofuel sources are more important than ever. One of the energy vectors of the future is microalgae, small aquatic organisms that convert sunlight into biomolecules with high energy value. Scientists throughout the world are researching the best strains of microalgae and developing the most efficient processing practices. This book shows how energy is potentially produced from microalgae and converted into sustainable biofuels.

Divided into fourteen parts, the book explores the microalgal production systems, process integration and process intensification applied to microalgal biofuels production, microalgae biorefineries for energy and coproducts production, life cycle assessment and exergy analysis of biofuels from microalgae, the bioeconomy of microalgal biofuels and the main fuel products obtained from microalgae (biodiesel, biohydrogen, bioethanol, biomethane, and volatile organic compounds). Additionally, a topic on recent patents on biofuels from microalgae is included, summarizing a range of technological routes and energy products, and outlines future developments.

Given the book's breadth of coverage and extensive bibliography, it offers an essential resource for researchers and industry professionals working in green energy and technology.

Santa Maria, Brazil — Eduardo Jacob-Lopes
Santa Maria, Brazil — Leila Queiroz Zepka
Rio Grande, Brazil — Maria Isabel Queiroz

Contents

Contributors

Mortaza Aghbashlo College of Agriculture and Natural Resources, University of Tehran, Karaj, Iran

Ihana Aguiar Severo Department of Food Science and Technology, Federal University of Santa Maria (UFSM), Santa Maria, RS, Brazil

Sergi Astals Advanced Water Management Centre, The University of Queensland, St. Lucia, QLD, Australia

Juliano Smanioto Barin Department of Food Science and Technology, Federal University of Santa Maria (UFSM), Santa Maria, RS, Brazil

Reinaldo Gaspar Bastos Center of Agricultural Sciences (CCA), Federal University of São Carlos (UFSCar), Araras, SP, Brazil

Carlos Ariel Cardona-Alzate Chemical Engineering Department, National University of Colombia, Manizales, Colombia

Alcinda Patrícia de Carvalho Lopes Laboratory for Process Engineering, Environment, Biotechnology and Energy (LEPABE), Chemical Engineering Department, University of Porto - Faculty of Engineering, Porto, Portugal; Laboratory of Separation and Reaction Engineering – Laboratory of Catalysis and Materials (LSRE-LCM), Chemical Engineering Department, University of Porto - Faculty of Engineering, Porto, Portugal

Michael K. Danquah Department of Chemical Engineering, Faculty of Engineering and Science, Curtin University, Sarawak, Malaysia

Debabrata Das Advance Technology Development Center, Indian Institute of Technology, Kharagpur, India; Department of Biotechnology, Indian Institute of Technology, Kharagpur, India

Mariany Costa Deprá Food Science and Technology Department, Federal University of Santa Maria, UFSM, Santa Maria, RS, Brazil

Andrés Donoso-Bravo Inria Chile, Santiago, Chile

Parag Gogate Chemical Engineering Department, Institute of Chemical Technology, Matunga, Mumbai, India

Pierre-Louis Gorry Processes and Technology Department and Doctoral Program in Natural Sciences and Engineering, Metropolitan Autonomous University Campus Cuajimalpa, Mexico City, Mexico

Eduardo Jacob-Lopes Department of Food Science and Technology, Federal University of Santa Maria (UFSM), Santa Maria, RS, Brazil

David Jeison Biochemical Engineering School, Pontificia Universidad Católica de Valparaíso, Valparaíso, Chile

Kailin Jiao College of Energy, Xiamen University, Xiamen, China

Saurabh Joshi Chemical Engineering Department, Institute of Chemical Technology, Matunga, Mumbai, India

Jiashuo Li Department of New Energy Science and Engineering, School of Energy and Power Engineering, Huazhong University of Science and Technology, Wuhan, China; State Key Laboratory of Coal Combustion, Huazhong University of Science and Technology, Wuhan, China

Lu Lin College of Energy, Xiamen University, Xiamen, China; Fujian Engineering and Research Center of Clean and High-Valued Technologies for Biomass, Xiamen University, Xiamen, China

Mariana Manzoni Maroneze Department of Food Science and Technology, Federal University of Santa Maria (UFSM), Santa Maria, RS, Brazil

Marcia Morales Processes and Technology Department and Doctoral Program in Natural Sciences and Engineering, Metropolitan Autonomous University Campus Cuajimalpa, Mexico City, Mexico

Cesar Mota Department of Sanitary and Environmental Engineering, Federal University of Minas Gerais, Belo Horizonte, MG, Brazil

Raúl Muñoz Department of Chemical Engineering and Environmental Technology, University of Valladolid, Valladolid, Spain

Mariana Ortiz-Sánchez Chemical Engineering Department, National University of Colombia, Manizales, Colombia

Sharadwata Pan School of Life Sciences Weihenstephan, Technical University of Munich, Freising, Germany

Fabiana Passos Department of Sanitary and Environmental Engineering, Federal University of Minas Gerais, Belo Horizonte, MG, Brazil

Kun Peng Department of New Energy Science and Engineering, School of Energy and Power Engineering, Huazhong University of Science and Technology, Wuhan, China

José Carlos Magalhães Pires Laboratory for Process Engineering, Environment, Biotechnology and Energy (LEPABE), Chemical Engineering Department, University of Porto - Faculty of Engineering, Porto, Portugal

Maria Isabel Queiroz School of Chemistry and Food, Federal University of Rio Grande (FURG), Rio Grande, RS, Brazil

Lucas Reijnders Institute for Biodiversity and Ecosystem Dynamics, University of Amsterdam, Amsterdam, The Netherlands

Daissy Lorena Restrepo-Serna Chemical Engineering Department, National University of Colombia, Manizales, Colombia

León Sánchez Doctoral Program in Biotechnology, Metropolitan Autonomous University Campus Iztapalapa, Iztapalapa, Mexico City, Mexico

Francisca Maria Loureiro Ferreira dos Santos Laboratory for Process Engineering, Environment, Biotechnology and Energy (LEPABE), Chemical Engineering Department, University of Porto - Faculty of Engineering, Porto, Portugal; Laboratory of Separation and Reaction Engineering – Laboratory of Catalysis and Materials (LSRE-LCM), Chemical Engineering Department, University of Porto - Faculty of Engineering, Porto, Portugal

Harshita Singh Advance Technology Development Center, Indian Institute of Technology, Kharagpur, India

Meisam Tabatabaei Agricultural Biotechnology Research Institute of Iran, Karaj, Iran; Biofuel Research Team, Karaj, Iran

Ahmad Farhad Talebi Faculty of Microbial Biotechnology, Semnan University, Semnan, Iran

Vítor Jorge Pais Vilar Laboratory of Separation and Reaction Engineering – Laboratory of Catalysis and Materials (LSRE-LCM), Chemical Engineering Department, University of Porto - Faculty of Engineering, Porto, Portugal

Roger Wagner Department of Food Science and Technology, Federal University of Santa Maria (UFSM), Santa Maria, RS, Brazil

Leila Queiroz Zepka Department of Food Science and Technology, Federal University of Santa Maria (UFSM), Santa Maria, RS, Brazil

Xianhai Zeng College of Energy, Xiamen University, Xiamen, China; Fujian Engineering and Research Center of Clean and High-Valued Technologies for Biomass, Xiamen University, Xiamen, China

Chapter 1
Energy from Microalgae: A Brief Introduction

Eduardo Jacob-Lopes, Leila Queiroz Zepka and Maria Isabel Queiroz

Abstract This chapter provides a brief overview of some of the major steps in the development of microalgae-based processes for renewable energy production. The chapter attempts to highlight the development and evolution of the key concepts and research in the field, preparing the reader for the following chapters, which will deepen the discussion on the subject.

Keywords Algae · Microalgae-based process · Algae products
Bioenergy · Biofuel

The race for renewables is underway. The growing trend toward the search for alternative and economically viable matrices is one of the main focuses of industrial biotechnology. Issues such as the global concerns of fossil fuels depletion, climate change, and increasing world population have become key determinants of the current energy imbalance (Staples et al. 2017).

Biofuels are considered the most likely sources of energy that can replace a sizeable amount of fossil fuels. Currently, biofuels are classified from first to fourth generation. The third generation biofuel production is mainly based on microalgae (Harun et al. 2010). Some microalgae are known to produce fairly high amounts of intra- and extracellular energy compounds that can be used for biofuels manufacture

E. Jacob-Lopes (✉) · L. Q. Zepka
Department of Food Science and Technology, Federal University of Santa Maria (UFSM), 97105-900 Santa Maria, RS, Brazil
e-mail: jacoblopes@pq.cnpq.br

L. Q. Zepka
e-mail: lqz@pq.cnpq.br

M. I. Queiroz
School of Chemistry and Food, Federal University of Rio Grande (FURG), 96201-900 Rio Grande, RS, Brazil
e-mail: queirozmariaisabel@gmail.com

E. Jacob-Lopes et al. (eds.), *Energy from Microalgae*, Green Energy and Technology, https://doi.org/10.1007/978-3-319-69093-3_1

(Singh et al. 2017). These organisms exhibit high photosynthetic efficiencies and yields up to twice that of terrestrial plants, and remain an attractive target for improving the sustainability of future bioenergy production (Chisti 2013).

Microalgae are a class of microorganisms that exhibit tremendously large biological diversity and metabolic plasticity (Cho et al. 2017). This terminology envelops a variety of prokaryotic and eukaryotic organisms. Some species can grow autotrophically and produce organic molecules while others are heterotrophic in nature, growing in the dark on complex organic material for energy and carbon source (Chew et al. 2017).

It is a consensus that the supply of sufficient energy qualities, with a minimum environmental impact, is among the main challengers of the energy world (Maroneze et al. 2016). However, the search for fossil energy substitutes that meet the requirements of energy sustainability in order to develop biofuels is not so recent. Microalgae are very promising candidates that can fill our energy hunger in a sustainable and environment-friendly manner. Two centuries ago, Rudolf Diesel, the inventor of the diesel engine, fueled the idea of the production of diesel from vegetable oil. This was the basis for using microalgae to generate energy (Barathiraja et al. 2017). Besides the energy concerns, the advantages in terms of environmental impact and sustainability have been considered. On the other hand, they also have the advantage of the parallel production of co-products and have the potential for the mitigation of pollutants, enabling the establishment of biorefineries in industrial integrated processes (Moreno-Garcia et al. 2017).

Regardless of the many possibilities of exploitation of energy from microalgae, today, one of the main interests in developing microalgae-based processes is because of the ability of these microorganisms to produce and accumulate lipids in their cells (Pereira et al. 2016). Microalgae oil consists of the neutral lipid triacylglycerol, which includes saturated and unsaturated fatty acids, which are stored in cytosolic and\or plastidic lipid bodies. The accumulation of such lipid bodies can be enhanced by abiotic stress, through to the adaptions of their biochemical metabolic pathways and cellular composition in response to external conditions including physiological inputs (Savchenko et al. 2017). In this sense, the possibility of lipids accumulation through the manipulation of environment culture conditions has a great potential for energy production. Biofuels from microalgae are no longer focused solely on achieving a high lipid yield and its conversion into biodiesel. Recent technology developments have been facilitated for the use of all algal metabolites. In addition, biodiesel, biohydrogen, bioethanol, bioethanol and, more recently, volatile organic compounds have been the main targets of the current exploitation of energy from microalgae (Santos et al. 2016; Zhu et al. 2017).

Independent of these potentialities, the single biggest and most critical barrier to the market deployment of commercially viable algae-based production remains the high cost of cultivating and harvesting the biomass feedstocks, currently a factor of 10–20, which is too high for commodity fuel production (Laurens et al. 2017).

Furthermore, in light of persisting low fossil fuel prices, the microalgae-based industry is forced to shift its focus from lower-value commodity biofuels and bioenergy products to higher-value (non-energy) products that can be profitable today. In this way, at least until oil prices return to near their pre-2014 levels, or carbon emissions reductions are rewarded through higher carbon pricing in a global climate disruption mitigation policy, primary strategies for bioenergy production from microalgae will need to rely on a multi-product biorefinery approach. As such, a biorefinery approach that generates multiple high-value products from microalgae will be essential to fully valorize biomass and enable the economically viable co-production of bioenergy. Industrial operations that were leading the commercial development of algae-based biofuels have been increasingly redirecting their commercial focus toward the production of higher-value food, feed, and specialty products (IEA 2017).

In fact, there are significant barriers currently impeding the commercialization and economic production of microalgae for fuel markets, in particular in supporting the resource demands for large-scale deployment. The barriers range from incomplete knowledge of microalgae biology to the challenges associated with the scale-up of the processes (Wijffels and Barbosa 2010).

Therefore, to accelerate the production of energy from microalgae, besides the robust increase of the productivities, the minimizing energy, water, nutrients, and land-use footprints need to be a primary objective of productions systems and future research and development (Pantel et al. 2017). The chapters presented in this book are intended to help provide a deeper understanding and insight into the promises and challenges for microalgal biofuels and bioenergy technologies to be substantial contributors to future fuel supplies.

References

Barathiraja, B., Sridharam, S., Sownya, V., Yuvaraj, L. D., & Praveenkumar, R. (2017). Microbial oil—A plausible alternate resource for food and fuel. *Bioresource Technology, 233,* 423–432.

Chew, K. W., Yap, J. Y., Show, P. L., Suan, N. H., Juan, J. C., Ling, T. C., et al. (2017). Microalgae: Biorefinery: High value products perspectives. *Bioresource Technology, 229,* 53–62.

Chisti, Y. (2013). Constraints to commercialization of algal fuels. *Journal of Biotechnology, 167,* 201–214.

Cho, D. H., Choi, J. W., Kang, Z., Kim, B. H., Oh, H. M., Kim, H. S., et al. (2017). Microalgal diversity fosters stable biomass productivity in open ponds treating wastewater. *Scientific Reports, 7,* 1–11.

Harun, R., Singh, M., Forde, G. M., & Danquah, M. K. (2010). Bioprocess engineering of microalgae to produce a variety of consumer products. *Renewable and Sustainable Energy Reviews, 14,* 1037–1047.

IEA Bioenergy. (2017). *State of technology review—Algae bioenergy*. 978-1-910154-30 4. http://www.ieabioenergy.com/publications/state-of-technology-review-algaebioenergy/.

Laurens, L. M. L., Chen-Glasser, M., & McMillan, J. D. (2017). A perspective on renewable bioenergy from photosynthetic algae as feedstock for biofuels and bioproducts. *Algal Research, 24,* 261–264.

Maroneze, M. M., Siqueira, S. F., Vendruscolo, R. G., Wagner, R., de Menezes, C. R., Zepka, L. Q., et al. (2016). The role of photoperiods on photobioreactors—A potential strategy to reduce costs. *Bioresource Technology, 219,* 493–499.

Moreno-Garcia, L., Adjalle, K., Barnabé, S., & Rhagavan, G. S. V. (2017). Microalgae biomass production for a biorefinery system: Recents advances and way towards sustainabillity. *Renewable and Sustainable Energy Reviews, 76,* 493–506.

Pantel, A., Gami, B., & Patel, P. (2017). Microalgae: Antiquity to era of integrated technology. *Renewable and Sustainable Energy Reviews, 71,* 535–547.

Pereira, H., Gangadhar, K. N., Schulze, P. S. C, Santos, T., Sousa, C.B., Schueler, L. M., et al. (2016). Isolation of a euryhaline microalgal strain, *Tetraselmis* sp. CTP4, as a robust feedstock for biodiesel production. *Scientific Reports, 6,* 1–11.

Santos, A. B., Fernandes, A. S., Wagner, R., Jacob-Lopes, E., & Zepka, L. Q. (2016). Biogeneration of volatile organic compounds produced by *Phormidium autumnale* in heterotrophic bioreactor. *Journal of Applied Phycology, 28*(3), 1561–1570.

Savchenko, O., Xing, J., Yang, X., Gu, Q., Shaheen, M., Huang, M., et al. (2017). Algal cell response to pulsed waved stimulation and its application to increase algal lipid production. *Scientific Reports, 7,* 1–13.

Singh, S. P., Pathak, J., & Sinha, R. P. (2017). Cyanobacterial factories for the production of green energy and value-added products: An integrated approach for economic viability. *Renewable and Sustainable Energy Reviews, 69,* 578–595.

Staples, M. D., Malina, R., & Barrett, S. R. H. (2017). The limits of bioenergy for mitigating global life-cycle greenhouse gas emissions from fossil fuels. *Nature Energy, 2,* 1–8.

Wijffels, R. H., & Barbosa, M. J. (2010). An outlook on microalgal biofuels. *Science, 329,* 796–799.

Zhu, L., Nugroho, Y. K., Shakeel, S. R., Li, Z., Martinkappi, B., & Hiltunen, E. (2017). Using microalgae to produce liquid transportation biodiesel: What is next? *Renewable Sustainable Energy Reviews, 78,* 391–400.

Chapter 2
Microalgal Production Systems with Highlights of Bioenergy Production

Mariana Manzoni Maroneze and Maria Isabel Queiroz

Abstract The purpose of this chapter is to provide an overview of the main systems of microalgae production with highlights of biofuel production. The large-scale production systems (raceway ponds, horizontal tubular photobioreactors, and heterotrophic bioreactors) and small-scale photobioreactors (vertical and flat-plate photobioreactors) will be presented and discussed with a special emphasis on the main factors affecting its efficiency, biomass productivities reported in the literature, scaling-up, costs of construction and operation, and commercial applications. Besides this, the recent developments in microalgae cultivation systems will be reviewed in their main aspects. Finally, the criteria for selecting an appropriate bioreactor for microalgae cultivation will be presented, as well as the pros and cons of each system will be discussed in this chapter.

Keywords Photobioreactor · Heterotrophic bioreactor · Biomass Energy · Biofuel

1 Introduction

Historically microalgae have been of interest since 1942, when Harder and von Witsch (1942) suggested that microalgae could be viable sources of lipids to be used as food or to produce biofuels. Since then, an increasing amount of research involving microalgae and their bioproducts has been performed. Currently, these microorganisms are considered one of the most promising sources for bioenergy production (Chisti 2016; Chew et al. 2017; Raslavičius et al. 2018).

M. M. Maroneze
Department of Food Science and Technology, Federal University of Santa Maria (UFSM), Santa Maria, RS 97105-900, Brazil

M. I. Queiroz (✉)
School of Chemistry and Food, Federal University of Rio Grande (FURG), Rio Grande, RS 96201-900, Brazil
e-mail: queirozmariaisabel@gmail.com

E. Jacob-Lopes et al. (eds.), *Energy from Microalgae*, Green Energy and Technology, https://doi.org/10.1007/978-3-319-69093-3_2

Compared with conventional oil seeds, the biofuels produced from microalgae have several advantages that include the higher productivities, the ability to use nonarable land for microalgal cultivation and possibility to use wastewater and gas flue as source of nutrients and carbon to promote growth (Jacob-Lopes et al. 2014; Collotta et al. 2017). Microalgae also can produce different types of biofuels, such as biodiesel, bioethanol, biohydrogen, syngas, biobutanol, and bioelectricity (Chang et al. 2017; Su et al. 2017a, b). Unfortunately, until now, the majority of economic analyses conclude that microalgae biofuels cannot compete with conventional fuels (Lundquist et al. 2010; Sun et al. 2011). On the other hand, the concept of biorefinery can be explored with the aim to improve economic aspects. This is possible because of the wide variety of high-value compounds that microalgae can produce, such as carotenoids, proteins, long-chain polyunsaturated fatty acids, vitamins, and phycobilins (Chew et al. 2017).

Industrialization of microalgae products requires large-scale culture systems, which generally are raceway ponds, closed photobioreactors (PBRs), or heterotrophic bioreactors. Open systems are much cheaper and easier to operate than closed systems, however, have many operational problems, such as contamination, evaporation, susceptibility to weather conditions, and extensive land requirements. On the other hand, closed systems can eliminate these limitations, but with a high capital cost, difficulty in scaling-up, and high shear stress. However, due to the high operational control and the high productivity provided by the PBRs, researchers have been invested heavily in the development of new photobioreactors designs, in order to reduce these limitations, and thus make microalgae-based processes viable (Chang et al. 2017).

This chapter discusses the systems of microalgae production in large-scale (raceway ponds, horizontal tubular photobioreactors, and heterotrophic bioreactors) and small-scale (vertical tubular photobioreactors and flat-plate photobioreactors), with emphasis on major factors that influence their efficiency, biomass productivities, costs of biomass production, scaling-up, and commercial applications. Moreover, recent developments in microalgae cultivation systems are presented. Finally, the advantages and disadvantages of all microalgae production systems discussed are compared, and the criteria for selecting an appropriate PBR are presented.

2 Large-Scale Microalgae Biomass Production

2.1 Raceway Ponds

The raceway ponds were first developed in the 1950s for treating wastewater and, since the 1960s, outdoor open raceways have been used in commercial production of microalgae and cyanobacteria (Chisti 2016). Currently, it is the most utilized system for commercial microalgae production, accounting for more than 95% of algae production worldwide, owing to their flexibility, low cost, and easy of scaling-up (Fernandez et al. 2013; Chang et al. 2017).

A raceway pond is a closed loop recirculation channel with a typical culture depth of about 0.25–0.30 m. The circulation mixes the nutrients, and cells are provided by the paddle wheels. The ponds are usually kept shallow because the algae need to be exposed to sunlight, and sunlight can only penetrate the pond water to a limited depth (Singh and Sharma 2012; Chang et al. 2017).

These ponds can be simply constructed in compacted soil with a 1- to 2-mm-thick plastic membrane. However, although it is cheaper, it is not commonly used for biomass production due to the high risk of contamination (Singh and Sharma 2012). To produce a biomass with high added value, the ponds are often made of concrete block walls and dividers lined with a plastic membrane to prevent seepage. Depending on the end use of the biomass, special care may be required to use liners that do not leach contaminating and inhibitory chemicals into the algal broth (Borowitzka 2005; Chisti 2016).

2.1.1 Major Factors Affecting the Raceway Pond Performance

Choice of Location

The choice of location of a raceway system has the greatest impact on biomass productivity. The factors to consider in the geographic location are average annual irradiance level, prevailing temperature, rainfall, land slope, potential nutrient sources, cost of the water, and land.

In terms of illumination, a minimum solar irradiation of 4.65 $kWh/m^2/d$ is required to sustain high growth rates (Benemann et al. 1982). According to Chisti (2013), in an ideal condition, the temperature should be around 25 °C, with a minimum of diurnal and seasonal variations. A geographic location with rainfall not more than 1000 mm of rain per year facilitates the microalgae cultivation, since that can minimize the dilution of algae stock in the ponds (Bennett et al. 2014). In an ideal situation, the land slope should not be greater than 2% to avoid significant earthmoving costs during pond construction, but the US Department of Energy (DOE) cites a 5% maximum slope (DOE 2010; Bennett et al. 2014).

Other factor that depends on the local climate is the evaporation, which is influenced by the level of irradiance, the wind velocity, the air temperature, and the absolute humidity. An average freshwater evaporation rate of 10 $L/m^2/d$ has been noted for some tropical regions. In this sense, freshwater needs to be added periodically to raceway to compensate the evaporation (Becker 1994; Chisti 2016).

Engineering Parameters

In raceways, the pond depth is one of the engineering parameters that has most influence on cultivation performance, because it is closely related to temperature control, mixing, and light utilization efficiency. In general, the biomass productivity is higher in cultivations with lower depth raceways, but this also depends on the microalgae species used and the dimensions.

In raceway ponds, the mixing serves several purposes such as periodic exposure of cells to sunlight, keeping cells into suspension, availability of the nutrient to algal cells, and removal of photosynthetically generated oxygen. In this sense, an ideal mixing supply can increase productivity by nearly 10 times. Conventionally, the mixing is conventionally measured by the Reynolds number (Re), which in an ideal situation is about 257,000, considering a 1.5 m wide channel with a broth depth of 0.3 m and a culture velocity of 0.3 m/s (Chisti 2016).

Carbon Supply and pH

The carbon is the major constituent of microalgal cells, with approximately 50% of the cell mass. All carbon is photosynthetically assimilated from CO_2 and, this assimilation is closely related to the pH of the medium, since that, if CO_2 is consumed rapidly and not replenished, the pH becomes alkaline. In raceway ponds, generally, the pH is instable, because the CO_2 absorption from atmosphere through the surface of a raceway is insufficient to support the high photosynthesis rate for a good part of the day (Chisti 2013). An alkaline pH results in generation of toxic ammonia from dissolved ammonium salts, lowers the affinity of algae for CO_2, and increases the flexibility of mother cells, delaying completion of the cell cycle (Juneja et al. 2013). For this reason, to obtain better productivities in raceway, it is necessary to engineer a supply of CO_2.

Gas diffusers are used in raceways to inject CO_2 in the form of fine bubbles. According to Li et al. (2014), the CO_2 concentration greater than 73 μmol/L at a pH of 8.0 is optimal for the normal growth of microalgae. To produce high-value compounds, commercial pure carbon dioxide has been extensively used in microalgal cultures. However, this entails in additional economic costs and reduces the economic viability and sustainability of the process. It is estimated that the cost of the carbon source in microalgae production ranges from 8 to 27% of the daily production cost (Li et al. 2014). Furthermore, in this type of system, between 35 and 70% of the pure CO_2 injected into a pond is lost to the atmosphere. As an alternative, flue gas can be used, which also could contribute to the mitigation of environmental problems (de Godos et al. 2014).

Oxygen Accumulation

The photosynthesis reaction produces stoichiometrically 1.9 tons of oxygen to produce 1 ton of microalgal biomass. So, when there is intense microalgal growth, an excess of oxygen is generated. At high concentrations of O_2, the productivity of microalgae reduces considerably due to photorespiration and photoinhibition effects (Raso et al. 2012).

In raceway ponds, the only mechanism commonly used for removal of oxygen from the medium is the agitation by the paddle wheel and is not particularly effective. Even with high surface areas, the oxygen removal is insufficient during periods of maximum photosynthetic activity. At these photosynthesis peaks, the performance of the system can decrease up to 35% due to the excess dissolved oxygen in the medium, which can reach 300% of the air saturation value. Moreover, the biochemical composition of microalgae biomass can be influenced by the oxygen level in the pond (Richmond 1990; Chisti 2013).

Culture Contamination

As the raceway ponds are open to the environment, they are easily contaminated by bacteria, viruses, fungi, and other microalgae species and by predators. An alternative to control the contamination is to place the lakes inside greenhouses with controlled environmental conditions, but for the production of biofuels on a large scale, this is economically unfeasible.

However, it is known that not many contaminants can survive under extreme conditions. In this way, the contamination can be avoided by cultivating some highly resistant microalgal strains at high pH or high salinity. The species with the best performance in commercial cultivation in raceways includes *Chlorella* sp., *Spirulina* sp., and *Dunaliella* sp., which are cultivated under stringent conditions that inhibit the growth of other microorganisms or other species of microalgae (Chang et al. 2017).

2.1.2 Biomass Production in Raceway Ponds

Although biomass productivities of 0.40 g/L/d or higher have already been reported (Wen et al. 2016), values much lower than these are typically found, as shown in Table 1. The reported productivities are specific for the reactor designs, operating conditions, local weather conditions, and algae species. For this reason, the productivities obtained with a specific system cannot be simply extrapolated to other growth conditions. The biomass productivities in raceways are considered low, but generally are compensated by high product prices and low construction and operating costs.

2.1.3 Cost of Construction and Operation of Raceway Ponds

In terms of cost of construction, the plastic-lined earthen are the raceway ponds with the best cost-benefit, as unlined earth ponds are not generally considered satisfactory for producing algal biomass (Chisti 2013). According to Chisti (2016), the cost estimated to produce a 100-ha plastic-lined pond of compacted earth was about US$ 144,830 per ha in 2014. This cost data can be corrected for inflation and

Table 1 Biomass productivities of different open raceway ponds located in different countries

Microalgae	Location	Total volume (L)	Culture depth (cm)	Productivity (g/L/d)	References
Nannochloropsis salina	Arizona, USA	780	25.4	0.013	Crowe et al. (2012)
Spirulina (Arthrospira)	La Mancha, MX	2360–603	15–20	0.144–0.151	Olguín et al. (2003)
Scenedesmus rubescens	Florida, USA	900	20	0.020	Lin and Lin (2011)
Scenedesmus acutus	Arizona, USA	2300	7.5	0.066	Eustance et al. (2015)
Spirulina platensis	Málaga, ES	135,000	30	0.027	Jiménez et al. (2003)
Scenedesmus sp.	Almería, ES	20,000	20	0.170	de Godos et al. (2014)

thus provides a reasonable estimate of the current cost (Chisti 2013). In this way, the capital cost estimated for 2017 is of US$ 149,598 per ha. This estimate includes the earthworks, the plastic lining, the carbon dioxide supply tubing, inlets and outlets, the baffles, the paddle wheel, and motor.

To produce dry microalgae biomass in outdoor commercial raceway ponds, Nosker et al. (2011) and Chisti (2007) estimated a cost of € 4.95 and US$ 3.80 per kg of dry weight, respectively. According to Nosker et al. (2011), the factors which influence production costs are irradiation conditions, mixing, photosynthetic efficiency of systems, culture media, and carbon dioxide costs. Thus, by optimizing these factors, the production cost can reduce up to € 0.68 per kg.

2.1.4 Scaling-up in Raceway Ponds

The microalgae cultivation in open ponds is already a consolidated and widely practiced method for large-scale cultivation, since that are easily scaled up. There have been records of large-scale cultivation in raceways since 1987, where two 1000 m^2 raceway ponds were used as a test facility between 1987 and 1990 in New Mexico. These tests were conducted to verify the potential of microalgal biomass production for low-cost biodiesel production and were considered technically feasible (Rawat et al. 2013).

Although already a widely used technology, cultivations in raceway ponds still pose many challenges in terms of economic viability for large-scale biofuel production. This is mainly due to the low biomass productivity presented in these systems the need for extensive areas of land and substantial costs for harvesting (Scott et al. 2010).

2.1.5 Commercial Microalgae Cultivation in Raceway Ponds

The commercialization of biofuel from microalgae is still in its early gestation and has lot of challenges to achieve cost-competitive fuels. Currently, the industrial microalgae biomass production is restricted to high value.

Raceway ponds are under operation worldwide to produce a diverse range of products. For example, Cyanotech Corporation, in Hawaii, has cultivations in raceways of *Spirulina platensis* and *Haematococcus pluvialis*, to produce Spirulina Pacifica® and BioAstin® Natural Astaxanthin, respectively. Nikken Sohonsha Corporation (Japan) produces more than 40 different products (healthcare products, medical products, cosmetics, dietary supplements, fertilizers, and animal feeds) from microalgae as *Chlorella*, *Dunaliella*, *Monodus,* and *Isochlysis*. Tianjin Norland Biotech (China) cultivates *Spirulina*, *Chlorella,* and *H. pluvialis* to produce *Spirulina* tablets, *Chlorella* tablets, astaxanthin, astaxanthin oil, and phycocyanin.

However, it is important to point out that many companies are working with pilot plant tests for biofuels production. Examples of companies that are using open systems in their tests are the LiveFuels (USA), OriginOil Inc. (USA), PatroSun (USA), Neste Oil (FI), Ingrepo (NL), and Aquaflow Bionomics (NZ) (Su et al. 2017a, b).

2.2 *Tubular Photobioreactor*

Recently, closed PBRs, especially tubular photobioreactors have been successfully used for commercial microalgal biomass production. Unlike open raceways, tubular photobioreactors permit a good control of culture conditions and high solar radiation availability and, consequently, a high biomass productivity, which makes this type of system potential for biofuel production and compounds of high commercial value (Kunjapur and Eldridge 2010; Abomohra et al. 2016).

A tubular photobioreactor consists of an array of straight transparent tubes that are usually made of plastic or glass and have a diameter of 0.1 m or less. These transparent tubes can be arranged in different patterns (e.g., straight, bent, or spiral) and orientations (e.g., horizontal, inclined, vertical, or helical) in order to maximize the sunlight capture (Huang et al. 2017). However, to increase the scale, the tubes are usually arrayed in a horizontal fence-like, which improves the land utilization, and also have a better angle for incident light (Junying et al. 2013).

Besides the solar array for algae growth, a tubular photobioreactor is also composed of a harvesting unit to separate algae from the suspension, a degassing column for gas exchange and cooling (heating) and a circulation pump (Wang et al. 2012). The microalgal culture flows through solar collector tubing and is recirculated by maintaining highly turbulent flow, which is produced using either a mechanical pump or a gentler airlift pump (Abomohra et al. 2016; Chang et al. 2017).

This type of photobioreactor can be illuminated by artificial or natural light. The artificial illumination is technically possible, but expensive compared with outdoor cultivations, which is just viable for commercial production of high added value products.

2.2.1 Major Factors Affecting the Tubular Photobioreactor Performance

Light Supply

In autotrophic microalgae production, the light availability is the most important factor that influences the cell productivity and is one of the most difficult to control in outdoor cultures, due to the variation in solar radiation during the day and during the change of season (Fernandez et al. 1997).

In terms of design, the light capture is influenced by the transparency of the materials and the surface/volume ratio. The most common materials used for PBR construction are glass, plexiglass, polyvinyl chloride (PVC), acrylic-PVC, and polyethylene. All these materials have transparency suitable for the microalgae cultivations. However, they all have their pros and cons and need to be evaluated according to the type of process and desired product. Glass is strong and transparent and very good material for the construction of laboratory-*scale PBRs*. However, it requires many connection parts for the construction of large-scale PBRs, which could be costly. For this reason, the plastic type is most suitable for large-scale tubular photobioreactor, mainly of polyethylene (Wang et al. 2012).

Temperature

As already mentioned, the optimal temperature for microalgae cultures is generally around 25 °C, and most microalgae species can tolerate temperatures between 16 °C and 35 °C. In closed PBR, generally, the volume is small because a thin optical thick mixing ness is applied for the sake of light transfer. Therefore, variations in temperature during the day/night cycle and the seasons' changes have significant effects on microalgal cultivation. In this sense, it is necessary to set up a cost-effective cooling system (Huang et al. 2017).

Several methods have been tested to prevent overheating of the microalgae cultivation. Among them are as follows: (i) shading of the tubes with dark-colored sheets (Torzillo 1997), (ii) cooling of the culture by spraying water on the surface of the photobioreactor (Becker 1994), (iii) submerging part of the photobioreactor or the entire culture on a large body of water (Becker 1994), and (iv) installing a heat exchanger for the photobioreactor (Watanabe et al. 2011). However, shading the PBR is inefficient because it greatly reduces the illumination and consequently in the yield of biomass. Water spraying is efficient for cooling, but entails an increase in the cultivation costs. On the other hand, the method of submersion besides efficient in the control of the temperature has been demonstrated to promote the average light intensity in the culture (Huang et al. 2017).

CO_2/O_2 Balance and Mixing

As explained in Sects. 2.1.1.3 and 2.1.1.4, the carbon dioxide and oxygen must be maintained in equilibrium and in moderate concentrations, since the excess of both causes damage to the cells. In this sense, the photobioreactor must contain a space for exhaust gases and an efficient mixing system, where promote turbulence and therefore mass transfer between the gas and liquid phases inside a photobioreactor (Wang et al. 2012).

In addition to its key role in the balance of gases and pH of the system, the mixing also is necessary to prevent sedimentation of algal cells, ensure that all cells of the population have uniform average exposure to light and nutrient and facilitate heat transfer and avoid thermal stratification. In tubular photobioreactor, the mixing is usually provided by aeration with CO_2-enriched gas bubbles or pumping, mechanical agitation, or a combination of these means. The choice depends on the scale of the system and the microalga species used, because some do not tolerate vigorous agitations (Suh and Lee 2003; Wang et al. 2012; Huang et al. 2017).

2.2.2 Biomass Productivity in Tubular Photobioreactor

The high biomass productivity is the greatest advantage of tubular photobioreactors, especially if compared to raceway ponds. Table 2 shows the biomass productivities in different tubular photobioreactors. The values range from 0.05 to 1.9 g/L/d. This variation is due to the type of geometric configuration used, microalgae species and operating and environmental conditions used in each study. If we compare these productivities values with those found in raceway ponds (Table 1), it is possible to see that except to the values found by Olaizola (2000), all the other productivities in tubular PRBs are greater than found in raceways.

2.2.3 Costs of Construction and Operation of Tubular Photobioreactors

The cost of PBRs has a major influence on production cost for large-scale biomass. The company AlgaeLink NV (Yerseke, The Netherlands) commercializes a horizontal serpentine PBR made of large-diameter transparent plastic tubes. For a system of 97 m^3, 1200 m^2 of occupied area, made of 2000 m long, 25 cm diameter PMMA tubes, according to Zitelli et al. (2013), the price was about € 194,000 in 2012. Through inflation calculation described in Chisti (2013), this price in 2017 is about € 202,798.

Table 2 Biomass productivities in tubular PBRs

Microalgae	Photobioreactor	Total volume (L)	Productivity (g/L/d)	References
Porphyridium cruentum	Airlift tubular	200	1.5	Camacho et al. (1999)
Phaeodactylon tricornutum	Airlift tubular	200	1.2	Fernández et al. (2001)
Phaeodactylon tricornutum	Airlift tubular	200	1.9	Grima et al. (2001)
Phaeodactylon tricornutum	Helical tubular	75	1.4	Ugwu et al. (2002)
Haematococcus pluvialis	Parallel tubular	25,000	0.05	Olaizola (2000)
Nannochloropsis gaditana	Fence-type tubular	340	0.59	San Pedro et al. (2014)

Norsker et al. (2011) calculated the cost for outdoor production of microalgae biomass in tubular photobioreactor and concluded that € 4.15 is the price for producing 1 kg of dry weight biomass, in a 100-ha plant. On the other hand, Grima (2009) found a cost of € 25 per kg of dry weight, in a horizontal tubular PBR of 4000 L. However, according to these authors, it is possible to reduce this cost up to € 0.5 per kg, through of process optimization.

2.2.4 Scaling-up

The scale-up of a tubular photobioreactor is not so simple in an open system, because it requires scaling-up of both the solar receiver and the airlift device. In principle, the volume of the solar receiver may be increased by increasing the diameter and the length of the tube. However, an increase in tube length can result in unacceptable concentrations of dissolved oxygen along the tubes. For this reason, in practice, only the tube diameter may be varied (Grima et al. 2001). Any change in tube diameter would imply a change in the light/dark cycle inside of photobioreactor. These cycles can improve the biomass productivity due to the ability of some species to store light energy to maintain their metabolism in the absence of light (Maroneze et al. 2016). Thus, the geometry of the PBR must be optimized according to the species used.

Grima et al. (2001) concluded that for *Phaeodactylum tricornutum,* the optimal photobioreactor (0.2 m^3) configuration and operations conditions were as follows: a solar receiver tube of 0.06 m diameter, 80 m long, connected to a 4 m tall airlift. Although not having a simple scaling-up, the tubular PBR is already quite widespread in large scale.

2.2.5 Commercial Applications of Horizontal Tubular PBRs

Currently, many biotech companies around the world are using tubular photobioreactors to produce microalgae biomass and several bioproducts. Among them are the Algaelink in the Netherlands that use horizontal and tubular photobioreactors for biomass and jet fuel production. The Heliae (USA) is using spiral tubular PBR to produce astaxanthin from *H. pluvialis.* In Cadiz, Spain, the Fitoplancton Marino SL uses a horizontal serpentine PBR cooled by immersion in a water pool to produce lyophilized microalgae biomass and slurries of several microalgae for aquaculture use (Torzillo et al. 2015; Chang et al. 2017).

2.3 *Microalgal Heterotrophic Bioreactors*

The heterotrophic bioreactors are a feasible alternative to overcome the light energy dependency that limits the scale-up and significantly complicates the design of photobioreactors (Vieira et al. 2012). Although restricted to a few microalgal species, the heterotrophic cultivation can be conducted in conventional reactor configurations such as stirred tank and bubble column reactors, which are relatively cheap, easily scalable, and generally present high kinetic performance (Queiroz et al. 2011; Perez-Garcia et al. 2011).

Moreover, to reduce the cost related to microalgae biofuel production, the organic carbon source and nutrients for the microalgae cultivation can be obtained from agro-industrial wastes (Queiroz et al. 2011; Francisco et al. 2015; Katiyar et al. 2017). In addition to meeting the demand for organic carbon, the use of wastewater in these cultivations also contributes to agro-industrial waste management (Maroneze et al. 2014).

On the other hand, the major limitations of these types of cultivation are the contamination and competition with other microorganisms that grow faster than the microalgae, inability to produce light-induced metabolites and inhibition of growth by excess organic substrate.

2.3.1 Major Factors Affecting Bioreactors in Heterotrophic Cultivations

Oxygen Supply

In aerobic bioprocesses, oxygen is a critical substrate for a cell metabolism that needs continuous supply, as it can easily become rate limiting due to its low solubility in water. According to Griffiths et al. (1960), independent of the organic substrate or the microalgae species, the biomass productivity is enhanced by higher levels of aeration.

In contrast, the aeration system energy requirement is a significant cost in bioreactors and also contributes to the carbon footprint of heterotrophic cultivations. So, for a viable biofuel production, a trade-off between the operating costs related to energy required for aeration and the productivity of the bioprocess (Santos et al. 2015). In this sense, Santos et al. (2015) concluded that for a heterotrophic bubble column bioreactor, the aeration of 0.5 VVM (volume of air per volume of medio per minute) is an equilibrium between kinetic performance and power requirements in bioreactor.

Mixing and Viscosity

Like in the cultivation systems already discussed, mixing is one of the most important operations in heterotrophic microalgal cultivation. This operation is necessary for uniformly distributing nutrients and for gas exchange. The adequate mixing can be provided by impellers and baffles or by aeration with airlift or bubble column systems (Perez-Garcia and Bashan 2015).

The viscosity of the medium is closely related to the mixing, where high viscosity in cultures requires higher impeller speed or airflow, which increases power consumption and operational costs. The viscosity comes mainly from the exogenous carbon source used, but is also increased with the high cell concentration and/or with the production of viscous cellular material.

2.3.2 Biomass Productivity in Heterotrophic Bioreactors

In terms of biomass production, the heterotrophic cultivations can present higher values of productivity, when compared to the other large-scale systems discussed in this chapter, as shown in Table 3, which summarizes the microalgal biomass productivities in heterotrophic cultivations with different carbon sources, bioreactors type, and microalgae species reported in the literature.

2.3.3 Costs in Heterotrophic Bioreactors

The production costs of the heterotrophic microalgae production depend on variables such as bioreactor, carbon source, microalgae strain, downstream processing operations, type and quality of the end product, among others. Tabernero et al. (2012) evaluated the production of microalgal biodiesel from *C. protothecoides* biomass grown heterotrophically. The cost estimated to produce one kilogram of biomass was US$ 1.29 per kg (corrected for 2017). This value was estimated for a biorefinery producing biomass in 465 continuously stirred bioreactors each of 150,000 L and producing 10 million/L/year of biodiesel.

A value below this (US$ 0.06/kg, corrected to 2017) was found by Roso et al. (2015) to produce *P. autumnale* biomass, in a techno-economic analysis of a

Table 3 Biomass microalgae productivity in heterotrophic cultivations

Microalgae	Bioreactor	Carbon source	Total volume (L)	Productivity (g/L/d)	References
Phormidium autumnale	Bubble column	Slaughterhouse wastewater	5	0.64	Roso et al. (2015)
Phormidium autumnale	Bubble column	Cassava starch	2	1.02	Francisco et al. (2014)
Phormidium autumnale	Bubble column	Cassava wastewater	2	6.68	Francisco et al. (2015)
Chlorella minutissima	Fermenter	Glycerol	7	0.44	Katiyar et al. (2017)
Chlorella protothecoides	Stirring tank	Glucose	5	7.40	Xiong et al. (2008)
Chlorella protothecoides	Fermenter	Cassava powder hydrolysate	5	7.66	Lu et al. (2010)
Chlorella pyrenoidosa	Stirred bioreactor	Food waste hydrolysate	2	3.45	Pleissner et al. (2013)
Aphanothece microscopica Nägeli	Bubble column	Fish processing wastewater	4.5	0.44	Queiroz et al. (2011)

simulated large-scale process to produce bulk oil and lipid-extracted algae in an agro-industrial biorefinery. These authors also found values of US$ 0.40/kg and US$ 0.07/kg (corrected to 2017) to produce bulk oil and lipid-extracted algae, respectively.

2.3.4 Scaling-up in Heterotrophic Bioreactors

Another differential of heterotrophic cultivation is the scaling-up relatively easy, and the bioreactors are available commercially for cultivation of several microorganisms with working volumes up to 100,000 L. Li et al. (2007) investigated the scale-up from 250 mL flasks to 11,000 L bioreactors of a heterotrophic cultivation with *C. protothecoides* to biodiesel production. The authors were successful in scaling-up and suggested that it is feasible to expand heterotrophic *Chlorella* cultivation for biodiesel production at the industry level.

According to Perez-Garcia and Bashan (2015), the practical aspects required for large-scale biofuel production from heterotrophic cultivation of microalgae are as follows: (i) the species must be robust and able to grow in the absence of light and under extreme conditions, such as high or low pH, high temperatures, or high salinity; (ii) the microalgae strain must also have a rapid growth to be able to compete with other heterotrophic microorganisms and thus avoid contamination; (iii) the exogenous carbon source must be inexpensive and easily found;

and (iv) the biofuel generated must present a quality standard required by the legislation and in quantity that makes the process economically viable.

2.3.5 Commercial Applications of Heterotrophic Bioreactors

The commercial production of microalgae via heterotrophic metabolism has been made by the Solazyme Inc., in Moema, São Paulo (Brazil). Solazyme's technology enables it to successfully convert a range of low-cost plant-based sugars into biofuels. The biofuels that this company has produced and tested are biodiesel (SoladieselBD®), renewable diesel (SoladieselRD®), aviation turbine fuel (Solajet™), and renewable jet fuel. Furthermore, Solazyme is producing renewable oils for the chemicals, nutrition and skin and personal care space utilizing today's existing industrial scale fermentation capacity.

3 Small-Scale Photobioreactors

3.1 Vertical Photobioreactors

Vertical reactors were among the first algal mass culture systems described in the literature (Cook 1950). These systems are compact, user-friendly bioreactors with a high ratio of surface area/volume, low contamination risk, and high biomass productivity. However, at present, these systems are not used as photobioreactors, except for investigational purposes, due to their difficult of scaling-up (Mirón et al. 1999).

Vertical photobioreactors consist of vertical tubes constructed with a transparent material (polyethylene or glass tubes) to allow the penetration of light. An air diffuser is located at the bottom of the reactor, where the sparged gas is converted into tiny bubbles. This sparging with gas mixture provides the constant agitation of the medium, mass transfer of CO_2 and also removes O_2 produced during photosynthesis. Based on their mode of liquid flow, vertical tubular photobioreactors can be divided into bubble column and airlift PBRs (Singh and Sharma 2012; Chang et al. 2017).

Bubble column photobioreactors are cylindrical vessel with height greater than twice the diameter, simply agitated by bubbling CO_2 and air from a sparger at the photobioreactor bottom without any special internal constructions and completely lack any moving parts (Singh and Sharma 2012; Koller 2015).

Airlift photobioreactors are cylindrical tubes with two interconnecting zones. One of the zones is called riser where gas mixture is sparged, whereas the other zone, the downcomer, does not receive the gas. The system can be with internal loop and external loop. In the first option, regions are separated either by a draft tube or a split cylinder. On the other hand, in the external loop, riser and downcomer are separated physically by two different tubes.

3.1.1 Major Factors Affecting the Vertical PBRs Performance

Light Available

In vertical photobioreactors, the illumination is accomplished externally, which can be natural or artificial. The light available plays an essential role for the good performance of any photosynthetic culture. To obtain sufficient illumination, both the airlift and the bubble column photobioreactors cannot exceed about 0.2 m in diameter; otherwise, the light availability will be reduced severely, mainly in the center of cylinder. Additionally, it must also be considered that the height of the cylinder should not exceed 4 m due to structural reasons (Huang et al. 2017). Furthermore, in a vertical tubular photobioreactor, the light availability also is influenced by aeration rates, gas holdup, and superficial velocity (Mirón et al. 1999).

Aeration Rate, Gas Holdup, and Superficial Gas Velocity

As in horizontal photobioreactors, the agitation of the system gives only by pneumatic path, the aeration is responsible for culture mixing. In an ideal aeration rate, the microalgae are kept in suspension, the light/dark cycle is minimized, the CO_2 is diffused homogeneously, and its excess is removed, thus maintaining the pH stable and the produced oxygen is removed. In general, in airlift photobioreactors, the mixing is better than bubble column and thus can sustain better biomass production of different microalgae (Fernandes et al. 2014).

Gas holdup is one of the most important parameters characterizing airlift and tubular photobioreactors. It is necessary to the hydrodynamic design in different industrial processes because it governs gas phase residence time and gas–liquid mass transfer. The gas holdup is defined as the volume of the gas phase divided by the total volume. This parameter is influenced mainly by the superficial gas velocity and the type of gas diffuser (Mirón et al. 1999).

Superficial gas velocity is the ratio of the volumetric gas flow rate and cross-sectional area of the reactor. The photosynthetic efficiency of the culture is affected by the dark zone that may exist at the center of the photobioreactor. This dark zone is totally dependent on the gas superficial velocity, which is the ratio of the volumetric gas flow rate and cross-sectional area of the reactor. According to Janssen et al. (2003), high gas velocity (>0.05 m/s) is recommended for increasing the photosynthetic efficiency.

Table 4 Biomass productivity in vertical photobioreactors

Microalgae	Photobioreactor	Total volume (L)	Productivity (g/L/d)	References
Haematococcus pluvialis	Bubble column	55	0.06	López et al. (2006)
Chlorella ellipsoidea	Bubble column	6–200	0.03–0.04	Wang et al. (2014)
Anabaena sp.	Bubble column	9	0.31	López et al. (2009)
Aphanotece microscopica Nägueli	Bubble column	3	0.77	Jacob-Lopes et al. (2009)
Scenedesmus obliquus	Bubble column	2	0.21	Maroneze et al. (2016)
Chlorella sp.	Airlift	100	0.21	Xu et al. (2002)
Isochrysis galbana	Airlift	1	0.60	Hu and Richmond (1994)
Chlorella sp.	Airlift	4	0.37	Chiu et al. (2009)
Chlorella sp.	Airlift	1.6	0.25	Lal and Das (2016)

3.1.2 Biomass Productivity in Vertical PBR

Productivities of microalgal biomass in vertical photobioreactors vary with the type of mode of liquid flow, dimensions, microalgae species, and operating and environmental conditions implemented. The values of productivities in these systems of production reported in the literature vary between 0.031 and 0.77 g/L/d, as shown in Table 4.

3.1.3 Costs in Vertical Photobioreactors

According to Wang et al. (2014), for a 20 L indoor bubble column PBR, the cost of biomass production was about US$ 431.39 per kg. On the other hand, in a 200 L outdoor bubble column photobioreactor, the cost to produce 1 kg of dry weight biomass was US$ 58.69. The estimated cost by the methodology proposed by Chisti (2013) in 2017 for biomass production is of approximately US$ 445.58 and US$ 60.63 per kg of dry weight biomass in a 20 L indoor bubble column photobioreactor and a 200 L outdoor bubble column photobioreactor, respectively.

3.1.4 Scaling-up in Vertical Photobioreactors

The vertical tubular photobioreactors are limited to laboratory and pilot scales, which is attributed to fragility of the material, gas transfer at the top regions of the system, temperature control, gas holdup, and a limited surface for illumination especially at up-scaled devices in case of algal species with high demands for illumination (Koller 2015). However, Mirón et al. (1999) affirm that such perceptions have never been substantiated and, according to their results, both bubble column and airlift photobioreactors are more suitable to scaling-up that horizontal PBRs.

A practical example of the difficulty of scale-up is the failure case of GreenFuel Technologies Corporation, at Arizona in 2007. Firstly, GreenFuel designs vertical inclined closed photobioreactors and installed pilot plant to recycle CO_2 emissions into microalgal biomass for biofuels production. With the success of the pilot plant, months later the company installed a photobioreactor at the same plant, but 100 times larger than its earlier test models. Due to incorrect scaling-up, the project of millions of dollars failed, its photobioreactors turned out to be twice as expensive as expected and the company had to fire nearly half its staff (Waltz 2009).

3.2 Flat-Plate Photobioreactors

Flat-plate photobioreactors have received much attention for microalgae biomass production due to their large illumination surface area (Ugwu et al. 2008). In this type of photobioreactor, a thin layer of culture is passed across a flat panel made of a transparent material, as glass, plexiglass, or polycarbonate (Faried et al. 2017). They can be oriented at different angles so as to modify the light intensity and use diffused and reflected light. Agitation can be provided either by bubbling air from its one side through perforated tube or by rotating it mechanically using a motor (Chang et al. 2017).

Flat-panel photobioreactors feature important advantages for biomass production of photoautotrophic microorganisms and may become a standard reactor type for the mass production of several algal species (Sierra et al. 2008). However, the capital and operational cost of such systems are still too high to produce microalgae biomass as feedstock for biofuels or other low-value products with currently available technologies (Li et al. 2014).

The construction of flat-plate reactors dates back to the early 1950s (Burlew 1953), since then, many different designs have been developed. Tredici and collaborators developed a rigid alveolar panel photobioreactor (Tredici et al. 1991; Tredici and Materassi 1992). Pulz and Scheibenbogen (1998) proposed a flat-plate PBR inner walls arranged to promote an ordered horizontal culture flow that was forced by a mechanical pump. Recently, Li et al. (2014) developed a flat-panel photobioreactor with internal bulk liquid flow and an external airlift with the purpose of developing a scalable industrial photobioreactor.

3.2.1 Major Factors Affecting the Flat-Plate Photobioreactor Performance

Light Supply

The flat-plate photobioreactors can be illuminated artificially or through sunlight. However, as the use of sunlight is much more economically feasible, and these systems have an excellent setting to capture sunlight, this is the most commonly used option.

The light absorption is totally dependent on the length of light path. In general, the biomass productivity is highest at the smallest light path and smallest at the longest light path PBR (Richmond and Cheng-Wu 2001). Other configuration that has an influence on light capture is the tilt angle of flat-plate photobioreactor. Throughout the year, the optimal tilt of the PBR that allows maximal incident light will change due to the position of the sun (Wang et al. 2012). Hu et al. (1998) described that as a general rule, the optimal angle for year-round biomass production is equal to the geographic latitude of the location.

Gas Balance and Mixing

A great advantage of flat-panel reactors is that they have a much shorter oxygen path than tubular reactors, so the accumulation of dissolved oxygen is low (Sierra et al. 2008; Chang et al. 2017). According to Sierra et al. (2008), in a flat-panel photobioreactor, an aeration of 0.25 VVM (volume of air per volume of liquid per minute) and a power supply of 53 W/m^3 are sufficient to maintain the balance of gases, mixing is ideally suited for most microalgal culture. Other authors reported even much higher aeration rates up to 2.0 VVM with positive effects (Alias et al. 2004; Wang et al. 2005).

Temperature

Microalgae cultivations in outdoor PBRs are exposed to seasonal and diurnal variation of temperature. These variations have a direct influence on the cellular growth and the chemical composition of the biomass, and therefore, for the development of an efficient and controlled process, the temperature must be maintained with the least possible variation.

Particularly, flat-plate photobioreactors are very susceptible to overheating due to its thin layer of cultivation and high light exposure. For this reason, the PBRs must have an efficient temperature control system. This control is usually done by water spraying (evaporative cooling) or alternatively, by using internal heat exchangers (Chang et al. 2017).

Table 5 Biomass productivities in flat-plate photobioreactor

Microalgae	Photobioreactor	Total volume (L)	Productivity (g/L/d)	References
Nannochloropsis sp.	Flat-plate	440	0.27	Cheng-Wu et al. (2001)
Spirulina platensis	Flat-plate inclined	6–50	0.3–4.3	Hu et al. (1996)
Chlorella vulgaris	Flat-plate airlift	30	0.16–0.95	Munkel et al. (2013)
Thermosynechococcus elongatus	Flat-plate airlift	10	2.9	Bergmann and Trösch (2016)
Nannochloropsis oculata	Short-light path flat-plate	–	12	Cuaresma et al. (2009)
Chlorella sorokiniana	Short-light path flat-plate	1	2.9–14.8	Tuantet et al. (2014)

3.2.2 Biomass Productivity in Flat-Plate Photobioreactor

Some values of biomass productivity reported in the literature are shown in Table 5, which range from 0.16–4.3 g/L/d. These values vary according to the species and parameters used for photobioreactor construction and cultivation. Due to the large light exposure surface area, high biomass productivity is found in these systems, however, are still limited to laboratory scale and pilot scale.

3.2.3 Costs in Flat-Plate PBRs

Tredici et al. (2016) evaluated the production cost of the microalga *Tetraselmis suecica* in a 1-ha plant made of "Green Wall Panel-II" (GWP®-II) photobioreactors located in Tuscany-Italy. The GWP® is flat disposable photobioreactor, designed and patented in 2004 and commercialized by Fotosintetica & Microbiologica S.r.l. Through a techno-economic analysis, they conclude that, for a 1-ha, the total capital investment is about € 1,661,777 and the total fixed capital per annum is of € 101,260. Also in this analysis, they found a cost of € 12.4 to produce 1 kg of biomass (dry weight). This cost can be reduced when the plant is installed in a region with more favorable climatic conditions. The authors related that in Tunisia, the cost of biomass production is of € 6.2 kg/in a 1-ha plant with the same PBR. Lower production costs (€ 5.96/kg) in a vertical flat-panel photobioreactor of commercial scale were found by Norsker et al. (2011), but if we update this value by calculating the inflation correction described in Chisti (2013), this value is of about € 6.36/kg.

3.2.4 Scaling-up in Flat-Plate Photobioreactors

The scale-up in flat-plate photobioreactors presents some challenges, which are usually caused by the large surface area of the photobioreactor. This type of design requires many modules and supports materials, shows difficulty in controlling culture temperature and is very susceptible to the fouling, which is the phenomena that occur when cells attach to the plastic walls, causing a reduction in light availability and an increased risk of contamination (Carvalho et al. 2014; Chang et al. 2017).

Despite the limitations, several commercial large-scale flat-plate photobioreactors have been developed. One example is the Green Wall Panel (GWP®) that has a concept of 'disposable panels' for large-scale applications. This system commercialized by Fotosintetica & Microbiologica S.r.l consists of vertical PBRs of 100-litre bags, made of a polyethylene foil enclosed in a rigid framework (Tredici et al. 2016). Other systems available commercially are the flat-plate airlift patented and produced by Subitec GmbH, in Germany. In this case, the photobioreactors are produced on scales varying from 6 to 180 L per unit.

4 Recent Developments in Microalgae Cultivation Systems

Recently, biofilm cultivation of microalgae emerged as a new biomass production strategy. These systems consist of a densely packed layer of microalgae that grow attached to a solid surface, which should be illuminated and should be frequently exposed to water containing nutrients. Among the advantages of the biofilm-based microalgae cultivation are the cost reduction related to microalgae harvesting, reduced light limitation, low footprint, low water consumption, and efficient CO_2 mass transfer. In contrast, the limitations of the system are the formation of gradients over the biofilm for pH, nutrients, and light (Gross et al. 2015; Hoh et al. 2015).

Another photobioreactor configuration that has attracted attention in recent years is the membrane photobioreactor, mainly for the cultivation of microalgae using wastewater. The membrane photobioreactor is a technology that integrates a conventional enclosed PBR with a submerged or side-stream membrane filtration process using microfiltration or ultrafiltration membranes for solid–liquid separation. These systems can operate in continuous mode, which increases the microalgal biomass production, they produce a high quality treated effluent with low levels of organic substances, pathogen, and suspended solids and are easy to operate and scale-up. However, only limited studies exist about these techniques and for a large-scale implementation, techno-economic analyses and environmental performance assessment are required to assess their viability (Billad et al. 2015; Luo et al. 2016).

Finally, hybrid photobioreactors have proved to be a promising technology for the mass production of microalgae compared with single PBRs. Hybrid photobioreactors are systems that combine different growth stages in two types of PBRs,

Table 6 Biomass productivities in emergent photobioreactors

Microalgae	Photobioreactor	Total volume (L)	Productivity (g/L/d)	References
Chlorella vulgaris	Biofilm photobioreactor	20.3	0.015	Tao et al. (2017)
Chlorella vulgaris	Biofilm photobioreactor	0.6–0.7	7.07	Pruvost et al. (2017)
Chlorella vulgaris	Membrane photobioreactor	25	0.06	Marbella et al. (2014)
Chlorella vulgaris	Membrane photobioreactor	10	0.04	Gao et al. (2014)
Chlorella vulgaris	Hybrid photobioreactor	1	0.05	Heidari et al. (2016)
Chlorella vulgaris	Hybrid photobioreactor	1.5	0.66	Jacob-Lopes et al. (2014)

closed and open, in which the disadvantage of one PBR is complemented by the other (Brennan and Owende 2010). These configurations aim to compensate the drawbacks caused by the limitation of surface/volume ratio and scale-up of open and closed conventional photobioreactors. These systems are based on a proper height/diameter ratio, generating configurations of reactors with heavy workloads in contrast to very long tubes or shallow ponds. The main advantages of these photobioreactors include low use of land area with high culture volume, low operating costs and are potential to scaling-up. On the other hand, this type of configuration is limited to the cultivation of microalgae species with the ability to store energy to sustain cell growth for periods in the dark, without affecting the rate of photosynthetic metabolism (Ramírez-Mérida et al. 2017).

The biomass productivities found in microalgae cultivation with these photobioreactors are shown in Table 6. All these systems are relatively new, and therefore, only a limited number of studies are found in the literature and are restricted to laboratory scale.

5 Criteria for the Selection of Microalgae Cultivation System

According to Chang et al. (2017), the main criteria to be considered in the choice of an ideal photobioreactor are as follows: (i) type and quality of the target product; (ii) tolerance of microalgal strains; and (iii) scale and performance versus cost.

The first criterion to be considered is the type and quality of the desired product. For the biofuel production, it is essential to produce a biomass rich in lipid or carbohydrate with a low cost to be competitive with conventional fossil fuels. In this case, a heterotrophic bioreactor integrated into a biorefinery system can be a good choice due to the high productivity, low cost, and low-land demand.

Additionally, these systems can operate in parallel in wastewater treatment, when they are used as a source of carbon and nutrients for algal growing. On the other hand, to produce light-induced metabolites and high-value products intended for human consumption, a closed PBR is more advisable (Li et al. 2014; Chang et al. 2017).

The characteristics of the microalgae strain that will be used must also be considered at the moment of the choice. Mainly in terms of adaptability and tolerance under outdoor conditions and shear forces and oxygen buildup generated by PBRs. In open ponds, strains must be able to compete with other microorganisms for nutrients and must have the ability to tolerate photoperiods and climate changes. In the case of closed photobioreactors, the strains must withstand strong shear forces generated by pumping or aeration and must be able to tolerate a possible excess of oxygen in the system (Brennan and Owende 2010; Chang et al. 2017).

When a biofuel is the target product, the most important issue is the cost of the biomass which will be processed to yield the fuel. For this, the systems must present a high kinetic performance at large-scale production. It is known that closed systems are significantly more efficient in biomass production compared with open systems. At the same time, most closed systems have a difficult and expensive scaling-up, and open systems can be scaled up easily and inexpensively to accommodate larger production rates. So, the choice of cultivation system must be based on the best trade-off between biomass productivity and production cost (Chang et al. 2017).

Table 7 Advantages and limitations of microalgae cultivation systems

Cultivation type	Cultivation system	Advantages	Limitations
Commercial			
	Raceway ponds	– Low investment – Low power consumption – Economical – Easy to clean – Easy maintenance	– Require large area of land – Cultures are easily contaminated – Low productivity – Limited to a few microalgal strains – Little control of culture conditions – Evaporation – Small illumination surface area
	Horizontal tubular photobioreactors	– High productivity – Large illumination surface area – Suitable for outdoor cultures – Relatively cheap	– Large land area demand – Poor mass transfer – Photoinhibition

(continued)

Table 7 (continued)

Cultivation type	Cultivation system	Advantages	Limitations
Lab-scale			
	Vertical photobioreactors	– High mass transfer – Good mixing – Potential for scalability – Easy to sterilize – Least land use – Reduced photoinhibition	– Small illumination area – Sophisticated construction materials – Support costs – Modest scalability
	Flat-plate photobioreactors	– High biomass productivity – Large illumination surface area – Suitable for outdoor cultures – Uniform distribution of light – Low power consumption	– Difficult to scale up – Difficult temperature control – Fouling – Photoinhibition – Shear damage from aeration
	Heterotrophic bioreactor	– High productivity	– Limited to a few microalgal strains
		– Low cost	– Susceptibility to contamination
		– Wastewater treatment	
		– Easy scaling-up	
Emergent			
	Biofilm photobioreactors	– Low cost of microalgae harvesting – Reduced light limitation – Low footprint – Low water consumption – Efficient CO_2 mass transfer	– Formation of gradients – Scaling-up
	Membrane photobioreactors	– High biomass productivity – High-quality treated effluent – Easy to operate	– Limited studies – Cost
	Hybrid photobioreactors	– Low use of land area – Low operating costs – High stability	– Limited to a few microalgal strains

In addition to considering all these factors, it is important to know all the advantages and limitations of each microalgae cultivation system. In this sense, Table 7 shows the pros and cons of all the systems presented in this chapter.

6 Final Considerations

The biofuels production from microalgae has been demonstrated to have broad potential of application, but these currently still remain at the exploratory stage. This chapter underlines several aspects involved in the microalgal production systems in order to help the development of biofuels from microalgae. Despite that a great deal of work has been done to develop systems for microalgae production, to date, there is no system without limitations. The main difficulties are related to the cost of construction and operation, scaling-up, contamination, and to a limited knowledge about the new cultivation systems. Therefore, to choose a system, trade-offs among productivity, costs, scaling-up, and value of final product should be carefully made.

References

Abomohra, A., Jin, W., Tu, R., Han, S., Eid, M., & Eladel, H. (2016). Microalgal biomass production as a sustainable feedstock for biodiesel: Current status and perspectives. *Renewable and Sustainable Energy Reviews, 64,* 596–606.

Alias, C. B., Lopez, M. C. G. M., Fernández, F. G. A., Sevilla, J. M. G., Sanchez, J. L. G., & Grima, E. M. (2004). Influence of power supply in the feasibility of *Phaeodactylum tricornutum* cultures. *Biotechnology and Bioengineering, 87,* 723–733.

Becker, E. W. (1994). *Microalgae-biotechnology and microbiology* (1st ed.). Cambridge: Cambridge University Press.

Benemann, J. R., Goebel, R. P., Augenstein, D. C., & Weissman, J. C. (1982). *Microalgae as a source of liquid fuels*. Final technical Report to U.S. DOE BER, viewed August 24, 2016, <https://www.osti.gov/scitech/biblio/6374113>.

Bennett, M. C., Turn, S. Q., & Chan, W. Y. (2014). A methodology to assess open pond, phototrophic, algae production potential: A Hawaii case study. *Biomass and Bioenergy, 66,* 168–75.

Bergmann, P., & Trösch, W. (2016). Repeated fed-batch cultivation of *Thermosynechococcus elongatus* BP-1 in flat-panel airlift photobioreactors with static mixers for improved light utilization: Influence of nitrate, carbon supply and photobioreactor design. *Algal Research, 17,* 79–86.

Billad, M. R., Arafat, H. A., & Vankelecom, I. F. J. (2015). Membrane technology in microalgae cultivation and harvesting: A review. *Biotechnology Advances, 32,* 1283–1300.

Borowitzka, M. A. (2005). Culturing microalgae in outdoor ponds. In R. A. Andersen (Ed.), *Algal culturing techniques* (pp. 205–218). Amsterdam: Elsevier Academic Press.

Brennan, L., & Owende, P. (2010). Biofuels from microalgae: A review of technologies for production, processing, and extractions of biofuels and co products. *Renewable and Sustainable Energy Reviews, 14,* 557–577.

Burlew, J. S. (1953). *Algal culture: From laboratory to pilot plant* (1st ed.). Washington: Carnegie Institution of Washington.

Camacho, R. F., Fernández, F. G. A., Pérez, J. A. S., Camacho, F. G., & Grima, E. M. (1999). Prediction of dissolved oxygen and carbon dioxide concentration profiles in tubular photobioreactors for microalgal culture. *Biotechnology and Bioengineering, 62,* 71–86.

Carvalho, J. C. M., Matsudo, M. C., Bezerra, R. P., Ferreira-Camargo, L. S., & Sato, S. (2014). Microalgae bioreactors. In R. Bajpai, A. Prokop, & M. Zappi (Eds.), *Algal biorefineries* (Vol. 1, pp. 83–126). Switzerland: Springer International Publishing.

Chang, J. S., Show, P. L., Ling, T. C., Chen, C. Y., Ho, S. H., Tan, C. H., et al. (2017). Photobioreactors. In C. Larroche, M. Sanroman, G. Du, & A. Pandey (Eds.), *Current developments in biotechnology and bioengineering: Bioprocesses, bioreactors and controls* (pp. 313–352). Atlanta: Elsevier.

Cheng-Wu, Z., Zmora, O., Kopel, R., & Richmond, A. (2001). An industrialsize flat glass reactor for mass production of *Nannochloropsis* sp. (*Eustigmatophyceae*). *Aquaculture, 195,* 35–49.

Chew, K. W., Yap, J. Y., Show, P. L., Suan, N. H., Juan, J. C., Ling, T. C., et al. (2017). Microalgae biorefinery: High value products perspectives. *Bioresource Technology, 229,* 53–62.

Chisti, Y. (2007). Biodiesel from microalgae. *Biotechnology Advances, 25,* 294–306.

Chisti, Y. (2013). Raceways-based production of algal crude oil In C. Posten & C. Walter (Eds.), *Microalgal biotechnology: Potential and production* (pp. 197–216). Berlin: de Gruyter.

Chisti, Y. (2016). Large-scale production of algal biomass: Raceway ponds. In F. Bux & Y. Chisti (Eds.), *Algae biotechnology: Products and processes* (pp. 21–40). New York: Springer.

Chiu, S. Y., Tsai, M. T., Kao, C. Y., Ong, S. C., & Lin, C. S. (2009). The air-lift photobioreactors with flow patterning for high-density cultures of microalgae and carbon dioxide removal. *Engineering in Life Sciences, 9,* 254–260.

Collotta, M., Champagne, P., Busi, L., & Alberti, M. (2017). Comparative LCA of flocculation for the harvesting of microalgae for biofuels production. *Procedia CIRP, 61,* 756760.

Cook, P. M. (1950). *Some problems in the large-scale culture of Chlorella* (pp. 53–75). Yellow Springs, OH: The Culture Foundation.

Crowe, B., Attalah, S., Agrawal, S., Waller, P., Ryan, R., Van Wagenen, J., et al. (2012). A comparison of *Nannochloropsis salina* growth performance in two outdoor pond designs: Conventional raceways versus the arid pond with superior temperature management. *International Journal of Chemical Engineering and Applications, 2012,* 9–21.

Cuaresma, M., Janssen, M., Vílchez, C., & Wijffels, R. H. (2009). Productivity of *Chlorella sorokiniana* in a short light-path (SLP) panel photobioreactor under high irradiance. *Biotechnology and Bioengineering, 104,* 352–359.

de Godos, I., Mendoza, J. L., Acién, F. G., Molina, E., Banks, C. J., Heaven, S., et al. (2014). Evaluation of carbon dioxide mass transfer in raceway reactors for microalgae culture using flue gases. *Bioresource Technology, 153,* 307–314.

Department of Energy (DOE). (2010). National algal biofuels technology roadmap, viewed August 24, 2016, <https://www1.eere.energy.gov/bioenergy/pdfs/algal_biofuels_roadmap.pdf>.

Eustance, E., Badvipour, S., Wray, J. T., & Sommerfeld, M. R. (2015). Biomass productivity of two Scenedesmus strains cultivated semi-continuously in outdoor raceway ponds and flat-panel photobioreactors. *Journal of Applied Phycology, 28,* 1471–1483.

Faried, M., Samer, M., Abdelsalam, E., Yousef, R. S., Attia, Y. A., & Ali, A. S. (2017). Biodiesel production from microalgae: Processes, technologies and recent advancements. *Renewable and Sustainable Energy Reviews, 79,* 893–913.

Fernandes, B. D., Mota, A., Ferreira, A., Dragone, D., Teixeira, J. A., & Vicente, A. A. (2014). Characterization of split cylinder airlift photobioreactors for efficient microalgae cultivation. *Chemical Engineering Science, 117,* 445–454.

Fernandez, F. G. A., Camacho, A. C., Pérez, J. A. S., Sevilla, J. M. F., & Grima, E. M. (1997). A model for light distribution and average solar irradiance inside outdoor tubular photobioreactors for the microalgal mass culture. *Biotechnology and Bioengineering, 55,* 701–714.

Fernandez, F. G. A., Sevilla, J. M. F., & Grima, E. M. (2013). Photobioreactors for the production of microalgae. *Reviews in Environmental Science and Bio/Technology, 12,* 131–151.
Fernández, F. G. A., Sevilla, J. M. F., Pérez, J. A. S., Grima, E. M., & Chisti, Y. (2001). Airlift-driven external-loop tubular photobioreactors for outdoor production of microalgae: Assessment of design and performance. *Chemical Engineering Science, 56,* 2721–2732.
Francisco, E. C., Franco, T. T., Wagner, R., & Jacob-Lopes, E. (2014). Assessment of different carbohydrates as exogenous carbon source in cultivation of cyanobacteria. *Bioprocess and Biosystems Engineering, 37,* 1497–505.
Francisco, E. C., Franco, T. T., Zepka, L. Q., & Jacob-Lopes, E. (2015). From waste-to-energy: The process integration and intensification for bulk oil and biodiesel production by microalgae. *Journal of Environmental Chemical Engineering, 3,* 482–487.
Gao, F., Yang, Z. H., Li, C., Wang, Y. J., Jin, W. H., & Deng, Y. B. (2014). Concentrated microalgae cultivation in treated sewage by membrane photobioreactor operated in batch flow mode. *Bioresource Technology, 167,* 441–446.
Griffiths, D. J., Thresher, C. L., & Street, H. E. (1960). The heterotrophic nutrition of *Chlorella vulgaris* (brannon no. 1 strain). *Annals of Botany, 24,* 1–11.
Grima, E. M. (2009). Algae biomass in Spain: A case study. In *First European Algae Biomass Association Conference & General Assembly*, Florence.
Grima, E. M., Fernández, J., Acién, F. G., & Chisti, Y. (2001). Tubular photobioreactor design for algal cultures. *Journal of Biotechnology, 92,* 113–131.
Gross, M., Jarboe, D., & Wen, Z. (2015). Biofilm-based algal cultivation systems. *Applied Microbiology and Biotechnology, 99,* 5781–5789.
Harder, R., & von Witsch, H. (1942). Ueber Massenkultur von Diatomeen. *Ber. Dtsch. Bot. Ges., 60,* 14–153.
Heidari, M., Kariminia, H. R., & Shayegan, J. (2016). Effect of culture age and initial inoculum size on lipid accumulation and productivity in a hybrid cultivation system of *Chlorella vulgaris*. *Process Safety and Environmental Protection, 104,* 111–122.
Hoh, D., Watson, S., & Kan, E. (2015). Algal biofilm reactors for integrated wastewater treatment and biofuel production: A review. *Chemical Engineering Journal, 287,* 466–473.
Hu, Q., Fairman, D., & Richmond, A. (1998). Optimal tilt angles of enclosed reactors for growing photoautotrophic microorganisms outdoors. *Journal of Fermentation and Bioengineering, 85,* 230–236.
Hu, Q., Guterman, H., & Richmond, A. (1996). A flat inclined modular photobioreactor for outdoor mass cultivation of photoautotrophs. *Biotechnology and Bioengineering, 51,* 51–60.
Hu, Q., & Richmond, A. (1994). Optimizing the population density in *Isochrysis galbana* grown outdoors in a glass column photobioreactor. *Journal of Applied Phycology, 6,* 391–396.
Huang, Q., Jiang, F., Wang, L., & Yang, C. (2017). Design of photobioreactors for mass cultivation of photosynthetic organisms. *Engineering, 3,* 318–329.
Jacob-Lopes, E., Scoparo, C. H. G., Lacerda, L. M. C. F., & Franco, T. T. (2009). Effect of light cycles (night/day) on CO_2 fixation and biomass production by microalgae in photobioreactors. *Chemical Engineering and Processing: Process Intensification, 48,* 306–310.
Jacob-Lopes, E., Zepka, L. Q., Merida, L. G. R., Maroneze, M. M., & Neves, C. (2014). Bioprocesso de conversão de dióxido de carbono de emissões industriais, bioprodutos, seus usos e fotobiorreator híbrido. BR n. PI2014000333.
Janssen, M., Tramper, J., Mur, L., & Wijffels, R. H. (2003). Enclosed outdoor photobioreactors: Light regime, photosynthetic efficiency, scale-up, and future prospects. *Biotechnology and Bioengineering, 81,* 193–210.
Jiménez, C., Cossío. B. R., & Niell, F. X. (2003). Relationship between physicochemical variables and productivity in open ponds for the production of *Spirulina*: A predictive model of algal yield. *Aquaculture, 221,* 331–45.
Juneja, A., Ceballos, R. M., & Murthy, G. S. (2013). Effects of environmental factors and nutrient availability on the biochemical composition of algae for biofuels production: A review. *Enegies, 6,* 4607–4638.

Junying, Z., Junfeng, R., & Baoning, Z. (2013). Factors in mass cultivation of microalgae for biodiesel. *Chinese Journal of Catalysis, 34,* 80–100.

Katiyar, R., Gurjar, B. R., Bharti, R. Q., Kumar, A., Biswas, S., & Pruthi, V. (2017). Heterotrophic cultivation of microalgae in photobioreactor using low cost crude glycerol for enhanced biodiesel production. *Renewable Energy, 113,* 1359–1365.

Koller, M. (2015). Design of closed photobioreactors for algal cultivation. In A. Prokop, R. K. Bajpai, & M. E. Zappi (Eds.), *Algal biorefineries volume 2: Products and refinery design* (pp. 139–186). Switzerland: Springer International Publishing.

Kunjapur, A. M., & Eldridge, R. B. (2010). Photobioreactor design for commercial biofuel production from microalgae. *Industrial and Engineering Chemistry Research, 49,* 3516–3526.

Lal, A., & Das, D. (2016). Biomass production and identification of suitable harvesting technique for *Chlorella* sp. MJ 11/11 and *Synechocystis* PCC 6803. *3 Biotech, 6,* 41–51.

Li, J., Stamato, M., Velliou, E., Jeffryes, C., & Agathos, S. N. (2014). Design and characterization of a scalable airlift flat panel photobioreactor for microalgae cultivation. *Journal of Applied Phycology, 27,* 75–86.

Li, X., Xu, H., & Wu, Q. (2007). Large-scale biodiesel production from microalga *Chlorella protothecoides* through heterotrophic cultivation in bioreactors. *Biotechnology and Bioengineering, 98,* 764–771.

Lin, Q., & Lin, J. (2011). Effects of nitrogen source and concentration on biomass and oil production of a *Scenedesmus rubescens* like microalga. *Bioresource Technology, 102,* 1615–1621.

Lopez, M. C. G., Del Rio Sanchez, E., Lopez, J. L. C., Fernandez, F. G. A., Sevilla, J. M. F., Rivas, J., et al. (2006). Comparative analysis of the outdoor culture of *Haematococcus pluvialis* in tubular and bubble column photobioreactors. *Journal of Biotechnology, 123,* 329–42.

López, C. V. G., Fernández, F. G. A., Sevilla, J. M. F., Fernández, J. F. S., García, M. C. F., & Grima, E. M. (2009). Utilization of the cyanobacteria *Anabaena* sp. ATCC 33047 in CO_2 removal processes. *Bioresource Technology, 100,* 5904–5910.

Lu, Y., Zhai, Y., Liu, M., & Wu, Q. (2010). Biodiesel production from algal oil using cassava (*Manihot esculenta* Crantz) as feedstock. *Journal of Applied Phycology, 22,* 573–578.

Lundquist, T. J., Woertz, I. C., Quinn, N. W. T., & Benemann, A. (2010). *Realistic technology and engineering assessment of algae biofuel production.* Berkeley: Energy Biosciences Institute, University of California.

Luo, Y., Le-Clech, P., & Henderson, R. K. (2016). Simultaneous microalgae cultivation and wastewater treatment in submerged membrane photobioreactors: A review. *Algal Research, 24,* 425–437.

Marbella, L., Bilad, M. R., Passaris, I., Discart, V., Bañadme, D., Beuckels, A., et al. (2014). Membrane photobioreactors for integrated microalgae cultivation and nutrient remediation of membrane bioreactors effluent. *Bioresource Technology, 163,* 228–235.

Maroneze, M. M., Barin, J. S., Menezes, C. R., Queiroz, M. I., Zepka, L. Q., & Jacob-Lopes, E. (2014). Treatment of cattle-slaughterhouse wastewater and the reuse of sludge for biodiesel production by microalgal heterotrophic bioreactors. *Scientia Agricola, 71,* 521–524.

Maroneze, M. M., Siqueira, S. F., Vendruscolo, R. G., Wagner, R., Menezes, C. R., Zepka, L. Q., et al. (2016). The role of photoperiods on photobioreactors—a potential strategy to reduce costs. *Bioresource Technology, 219,* 493–499.

Mirón, A. S., Gómez, A. C., Camacho, F. G., Grima, E. M., & Chisti, Y. (1999). Comparative evaluation of compact photobioreactors for large-scale monoculture of microalgae. *Journal of Biotechnology, 70,* 249–270.

Münkel, R., Schmid-Staiger, U., Werner, A., & Hirth, T. (2013). Optimization of outdoor cultivation in flat panel airlift reactors for lipid production by *Chlorella vulgaris. Biotechnology and Bioengineering, 110,* 2882–2893.

Norsker, N. H., Barbosa, M. J., Vermuë, M. H., & Wijffels, R. H. (2011). Microalgal production-a close look at the economics. *Biotechnology Advances, 29,* 24–27.

Olaizola, M. (2000). Commercial production of astaxanthin from *Haematococcus pluvialis* using 25,000-liter outdoor photobioreactors. *Journal of Applied Phycology, 12,* 499–506.

Olguín, E., Galicia, S., Mercado, G., & Pérez, T. (2003). Annual productivity of *Spirulina* (*Arthrospira*) and nutrient removal in a pig wastewater recycling process under tropical conditions. *Journal of Applied Phycology, 15,* 249–257.

Perez-Garcia, O., & Bashan, Y. (2015). Microalgal heterotrophic and mixotrophic culturing for bio-refining: From metabolic routes to techno-economics. In A. Prokop, R. K. Bajpai, & M. E. Zappi (Eds.), *Algal biorefineries volume 2: Products and refinery design* (pp. 61–132). Switzerland: Springer International Publishing.

Perez-Garcia, O., Escalante, F. M. E., de-Bashan, L. E., & Bashan, Y. (2011). Heterotrophic cultures of microalgae: Metabolism and potential products. *Water Research, 45,* 11–36.

Pleissner, D., Lam, W. C., Sun, Z., & Lin, C. S. K. (2013). Food waste as nutrient source in heterotrophic microalgae cultivation. *Bioresource Technology, 137,* 139–146.

Pruvost, J., Le Borgne, F., Artu, A., & Legrand, J. (2017). Development of a thin-film solar photobioreactor with high biomass volumetric productivity (AlgoFilm©) based on process intensification principles. *Algal Research, 21,* 120–137.

Pulz, O., & Scheibenbogen, K. (1998). Photobioreactors: Design and performance with respect to light energy input. *Advances in Biochemical Engineering/Biotechnology, 59,* 123–152.

Queiroz, M. I., Hornes, M. O., Silva-Manetti, A. G., & Jacob-Lopes, E. (2011). Single-cell oil production by cyanobacterium *Aphanothece microscopica Nägeli* cultivated heterotrophically in fish processing wastewater. *Applied Energy, 88,* 3438–3443.

Ramírez-Mérida, L. G. R., Zepka, L. Q., & Jacob-Lopes, E. (2017). Current production of microalgae at industrial scale. In J. C. M. Pires (Ed.), *Recent advances in renewable energy* (pp. 242–260). Sharjah: Bentham Science Publishers.

Raslavičius, L., Striūgas, N., & Felneris, M. (2018). New insights into algae factories of the future. *Renewable and Sustainable Energy Reviews, 81,* 643–654.

Raso, S., van Genugten, B., Vermuë, M., & Wijffels, R. H. (2012). Effect of oxygen concentration on the growth of *Nannochloropsis* sp. at low light intensity. *Journal of Applied Phycology, 24,* 863–871.

Rawat, I., Kumar, R. R., Mutanda, T., & Bux, F. (2013). Biodiesel from microalgae: A critical evaluation from laboratory to large scale production. *Applied Energy, 103,* 444–467.

Richmond, A. (1990). Large scale microalgal culture and applications. In F. E. Round & D. J. Chapman (Eds.), *Progress in phycological research* (pp. 269–330). Britol: Biopress Ltd.

Richmond, A., & Cheng-Wu, Z. (2001). Optimization of a flat plate glass reactor for mass production of *Nannochloropsis* sp. outdoors. *Journal of Biotechnology, 85,* 259–269.

Roso, G. R., Santos, A. M., Zepka, L. Q., & Jacob-Lopes, E. (2015). The econometrics of production of bulk oil and lipid extracted algae in an agroindustrial biorefinery. *Current Biotechnology, 4,* 547–553.

San Pedro, A., González-López, C. V., Acién, F. G., & Grima, E. M. (2014). Outdoor pilot-scale production of *Nannochloropsis gaditana*: Influence of culture parameters and lipid production rates in tubular photobioreactors. *Bioresource Technology, 169,* 667–676.

Santos, A. M., Deprá, M. C., Santos, A. M., Zepka, L. Q., & Jacob-Lopes, E. (2015). Aeration energy requirements in microalgal heterotrophic bioreactors applied to agroindustrial wastewater treatment. *Current Biotechnology, 4,* 249–254.

Scott, S. A., Davey, M. P., Dennis, J. S., Horst, O., Howe, C. J., Lea-Smith, D. J., et al. (2010). Biodiesel from algae: Challenges and prospects. *Current Opinion in Biotechnology, 21,* 277–286.

Sierra, E., Acién, F. G., Fernández, J. M., García, J. L., González, C., & Molina, E. (2008). Characterization of a flat plate photobioreactor for the production of microalgae. *Chemical Engineering Journal, 138,* 136–147.

Singh, R. N., & Sharma, S. (2012). Development of suitable photobioreactor for algae production —a review. *Renewable and Sustainable Energy Reviews, 16,* 2347–2353.

Su, H., Zhou, X., Xia, X., Sun, Z., & Zhang. Y. (2017a). Progress of microalgae biofuel's commercialization. *Renewable and Sustainable Energy Reviews, 74,* 402–411.

Su, Y., Song, K., Zhang, P., Su, Y., Cheng, J., & Chen, X. (2017b). Progress of microalgae biofuel's commercialization. *Renewable and Sustainable Energy Reviews, 74,* 402–411.

Suh, I. S., & Lee, C. G. (2003). Photobioreactor engineering: Design and performance. *Biotechnology and Bioprocess Engineering, 8,* 313–321.

Sun, A., Davis, R., Starbuck, M., Ben-Amotz, A., Pate, R., & Piencos, P. T. (2011). Comparative cost analysis of algal oil production for biofuels. *Energy, 36,* 5169–5179.

Tabernero, A., Martín del Valle, E. M., & Galán, M. A. (2012). Evaluating the industrial potential of biodiesel from a microalgae heterotrophic culture: Scale-up and economics. *Biochemical Engineering Journal, 63,* 104–115.

Tao, Q., Gao, F., Qian, C. Y., Guo, X. Z., Zheng, Z., & Yang, Z. H. (2017). Enhanced biomass/biofuel production and nutrient removal in an algal biofilm airlift photobioreactor. *Algal Research, 21,* 9–15.

Torzillo, G. (1997). Tubular bioreactors. In A. Vonshak (Ed.), *Spirulina platensis (Arthrospira): Phisiology, cell-biology and biotechnology* (1st ed., pp. 101–115). London: Taylor and Francis.

Torzillo, G., Zittelli, G. C., & Chini Zittelli, G. (2015). Tubular photobioreactors. In A. Prokop, R. K. Bajpai, & M. E. Zappi (Eds.), *Algal biorefineries volume 2: Products and refinery design* (pp. 187–212). Switzerland: Springer International Publishing.

Tredici, M. R., Carlozzi, P., Zittelli, G. C., & Materassi, R. (1991). A vertical alveolar panel (VAP) for outdoor mass cultivation of microalgae and cyanobacteria. *Bioresource Technology, 38,* 153–159.

Tredici, M. R., & Materassi, R. (1992). From open ponds to vertical alveolar panels: The Italian experience in the development of reactors for the mass cultivation of photoautotrophic microorganisms. *Journal of Applied Phycology, 4,* 221–231.

Tredici, M. R., Rodolfi, L., Biondi, N., Bassi, N., & Sampietro, G. (2016). Techno-economic analysis of microalgal biomass production in a 1-há Green Wall Panel (GWP®) plant. *Algal Research, 19,* 253–263.

Tuantet, K., Temmink, H., Zeeman, G., Janssen, M., Wijffels, R. H., & Buisman, C. J. N. (2014). Nutrient removal and microalgal biomass production on urine in a short light-path photobioreactor. *Water Research, 55,* 162–174.

Ugwu, C. U., Aoyagi, H., & Uchiyama, H. (2008). Photobioreactors for mass cultivation of algae. *Bioresource Technology, 99,* 4021–4028.

Ugwu, C. U., Ogbonna, J. C., & Tanaka, H. (2002). Improvement of mass transfer characteristics and productivities of inclined tubular photobioreactors by installation of internal static mixers. *Applied Microbiology and Biotechnology, 58,* 600–607.

Vieira, J. G., Manetti, A. G. S., Jacob-Lopes, E., & Queiroz, M. I. (2012). Uptake of phosphorus from dairy wastewater by heterotrophic cultures of cyanobacteria. *Desalination and Water Treatment, 40,* 224–230.

Waltz, E. (2009). Biotech's green gold? *Nature Biotechnology, 27,* 15–18.

Wang, B., Lan, C. Q., & Horsman, M. (2012). Closed photobioreactors for production of microalgal biomasses. *Biotechnology Advances, 30,* 904–912.

Wang, S. K., Hu, Y. R., Wang, F., Stiles, M. R., & Liu, C. Z. (2014). Scale-up cultivation of *Chlorella ellipsoidea* from indoor to outdoor in bubble column bioreactors. *Bioresource Technology, 156,* 117–122.

Wang, C. H., Sun, Y. Y., Xing, R. L., & Sun, L. Q. (2005). Effect of liquid circulation velocity and cell density on the growth of *Parietochloris incisa* in flat plate photobioreactors. *Biotechnology and Bioprocess Engineering, 10,* 103–108.

Watanabe, Y., de la Noue, J., & Hall, D. O. (2011). Photosynthetic performance of a helical tubular photobioreactor incorporating the cyanobacterium *Spirulina platensis*. *Biotechnology and Bioengineering, 47,* 261–269.

Wen, X., Du, K., Wang, Z., Peng, X., Luo, L., Tao, H., et al. (2016). Effective cultivation of microalgae for biofuel production: A pilot-scale evaluation of a novel oleaginous microalga *Graesiella* sp. WBG-1. *Biotechnology for Biofuels, 9,* 123–135.

Xiong, W., Li, X., Xiang, J., & Wu, Q. (2008). High-density fermentation of microalga *Chlorella protothecoides* in bioreactor for microbial-diesel production. *Applied Microbiology and Biotechnology, 78,* 29–36.

Xu, Z., Baicheng, Z., Yiping, Z., Zhaoling, C., Wei, C., & Fan, O. (2002). A simple and low-cost airlift photobioreactor for microalgal mass culture. *Biotechnology Letters, 24,* 1767–1771.
Zitelli, G. C., Rodolfi, L., Bassi, N., Biondi, N., & Tredici, M. R. (2013). Photobioreactors for biofuel production. In M. A. Borowitzka & N. R. Moheimani (Eds.), *Algae for biofuels and energy* (pp. 115–131). Dordrecht: Springer.

Chapter 3
Process Integration Applied to Microalgal Biofuels Production

Alcinda Patrícia de Carvalho Lopes,
Francisca Maria Loureiro Ferreira dos Santos,
Vítor Jorge Pais Vilar and José Carlos Magalhães Pires

Abstract The rapid development of modern society has resulted in an increased demand for energy and, consequently, an increased use of fossil fuel reserves, compromising the energy sector sustainability. Moreover, the use of this source of energy led to the accumulation of greenhouse gases (GHGs) in atmosphere, which are associated with climate change. In this context, European Union has established new directives regarding GHG emissions and the renewable energy use. Microalgae may have an important role in the achievement of these goals. These photosynthetic microorganisms have a high growth rate, are able to capture CO_2, the biomass can be used to produce biofuels, constituting an undeniable economic potential. Microalgae may also be a source of low carbon fuel, being one of the most studied biofuels feedstock. They are considered a sustainable energy resource, able to reduce significantly the dependence on fossil fuel. They can grow on places that are unsuitable for agriculture, not competing with land for food production. The use of wastewater as microalgal culture medium will reduce the required amount of freshwater and nutrients, achieving simultaneously an effluent with low nutrient concentrations. An important step to increase the competitiveness (promoting simultaneously the environmental sustainability) of microalgal biofuels regarding fossil fuels is the optimization of culture parameters using wastewater as culture medium. Thus, this chapter aims to present the recent studies regarding the integration of wastewater treatment and microalgal cultivation for biomass/biofuel production.

Keywords Biofuel · Microalgae · Process integration · Wastewater treatment
Sustainability

A. P. de Carvalho Lopes · F. M. L. F. dos Santos · J. C. M. Pires (✉)
Laboratory for Process Engineering, Environment, Biotechnology and Energy (LEPABE),
Chemical Engineering Department, University of Porto - Faculty of Engineering,
Rua Dr. Roberto Frias, 4200-465 Porto, Portugal
e-mail: jcpires@fe.up.pt

A. P. de Carvalho Lopes · F. M. L. F. dos Santos · V. J. P. Vilar
Laboratory of Separation and Reaction Engineering – Laboratory of Catalysis and Materials
(LSRE-LCM), Chemical Engineering Department, University of Porto - Faculty of
Engineering, Rua Dr. Roberto Frias, 4200-465 Porto, Portugal

E. Jacob-Lopes et al. (eds.), *Energy from Microalgae*, Green Energy
and Technology, https://doi.org/10.1007/978-3-319-69093-3_3

1 Introduction

The increase of world population is associated with the increase of energy consumption to levels that can compromise the economic growth. Energy is mainly supplied by fossil fuels, which price volatility and sustainable issues (air pollution and climate change) are the main drawbacks. Concerning climate change, the desired balance between CO_2 emissions and sinks (controlling the increase of atmospheric CO_2 concentration) may be achieved through three political strategies (Pires 2017): (i) energy efficiency enhancement; (ii) renewable energy development; and (iii) forest protection. In this context, biofuels have a huge potential to reduce CO_2 emissions to atmosphere (clean energy), as they can substitute fossil fuel energy products without significant technological changes. However, biofuel must be produced from non-edible feedstocks to avoid competition with human food market.

Microalgal culture has attracted the attention of the scientific community due to the high biomass productivity that can be achieved. High growth rates and ability to fix CO_2 are important characteristics to be considered one of the most promising alternatives for biofuel production (Chisti 2007). In addition, they can grow in places that are unsuitable for agriculture, not competing for land with food production practices. However, the cost of microalgal production is still high, being the industrial-scale microalgal culture limited to high-value products. The increase of nutrients (nitrogen and phosphorus) price in the last years is one of the significant contributions to the production cost (Pires et al. 2013). Thus, to obtain microalgal biomass at low cost (to be used for biofuel production), the integration of processes must be performed. To reduce nutrients requirements, microalgae can be cultivated in nutrient-rich wastewater. At the same time, this process integration also reduces the need of freshwater and promotes the treatment of these effluents. Therefore, this chapter aims to present technological issues related to the integration of wastewater treatment and microalgal cultivation for biomass/biofuel production. Recent advances and challenges are also discussed.

2 Microalgae

Microalgae may be classified as prokaryotic or eukaryote organism (Richmond 2004). With respect to the prokaryotic domain, cyanobacteria (also called blue-green algae) are the only ones belonging to this group. On the other hand, in the eukaryotic domain, there are several classes of algae and the most relevant are the following: green algae (*Chlorophyceae*), Golden algae (*Chrysophyceae*) and diatoms (*Bacillariophyceae*). Microalgae can be found more often in the water—freshwater, seawater or brackish water (Lam et al. 2017; Lee 2008). However, they can also be found in all other terrestrial environments, such as snow or hot springs. In most habitats, they act as primary producers in the food chain, synthesizing organic matter from solar energy, carbon dioxide, water and nutrients (e.g. nitrogen and phosphorus) through photosynthesis. In addition, they also produce the oxygen required for metabolism of consumer organisms.

Microalgae are an extremely diverse group of organisms. For example, they contain a significant amount of lipids in the form of fatty acids, which can be extracted for subsequent production of biodiesel (Lam et al. 2017).

2.1 *Cellular Metabolism*

The algae are able to grow with different cellular metabolisms, focusing on the main forms of nutrition, including autotrophy and heterotrophy (Richmond 2004). Therefore, microalgae may grow based on four types of cell metabolisms: autotrophy, heterotrophy, mixotrophy and photoheterotrophy.

Autotrophic organisms obtain energy and electrons (necessary for CO_2 reduction) through the absorption of solar energy and substrates oxidation (mostly water), respectively (Richmond 2004). On the other hand, the heterotrophic organisms use only organic compounds as carbon and energy source (Abreu et al. 2012). With regard to growth under mixotrophy, it is equivalent to grow under autotrophic and heterotrophic conditions, since both organic compounds and CO_2 can be assimilated by microalgae depending on the growing conditions (Richmond 2004). Thus, mixotrophic microorganisms synthesize compounds characteristic of both types of metabolisms, showing high production rates (Cerón García et al. 2005). Concerning photoheterotrophic metabolism, organisms require light energy as a source of energy and organic compounds as carbon source (Richmond 2004).

2.2 *Microalgal Growth Conditions*

The growth of the culture as well as the biochemical composition of biomass is not only determined by the microalgae species in cultivation (Rocha 2012). The medium composition, pH, temperature and light intensity are some of the parameters that can influence the growth and biochemical composition of microalgae.

2.2.1 Nutrients

According to Chisti (2007), the molecular formula of the microalgal biomass is $CO_{0.48}H_{1.83}N_{0.11}P_{0.01}$. Thus, the most important nutrients for autotrophic growth (known as macronutrients) are the carbon (C), nitrogen (N) and phosphorus (P) (Richmond 2004).

Carbon is the macronutrient needed in high concentrations, since it is the main constituent of all organic substances synthesized by the cells, such as proteins, carbohydrates, nucleic acids, vitamins and lipids (Richmond 2004). Microalgae have inorganic carbon assimilation processes: diffusion (5.0 < pH < 7.0) and active transport (pH > 7.0) (Gonçalves et al. 2017). In order to achieve high autotrophic production rates, CO_2 and bicarbonates (HCO_3^-) supply is the most important (Richmond 2004). For certain species of microalgae that grow in mixotrophic conditions, organic compounds (e.g. sugars, acids and alcohols) can be used as carbon source.

Nitrogen has also an important role, since it is a basic element for the formation of proteins, nucleic acids, vitamins and photosynthetic pigments (Richmond 2004). The assimilation mechanism of nitrate and ammonium (NH_4^+) by microalgae is active transport (Gonçalves et al. 2017). Nitrogen is mainly provided in the form of nitrate (NO_3^-), but sometimes ammonium (NH_4^+) and urea can also be used (Richmond 2004). Silva et al. (2015) evaluated the preferred source of nitrogen (NO_3^- and NH_4^+) for two species of microalgae (*Chlorella vulgaris* and *Pseudokirchneriella subcapitata*). The authors concluded that the ammonium was preferred source of nitrogen for microalgae *C. vulgaris*, since its assimilation by the microalgae involves lower energy consumption (Jia et al. 2016). When the microalgae are limited by nitrogen a discoloration of the cells usually occurs (reduction of chlorophylls and carotenoids increase) and a build-up of organic compounds such as polysaccharides and some oils (Becker 1994). Goiris et al. (2015) studied the impact of nutrient limitation in the production of antioxidants in three species of microalgae (*Phaeodactylum tricornutum*, *Tetraselmis suecica* and *C. vulgaris*). The content of chlorophyll *a* in biomass was significantly lower when the microalgae were limited by nitrogen.

Phosphorus is essential nutrient for growth and for many cellular metabolic activities, such as energy transfer, synthesis of nucleic acids, deoxyribonucleic acid (DNA), among others (Richmond 2004). Similarly to nitrogen, phosphorus is also assimilated by the microalgae through active transport (Gonçalves et al. 2017). This chemical element is preferentially added in the form of orthophosphate (PO_4^{3-}), and its absorption is energy dependent (Richmond 2004). The supply of phosphorus also influences the composition of the produced biomass (Borowitzka 1988). The content of lipids and carbohydrates is especially affected by internal and external phosphorus supply. The N:P ratio in the culture medium is also important, as it influences not only the productivity, but also the dominant species in culture (Richmond 2004). In 1934, Alfred C. Redfield estimated the N:P ratio of 16:1 (known as Redfield ratio) through the elemental composition of microalgal cells. However, several studies have tested different ratios (Martin et al. 1987; Minster and Boulahdid 1987; Shaffer et al. 1999; Takahashi et al. 1985). Silva et al. (2015) evaluated the effect of N:P ratio on the growth of microalgae *C. vulgaris* and *P. subcapitata*. The N:P ratios of 8:1, 16:1 and 24:1 were evaluated. The N:P ratio of 8:1 was the one that more favoured the growth of microalgae *C. vulgaris*.

In addition to C, N and P, other nutrients are also important for cell growth, such as the sulphur (S), potassium (K), sodium (Na), iron (Fe), magnesium (Mg) and calcium (Ca) (Richmond 2004). In addition to these, other trace elements (micronutrients) are important, such as boron (B), copper (Cu), manganese (Mn), zinc (Zn), molybdenum (Mo), cobalt (Co), vanadium (V) and selenium (Se).

2.2.2 PH

During the photosynthetic CO_2 fixation, the hydroxide ion (OH^-) accumulates in the growing medium, leading to a gradual increase of pH (Richmond 2004). This shifts the chemical equilibrium of the inorganic carbon present in the medium towards the formation of carbonates (CO_3^{2-}). However, they are not the preferred carbon source for microalgae (Lower 1999). On the other hand, a decrease on the solution pH shifts the chemical equilibrium towards the formation of CO_2, which is one of the preferred carbon sources for microalgae. Nevertheless, this process can lead to the release of CO_2 into the atmosphere, decreasing the concentration of this nutrient extremely important for the cultivation of microalgae.

With regard to nitrogen, when it is provided in the form of ammonium, an increase on the solution pH can result in a decrease on the concentration of nitrogen available for microalgae (Guštin and Marinšek-Logar 2011; Cai et al. 2013). High pH values move the chemical equilibrium of ammonium for the production of ammonia that can be released into the atmosphere due to the aeration of the culture, reducing the availability of nitrogen for microalgae.

The concentration of phosphorus in culture medium can also be influenced by elevated pH, as it can lead to precipitation of phosphate (in the forms of calcium phosphate, iron phosphate and aluminium phosphate) and therefore limit the amount of phosphorus available for microalgae (Wang and Nancollas 2008; Cai et al. 2013).

The pH can directly affect the microalgae, as the pH of microalgal cytoplasm is neutral or slightly alkaline, and enzymes are pH-sensitive and may be inactive in acidic conditions (Chiranjeevi and Mohan 2016). Therefore, extreme pH conditions can cause the disruption of many cellular processes, which may lead to the collapse of culture (Jia et al. 2016).

Tripathi et al. (2015) studied the effect of pH on the growth of *Scenedesmus* sp. microalgae in a range of 7–10 and concluded that the optimal pH for this species was 8. Munir et al. (2015) evaluated the pH effect on the growth of two species of microalgae (*Spirogyra* sp. and *Oedogonium* sp.) in a range of 6.5–9.0, achieving the highest growth at pH 7.5 for both species.

2.2.3 Light Intensity and Temperature

The light energy received by microalgae is a function of the photon flux density that reaches the surface of the culture (Richmond 2004). The cells absorb only a fraction of the photon flux, which is influenced by several factors, such as (i) cell density;

(ii) the optical properties of the cells; (iii) the optical path length of photobioreactor; and (iv) the mixing degree. The photons that are not absorbed by the photosynthetic reaction centres of the cells can be reflected or the associated energy is dissipated in the form of heat. A basic aspect of the interaction of light and temperature is the fact that the optimum temperature for photosynthesis increase with increasing light intensity. Gonçalves et al. (2016) evaluated the effect of light and temperature on the growth of microalgae (*C. vulgaris*, *P. subcapitata*, *Synechocystis salina* and *Microcystis aeruginosa*) and nutrients uptake. In the case of *C. vulgaris*, these authors found that the optimum temperature for growth was 25 °C and the optimum daily irradiance was 208 μmol/m^2/s.

Figure 1 presents the variation of photosynthetic rate with the luminous intensity. With the increase in the light intensity, the photosynthetic rate can reach a value corresponding to the saturation (Richmond 2004). A further increase in the light intensity will not result in an increase of growth rate, and it can become unfavourable, manifesting itself by a decrease in growth rate and culminating in photoinhibition and/or the death of culture in extreme cases.

2.3 Microalgal Culture Technology

Currently, microalgal cultivation technologies can be divided in two classes: open systems (raceway ponds, lagoons, among others) and closed systems/ photobioreactors (tubular, bubble column and airlift) (Dasgupta et al. 2010; Kochen 2010).

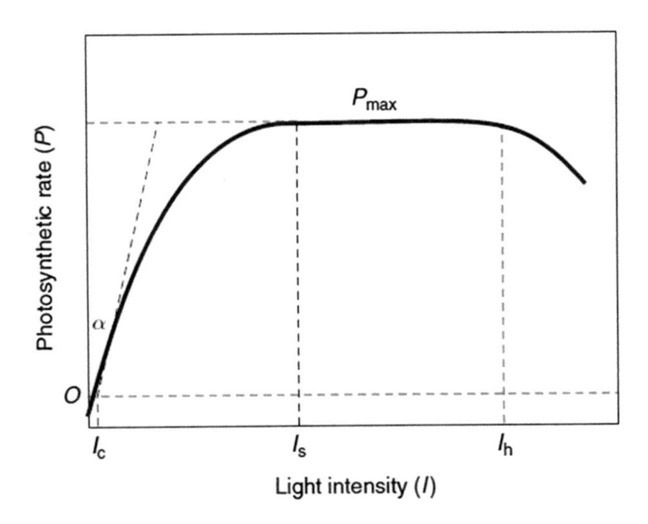

Fig. 1 Variation of photosynthetic rate with light intensity (adapted from Richmond 2004)

2.3.1 Open Systems

Raceway ponds systems consist of a closed-circuit recirculation channel, usually with 0.3 m depth (Benemann and Oswald 1996; Craggs et al. 2012). Mixing and circulation are produced by a paddlewheel, and flow is guided in the curves by deflectors placed in the flow channel, as shown in Fig. 2. During the day, the culture is fed continuously in front of paddlewheel, where run-off begins (Chisti 2007). The paddlewheel operates continuously to prevent sedimentation of biomass.

In open systems, cooling is achieved only by evaporation (Chisti 2007). The temperature varies seasonally and throughout the day, and a significant loss of water by evaporation can be observed. As an open system, the use of CO_2 is less efficient than in closed systems (due to loss of this compound to the atmosphere), also representing a significant cost in the production of microalgae. In addition, the open systems are more susceptible to contamination from other algae or microorganisms. Besides the requirement of large cultivation areas and the limitation of cultivation period (due to contamination problems), these systems are associated to low productivity and inefficient mixture that does not prevent the existence of optically dark areas. However, raceway ponds have a lower production cost, as compared to the closed photobioreactors (Chisti 2007; Harun et al. 2010; Pushparaj et al. 1997). Additionally, the raceway ponds have the advantage of being easily cleaned (removal of biofilm accumulated on the channels walls) (Chisti 2007).

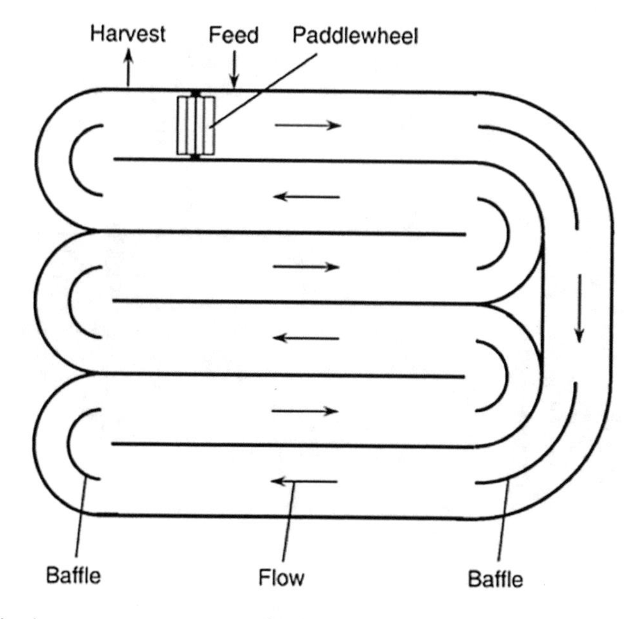

Fig. 2 Cultivation system—raceway pond (adapted from Walter 2011)

2.3.2 Closed Systems

The photobioreactors allow the culture of a single species of microalgae for long periods of cultivation (Molina Grima et al. 1999). Among the various types of photobioreactors for monocultures cultivation, the tubular photobioreactors are the most suitable for large-scale production of microalgae.

The tubular photobioreactors consist of a set of transparent tubes, usually glass or plastic (Chisti 2007). The tubes have typically a diameter less than or equal to 0.1 m. This parameter is limited in order to allow the penetration of light inside the tube, thus guaranteeing light availability to the whole culture. Culture circulates within the tubes, passing by a reservoir (degassing column) and returns again to the tubes, as shown in Fig. 3. There are other variants of photobioreactors; however, they are not usually applied (Carvalho et al. 2006; Molina Grima et al. 1999; Pulz 2001; Tredici 2002).

Some of the advantages of photobioreactors are: (i) the control of cultivation conditions (pH, agitation, concentration of CO_2 and oxygen—O_2) is facilitated; (ii) the reduction of water and CO_2 losses; (iii) the possibility of operating with high cell concentrations and volumetric productivities; and (iv) the reduction of contamination by other microorganisms (Li et al. 2008). However, the photobioreactors have some drawbacks, among them the overheating of the culture, the accumulation of O_2 and the high costs of construction.

2.4 *Microalgal Applications*

The microalgae biomass can be used to generate added-value products. The biomass applicability varies depending on the used microalgae species (Spolaore et al. 2006). Currently, there are numerous commercial applications for microalgae, such

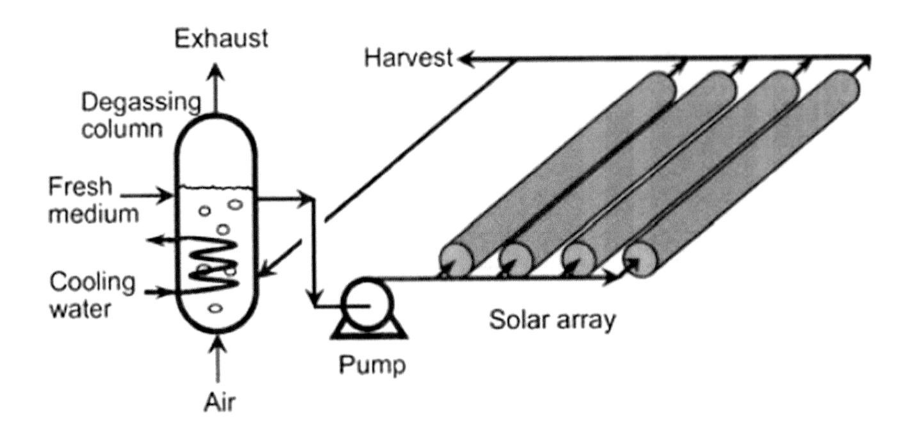

Fig. 3 Tubular photobioreactor (adapted from Chisti 2007)

as: (i) use of microalgae to increase the nutritional value of food and animal feed due to its chemical composition; (ii) extraction of high-value products from microalgae that can be incorporation in cosmetic products; (iii) production of biofuels; (iv) CO_2 capture; and (v) use of microalgae for the depuration of wastewaters.

Gouveia and Empis (2003) concluded that biomass of *C. vulgaris* and *H. pluvialis* was a relatively concentrated natural source of carotenoids, which are natural pigments that exhibit antioxidant capacity. The main carotenoids, found in microalgae, with commercial interest are the β-carotene, lutein and astaxanthin (Mostafa 2012). Besides these compounds, microalgae may be used for the production of biofuels and other bioproducts: proteins, cosmetics, pharmaceutical products, among others.

In the context of environmental applications, microalgae may be used for wastewater treatment in WWTPs (Hoffmann 1998; Oswald 2003). The discharge of wastewaters with high amounts of N and P can cause severe eutrophication of watercourses at downstream (Correll 1998). Thus, the removal of N and P based on microalgae can be quite efficient, cheaper and ecologically safer than physical and chemical treatments currently used (Hoffmann 1998).

Microalgae can also be used in biofixation of atmospheric CO_2 (or from industrial gaseous effluents) through photosynthesis, thus contributing for the reduction of this important greenhouse gas (Nascimento et al. 2015; Sheehan et al. 1998). Microalgae can capture about 1.7–2.4 tons of CO_2 per ton of biomass.

3 Wastewater Treatment by Microalgae

Microalgae can play an important role in the treatment of wastewater, particularly at the level of nutrients removal and reduction of WWTPs operating costs.

3.1 Nutrient Removal

Urban wastewaters are rich in carbon, nitrogen, phosphorus and other minerals, which have to be removed before effluent discharge in water bodies (Cabanelas et al. 2013). An excess of organic carbon and nutrients released into rivers and lakes can lead to decreased dissolved oxygen, toxicity of aquatic life and to eutrophication.

In natural aquatic systems, microalgae assimilate large amounts of nutrients and metals during their growth. Microalgae can digest inorganic sources of nitrogen such as ammonium, nitrite and nitrate (Jia et al. 2016).

The use of microalgae in the wastewater treatment plant was first proposed by Oswald and Gotass (1957) and in recent decades has received a lot of attention. The premise of this approach is that the mixotrophic systems can be designed to reduce the organic carbon, as well as the nutrients in urban wastewater to values lower than discharge limits (McCarty et al. 2011). Microalgae can be a good approach for the tertiary treatment of urban wastewaters because they require large amounts of nitrogen and phosphorus to their growth, including for the synthesis of proteins (40–60% by dry weight), nucleic acids and phospholipids (Silva-Benavides and Torzillo 2011). The wastewater treatment based on microalgae can remove N and P more efficiently than the traditional activated sludge treatment (Lau et al. 1995; Lavoie and De La Noüe 1983; Tam and Wong 1989). In addition to the removal of these nutrients, microalgae have the ability to remove heavy metals from wastewater (Rai et al. 1981). Finally, the microalgae can perform a disinfectant effect in the effluent due to the pH increase inherent to photosynthesis (De La Noue and De Pauw 1988). The mentioned advantages make this system an excellent alternative to the traditional technologies employed for wastewater.

Nutrient removal efficiencies are dependent on the wastewater composition and environmental conditions, such as light intensity, the N:P ratio, the light/dark cycle and microalgal species (Aslan and Kapdan 2006). The most studied microalgal species for the treatment of urban wastewaters are *Chlorella*, *Scenedesmus*, *Phormidium*, *Botryococcus*, *Chlamydomonas* and *Spirulina* (Chinnasamy et al. 2010; Kong et al. 2010; Olguín 2003; Wang et al. 2010). Taking into account the potential of microalgae for wastewater treatment, Table 1 presents several studies that demonstrate the viability of microalgal cultures in the nutrients removal from different types of wastewater.

3.2 Limitations of Conventional Treatments

The consortium of microorganisms present in activated sludge systems require phosphorus for their growth, which results in partial removal of phosphate during the secondary treatment (Yau 2016). However, to achieve discharge limits of 1 mg P/L normally is required the use of inorganic coagulants (such as lime, aluminium sulphate and iron chloride). Besides the increase on the treatment cost, the addition of these coagulants is less environmentally sustainable than the removal of phosphorus by microalgae.

Another limitation is the main by-product generated in biological treatment: activated sludge waste. To treat 1 million litres of wastewater, the biological treatment produces about 70–100 kg of activated sludge in dry basis (Athanasoulia et al. 2012). Consequently, the treatment and disposal of this waste requires a considerable deposition area and a high-energy expense. In addition, the mechanical aeration (necessary in biological treatment) can cause the release of volatile contaminants into the atmosphere (Jia et al. 2016). The role of microalgae in this step could reduce or even prevent the release of these contaminants, since microalgae would produce oxygen and thus reduce the need for aeration.

Finally, greenhouse gases (such as methane—CH_4, N_2O and CO_2) are released to the atmosphere in the biological treatment (Campos et al. 2016). Conventional

Table 1 Nutrient removal from wastewater with microalgal cultures

Microalgae / Wastewater	Experimental setup	Removal efficiency (%)			References
		$[NH_4^+–N]_i$ (mg/L)	$[NO_3^-–N]_i$ (mg/L)	$[TP]_i$ (mg/L)	
B. braunii	Fermenter BioFlo; *V* = 9 L; *T* = 25 °C; LI = 3500 lx; LDR = 12:12; CT = 14 d		80	~100	Sydney et al. (2011)
Domestic			390	385 [P–PO_4^{3-}]	
C. vulgaris	Bubble column PBR; *V* = 2 L; *T* = 30 °C; LI = 3000 lx; CT = 14 d	97		96	Feng et al. (2011)
Synthetic		20		4	
Chlorella sp.	Coil bioreactor; *V* = 25 L; *T* = 25 ± 2 °C; LI = 50 μmol/m²/s; CT = 14 d	94		81	Li et al. (2011)
Municipal		83		212	
N. oleoabundans	Bubble column PBR; *V* = 400 mL; *T* = 30 °C; LI = 1280 lm; CT = 7 d		99	100	Wang and Lan (2011)
Synthetic			140	47	

CT cultivation time; *LDR* light:dark ratio; *LI* light intensity; *T* temperature; *V* volume treatments have no capacity to carry out the CO_2 capture and thus prevent its release to the atmosphere.

3.3 *Benefits of Microalgae in WWTPs*

In the activated sludge treatment, it is estimated that 1 kg of BOD removal consumes about 1 kWh of electricity for aeration (implying 1 kg of CO_2 emissions in the electricity production) and produces about 0.45 kg of biomass residues (Oswald 2003). On the other hand, the removal of 1 kg of BOD by microalgae (photosynthetic pathway), in mixotrophic systems, does not require energy input and it can produce enough biomass to generate methane, which will produce 1 kWh of electricity. The wastewater treatment based on microalgae is an ecological process, without secondary pollution, and it allows the efficient recycling of nutrients (Mulbry et al. 2008; Muñoz and Guieysse 2006; Pizarro et al. 2006).

The microalgae biomass resulting from wastewater treatment systems can give rise to products with commercial interest, such as fertilizer, animal feed, fine chemicals, biofuels, among others, thereby reducing the total cost of the treatment plant (De La Noüe et al. 1992).

Finally, the CO_2 biofixation by microalgae is an environmentally friendly method to remove carbon from the atmosphere (Singh and Yadav 2015). Microalgae have been described as having high capacity to fix CO_2 when compared with land plants (Chen et al. 2013). They may fix the CO_2 released by the activated sludge, avoiding its release to the atmosphere.

The interest in the cultures of microalgae is that conventional treatment processes present some important disadvantages, such as: (i) variable efficiency depending on the nutrient to be removed; (ii) expensive treatment; (iii) chemical processes leading to secondary pollution; and (iv) nutrient loss with possible value (N and P) (De La Noüe et al. 1992).

Thus, the increase in global warming, scarcity of fossil fuels and the need to mitigate emissions of greenhouse gases, the study of the feasibility of biological wastewater treatment based on microalgae (associated with the production of biofuels) is of utmost importance (Rawat et al. 2011).

3.4 *Reduction of Operating Costs*

The introduction of microalgae in wastewater treatment processes can reduce the costs associated to aeration and coagulation and at the same time can obtain biomass with high commercial value (Christenson and Sims 2011). The microalgae-based treatment system is a less expensive and ecologically safer technology when compared with physical and chemical processes, with the additional benefits of resources recovery and recycling. The biological nitrification/

denitrification system, the most common process used for nitrogen removal, has as end product the nitrogen gas (N_2), while the treatment with microalgae retains the nitrogen compounds on biomass, adding value.

The aerobic photosynthetic pathway is especially interesting, as it allows to reduce the operating costs associated with the aeration of the biological treatment (Borowitzka and Borowitzka 1988), which represents more than 50% of the energy needs in a WWTP (EPA Office of Water 2006). Recent studies have shown that microalgae may also support the aerobic degradation of several hazardous contaminants (Muñoz and Guieysse 2006; Safonova et al. 2004).

4 Microalgal Limitations in Wastewater Treatment

As in all treatment systems, microalgal culture also presents some disadvantages/ limitations in wastewater treatment.

4.1 Temperature Variability

The productivity of microalgae increases with increasing temperature up to an optimum value, above which the respiration and photorespiration of microalgae reduce overall productivity (Park et al. 2011). Thus, water temperature too high or too low can have a negative effect on the growth of microalgae and it can cause growth inhibition. The ideal temperature, measured under maximum growth conditions (optimal conditions of nutrients and light), varies from species to species; however, for many species it is between 28 and 35 °C. The optimum temperature changes when the growth is limited by nutrients and/or light. Moreover, microalgal growth decreases when they are subject to sudden changes of temperature (Larsdotter 2006). For example, the exposure of a species at a temperature of 10 °C, when they were adapted to higher values, resulted in a reduction of about 50% of the chlorophyll *a* in 15 h. In addition, a high luminous intensity associated with low temperatures is also another factor that causes growth inhibition.

Taking into account the variability of temperature during the day and the year, it is expected that the efficiency of wastewater treatment based on microalgae can be substantially affected. Therefore, the seasonal temperature variability is one of the main limitations in the introduction of microalgae in WWTPs, since it will be difficult to keep the cultures within a range of acceptable temperatures throughout the year.

4.2 *Light/Dark Cycle*

The quality, intensity and light period are very important parameters in microalgae production (Cardinale 2011). In outdoor cultivation systems, solar radiation is the only source of light and its availability is therefore dependent on the geographical location, climate and seasonality (Novoveská et al. 2016). Light regimes (to which cultures are submitted) are considered an important factor in the productivity and efficiency of photosynthetic reactions (Sicko-Goad and Andresen 1991; Toro 1989).

Lee and Lee (2001) evaluated the effect of the light/dark cycle in the treatment of wastewater by *Chlorella kessleri*. This study showed that the amount of nitrate removed was 31.6 mg NO_3/L after 3 days under continuous light conditions, and 14.0 mg NO_3/L with the light/dark cycle of 12:12. However, the removal of organic carbon and phosphate was higher for the culture under conditions of light/dark cycle of 12:12. Therefore, *C. kessleri* could grow heterotrophically during the dark periods, once the microalgae are able to metabolize the organic carbon to their growth without photosynthesis. With regard to the species (*C. vulgaris*), Santos et al. (2009) studied the effect of two light/dark cycles: 24:0 and 12:12. This study demonstrated that the cycle of continuous light presented a higher growth rate.

Depending on the types of effluent (primary, secondary or tertiary), the light/dark cycle may have different impacts on the treatment efficiency, since the primary and secondary effluents have high concentrations of organic carbon. On the other hand, the tertiary effluent has low amounts of organic carbon, which can limit the heterotrophic growth of microalgae and, consequently, the treatment efficiency.

4.3 *Competition with the Microflora Present in Wastewater*

The coexistence of microalgae and bacteria is a biological process that occurs by the interaction of two distinct processes: photosynthesis of microalgae and bacterial

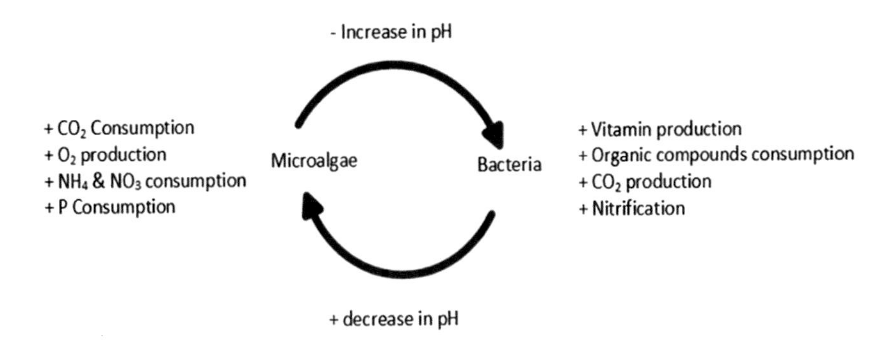

Fig. 4 Nutrient exchange in microalgae-bacteria consortium (adapted from Anbalagan 2016)

respiration (Anbalagan 2016). These two processes occur simultaneously in wastewater with nutrient exchange, as shown in Fig. 4.

Sforza et al. (2014) evaluated the effect of the wastewater native microflora on the growth of microalgae *Chlorella protothecoides*. The obtained results indicated that significant differences were not detected in the growth of microalgae, suggesting that the presence of the native microflora in wastewater does not influence its growth.

4.4 Biomass Harvesting

Despite the numerous advantages of wastewater bioremediation by microalgae, there are also some obstacles that limit their application on a large scale, such as harvesting of biomass. Currently, biomass harvesting is quite expensive, since this step is about 20–30% of the biomass production cost (Molina Grima et al. 2003). The microalgae separation from effluent remains the main obstacle for the bioremediation of wastewater in WWTPs, in part due to the small size of the microalgae. The size of eukaryotic unicellular microalgae usually varies between 3 and 30 μm (Molina Grima et al. 2003), and the size of the cyanobacteria varies between 0.2 and 2 μm (Chorus and Bartram 1999). Moreover, the fact that cultures are relatively diluted (200–600 mg/L) (Uduman et al. 2010) with densities close to the water also affects the harvesting process. Finally, the negative charge of microalgae maintains cells in suspension (Danquah et al. 2009a). Until now, there is no method for microalgal harvesting that is economically viable and efficient (Barros et al. 2015). Biomass harvesting techniques applied to microalgae include coagulation/flocculation, auto and bioflocculation, gravitational sedimentation, flotation, electrical processes, filtration and centrifugation. However, none of these techniques meets the ideal conditions for use in large scale (cost vs. efficiency) (Christenson and Sims 2011). Cost reduction of harvesting is therefore considered to be a key factor for the development of sustainable production in large scale of microalgal biomass. An ideal process should be effective for most species of microalgae and it must allow the obtaining of high concentrations of biomass (Danquah et al. 2009b). In addition, the harvesting process must introduce moderate costs of operation, maintenance and energy.

4.5 Wastewater Characteristics

The amount and quality of light penetration affect the photosynthetic process of organisms that use sunlight as an energy source (Butler et al. 2017). Thus, as photosynthetic organisms, the colour of wastewater, as well as the amount of particles in suspension should be factors to be used for the cultivation of microalgae (Yaakob and Fakir 2011). The high content of particulates in wastewaters can affect microalgae growth due to shadowing effects. In addition, the microorganisms also contribute to the turbidity of the water, limiting even more the depth of light penetration. Taking into account these factors that limit light penetration, photosynthesis occurs only in the superficial layers of the culture (USEPA 2011), influencing the overall treatment efficiency. In order to maximize the light penetration, the mixing degree inside the photobioreactors is an important factor, as all cells can be exposed to light in a turbulent regime for at least a short period of time, being possible to achieve high productivity (Yaakob and Fakir 2011).

5 Biofuels Production with Microalgae Cultivated in Wastewater: Recent Advances and Challenges

The production of microalgal biofuels has two major challenges: (i) production costs reduction and (ii) identification of the harvesting process. The integration of biomass production and wastewater treatment reduces the requirements of nutrients and freshwater. Studies with real wastewaters should be performed to evaluate the nutrient removal efficiencies (wastewater treatment efficiency) and biomass productivities (possible growth inhibition). With the achieved biomass, the potential for production of different biofuels (biodiesel, bioethanol, biogas, between others) should be assessed. Table 2 shows some recent studies focusing on biofuel production with microalgae cultivated in wastewater. Prandini et al. (2016) evaluated the growth of microalgae *Scenedesmus* sp. in piggery wastewater and bubbled swine wastewater-derived biogas (for biogas filtration). Microalgal culture was able to assimilate N–NH_3, P–PO_4^{3-} and CO_2 at a rate of 21 ± 1, 4 ± 3 and 219 ± 5 mg/L/d, respectively. H_2S in biogas (up to 3000 ppm) was not inhibitory and it was completely removed. Hernandez et al. (2016) tested a consortium of microalgae composed by *Chlamydomonas subcaudata*, *Anabaena* sp. and *Nitzschia* sp. for treatment of slaughterhouse wastewater in two high-rate algal ponds—HRAPs (indoor and outdoor) during 115 d. High removal efficiencies of chemical oxygen demand and soluble phosphorus were achieved in both HRAPs. The maximum productivity was 12.7 g/m^2/d. High quality of free fatty acids (FFA) was achieved in a ratio of 142 mg FFA/g. Biogas production was also assessed, resulting in 195 mL CH_4/g. Lutzu et al. (2016) evaluated the potential of brewery wastewater as microalgal culture medium. Adjustments in nitrogen and phosphorus concentrations were needed to improve biomass and lipid productivities. The chemical analysis of the fatty acids methyl esters showed that high fractions (67.24%) are unsaturated ones and they are composed mainly by C16–C18. Concerning the wastewater treatment, high removal efficiencies were achieved for nitrogen and phosphorus (>99%) and a significant reduction of chemical oxygen demand was observed (65%). Despite the recent studies reported in the literature, further researches are still needed. Due to the natural variability of wastewater composition, microalgal culture should be tested under environmental stresses in

Table 2 Biofuel production with microalgae cultivated in wastewater

Biofuel/ microalgae	Experimental setup	*P* (mg/L/d)/BFC	Reference
Biogas/ *Scenedesmus* sp.	Swine wastewater; $V = 16.9$ L; $T = 22 \pm 2$ °C; LI = 148.5 μmol/m^2/s; LDR = 12:12 and 24:0	$P_x = 142.0$	Prandini et al. (2016)
Biogas and lipids/mix of microalgae	Slaughterhouse wastewater; HRAP; $V = 75$ L; $T = 20$–25 °C; LI = 63–760 μmol/m^2/s; HRT = 10–15 d	$P_x = 1.05$–2.56; $P_L = 13$–15/142 mg FFA/g; 195 mL CH_4/g	Hernandez et al. (2016)
Biodiesel/ *Chlorella* sp.	Domestic wastewater; HRAP; $T = 20.2 \pm 5.7$ °C; pH = 12; LDR = 8:16	P_x = n.a./SFA = 46–67%; UFA = 10–40%	Drira et al. (2016)
Biodiesel/ *Scenedesmus dimorphus*	Brewery wastewater; bubble column PBR; $V = 250$ mL; LI = 100 μmol/m^2/s; LDR = 24:0; CT = 12 d	P_x = n.a./SFA = 31–37%; UFA = 62–68%	Lutzu et al. (2016)
Biodiesel/ *Scenedesmus obliquus*	Municipal wastewater; Erlenmeyer flasks; $V = 1$ L; $T = 25 \pm 1$ °C; LI = 100 μmol/m^2/s; LDR = 12:12	$P_x = 4.8$–7.5; $P_L = 0.44$–1.98/ SFA = 27–41%; UFA = 10–26%	Han et al. (2016)
Biogas/ *Chlorella vulgaris*	Domestic wastewater; bubble column PBR; $V = 40$ L; pH = 6; LI = 150 μmol/m^2/s; LDR = 24:0; CT = 21 d	P_x = n.a./223–408 mL CH_4/g	Calicioglu and Demirer (2016)
Biomass/ *Acutodesmus dimorphus*	Industrial wastewater; Erlenmeyer flasks; $V = 1$ L; $T = 35$ °C; LI = 60 μmol/m^2/s; LDR = 12:12; CT = 8 d	$P_x = 210$; LC = 25.05%	Chokshi et al. (2016)
Lipids/ *Scenedesmus obliquus*	Domestic wastewater; raceway pond; $V = 533$ L; pH = 8; CT = 5 d	$P_x = 87.3$; LC = 33.6%	Arbib et al. (2017)

BFC biofuel characteristics; *FFA* free fatty acids; *HRAP* high-rate algal ponds; *HRT* hydraulic retention time; *LDR* light–dark ratio; *LC* lipids content; *LI* light intensity; *n.a.* not available; *P* productivity; P_L lipid productivity; P_X biomass productivity; *SFA* saturated fatty acids; *T* temperature; *UFA* unsaturated fatty acids; *V* volume

order to evaluate their tolerance capacity. With a less-controlled environment, the development of an innovative, efficient and cost-effective harvesting process is highly required. In addition, studies with life cycle assessment for economic viability, carbon footprint and sustainability should be performed.

6 Conclusions

The continuous growth of population and energy needs of the industry and transport sectors increased interest in the use of renewable energy sources. Besides not being renewable sources, fossil fuel energy (oil, coal and natural gas) emits considerable amounts of greenhouse gases to the atmosphere leading to increased global warming. Therefore, the production of biofuels from microalgal biomass is considered a source of sustainable energy, since the cultivation of biomass can be integrated with wastewater treatment. The presence of large amounts of C, N and P (macronutrients for microalgal growth) in urban wastewaters allows that this kind of effluents may be used as cultivation media for microalgal culture. Consequently, the cultivation of microalgae in wastewater treatment plants can play a dual role, since it allows the removal of nutrients from effluent and the production of biomass for subsequent production of biofuels. Bioremediation of wastewater is an ecological process and no secondary pollution, since biomass produced is reused and enables the efficient recycling of nutrients. In addition, the cultivation of microalgae using wastewater as culture medium presents numerous advantages, such as: (i) reduced need for aeration; (ii) higher consumption of P than in the biological treatment; and (iii) biofixation capacity of CO_2 by the microalgae. However, there are still some obstacles need to be overcome: the effect of temperature variability, light/dark cycle, competition with the microflora and wastewater chemical composition.

Acknowledgements This work was financially supported by: Project POCI-01-0145-FEDER-006939 (LEPABE), Project POCI-01-0145-FEDER-006984 (Associate Laboratory LSRE-LCM) and Project AlProcMat@N2020-NORTE-01-0145-FEDER-000006—funded by FEDER funds through COMPETE2020—Programa Operacional Competitividade e Internacionalização (POCI)—and by national funds through FCT—Fundação para a Ciência e a Tecnologia. V. J. P. Vilar acknowledges the FCT Investigator 2013 Programme (IF/00273/2013). J. C. M. Pires acknowledges the FCT Investigator 2015 Programme (IF/01341/2015).

References

Abreu, A. P., Fernandes, B., Vicente, A. A., Teixeira, J., & Dragone, G. (2012). Mixotrophic cultivation of *Chlorella vulgaris* using industrial dairy waste as organic carbon source. *Bioresource Technology, 118,* 61–66.

Anbalagan, A. (2016). *Indigenous microalgae-activated sludge cultivation system for wastewater treatment*. Mälardalen University.

Arbib, Z., De Godos, I., Ruiz, J., & Perales, J. A. (2017). Optimization of pilot high rate algal ponds for simultaneous nutrient removal and lipids production. *Science of the Total Environment, 589,* 66–72.

Aslan, S., & Kapdan, I. K. (2006). Batch kinetics of nitrogen and phosphorus removal from synthetic wastewater by algae. *Ecological Engineering, 28,* 64–70.

Athanasoulia, E., Melidis, P., & Aivasidis, A. (2012). Optimization of biogas production from waste activated sludge through serial digestion. *Renewable Energy, 47,* 147–151.

Barros, A. I., Gonçalves, A. L., Simões, M., & Pires, J. C. M. (2015). Harvesting techniques applied to microalgae: A review. *Renewable and Sustainable Energy Reviews, 41,* 1489–1500.

Becker, E. W. (1994). *Microalgae: Biotechnology and microbiology*. U. K.: Cambridge University Press.

Benemann, J. R., & Oswald, W. J. (1996). *Systems and economic analysis of microalgae ponds for conversion of CO_2 to biomass*. Final report. Berkeley, CA (United States): Department of Civil Engineering, California University.

Borowitzka, M. A. (1988). Fats, oils and carbohydrates. In M. A. Borowitzka & L. J. Borowitzka (Eds.), *Micro-algal biotechnology*. Cambridge: Cambridge University Press.

Borowitzka, M. A., & Borowitzka, L. J. (1988). *Micro-algal biotechnology*. UK: Cambridge University Press.

Butler, E., Hung, Y.-T., Suleiman Al Ahmad, M., Yeh, R. Y.-L., Liu, R. L.-H., & Fu, Y.-P. (2017). Oxidation pond for municipal wastewater treatment. *Applied Water Science, 7,* 31–51.

Cabanelas, I. T. D., Arbib, Z., Chinalia, F. A., Souza, C. O., Perales, J. A., Almeida, P. F., et al. (2013). From waste to energy: Microalgae production in wastewater and glycerol. *Applied Energy, 109,* 283–290.

Cai, T., Park, S. Y., & Li, Y. B. (2013). Nutrient recovery from wastewater streams by microalgae: Status and prospects. *Renewable and Sustainable Energy Reviews, 19,* 360–369.

Calicioglu, O., & Demirer, G. N. (2016). Biogas production from waste microalgal biomass obtained from nutrient removal of domestic wastewater. *Waste and Biomass Valorization, 7,* 1397–1408.

Campos, J. L., Valenzuela-Heredia, D., Pedrouso, A., Belmonte, M., & Mosquera-Corral, A. (2016). Greenhouse gases emissions from wastewater treatment plants: Minimization, treatment, and prevention. *Journal of Chemistry, 2016,* 1–12.

Cardinale, B. J. (2011). Biodiversity improves water quality through niche partitioning. *Nature, 472,* 86–89.

Carvalho, A. P., Meireles, L. A., & Malcata, F. X. (2006). Microalgal reactors: A review of enclosed system designs and performances. *Biotechnology Progress, 22*(6), 1490–1506.

Cerón García, M., Sánchez Mirón, A., Fernández Sevilla, J. M., Molina-Grima, E., & García Camacho, F. (2005). Mixotrophic growth of the microalga Phaeodactylum tricornutum: Influence of different nitrogen and organic carbon sources on productivity and biomass composition. *Process Biochemistry, 40,* 297–305.

Chen, C. Y., Kao, P. C., Tsai, C. J., Lee, D. J., & Chang, J. S. (2013). Engineering strategies for simultaneous enhancement of C-phycocyanin production and CO_2 fixation with *Spirulina platensis*. *Bioresource Technology, 145,* 307–312.

Chinnasamy, S., Bhatnagar, A., Hunt, R. W., & Das, K. C. (2010). Microalgae cultivation in a wastewater dominated by carpet mill effluents for biofuel applications. *Bioresource Technology, 101,* 3097–3105.

Chiranjeevi, P., & Mohan, S. V. (2016). Critical parametric influence on microalgae cultivation towards maximizing biomass growth with simultaneous lipid productivity. *Renewable Energy, 98,* 64–71.

Chisti, Y. (2007). Biodiesel from microalgae. *Biotechnology Advances, 25,* 294–306.

Chokshi, K., Pancha, I., Ghosh, A., & Mishra, S. (2016). Microalgal biomass generation by phycoremediation of dairy industry wastewater: An integrated approach towards sustainable biofuel production. *Bioresource Technology, 221,* 455–460.

Chorus, I., & Bartram, J. (1999). *Toxic cyanobacteria in water: a guide to their public health consequences, monitoring, and management*. London: E & FN Spon.

Christenson, L., & Sims, R. (2011). Production and harvesting of microalgae for wastewater treatment, biofuels, and bioproducts. *Biotechnology Advances, 29,* 686–702.

Correll, D. L. (1998). The role of phosphorus in the eutrophication of receiving waters: A review. *Journal of Environmental Quality, 27*(2), 261–266.

Craggs, R., Sutherland, D., & Campbell, H. (2012). Hectare-scale demonstration of high rate algal ponds for enhanced wastewater treatment and biofuel production. *Journal of Applied Phycology, 24,* 329–337.

Danquah, M. K., Ang, L., Uduman, N., Moheimani, N., & Forde, G. M. (2009a). Dewatering of microalgal culture for biodiesel production: Exploring polymer flocculation and tangential flow filtration. *Journal of Chemical Technology and Biotechnology, 84,* 1078–1083.

Danquah, M. K., Gladman, B., Moheimani, N., & Forde, G. M. (2009b). Microalgal growth characteristics and subsequent influence on dewatering efficiency. *Chemical Engineering Journal, 151,* 73–78.

Dasgupta, C. N., Jose Gilbert, J., Lindblad, P., Heidorn, T., Borgvang, S. A., Skjanes, K., et al. (2010). Recent trends on the development of photobiological processes and photobioreactors for the improvement of hydrogen production. *International Journal of Hydrogen Energy, 35,* 10218–10238.

De La Noue, J., & De Pauw, N. (1988). The potential of microalgal biotechnology: A review of production and uses of microalgae. *Biotechnology Advances, 6,* 725–770.

De La Noüe, J., Laliberté, G., & Proulx, D. (1992). Algae and waste water. *Journal of Applied Phycology, 4,* 247–254.

Drira, N., Piras, A., Rosa, A., Porcedda, S., & Dhaouadi, H. (2016). Microalgae from domestic wastewater facility's high rate algal pond: Lipids extraction, characterization and biodiesel production. *Bioresource Technology, 206,* 239–244.

EPA Office of Water. (2006). *Wastewater management fact sheet, energy conservation.*

Feng, Y., Li, C., & Zhang, D. (2011). Lipid production of *Chlorella vulgaris* cultured in artificial wastewater medium. *Bioresource Technology, 102,* 101–105.

Goiris, K., Van Colen, W., Wilches, I., León-Tamariz, F., De Cooman, L., & Muylaert, K. (2015). Impact of nutrient stress on antioxidant production in three species of microalgae. *Algal Research, 7,* 51–57.

Goncalves, A. L., Pires, J. C. M., & Simoes, M. (2016). The effects of light and temperature on microalgal growth and nutrient removal: An experimental and mathematical approach. *RSC Advances, 6,* 22896–22907.

Gonçalves, A. L., Pires, J. C. M., & Simões, M. (2017). A review on the use of microalgal consortia for wastewater treatment. *Algal Research, 24,* Part B, 403–415.

Gouveia, L., & Empis, J. (2003). Relative stabilities of microalgal carotenoids in microalgal extracts, biomass and fish feed: Effect of storage conditions. *Innovative Food Science & Emerging Technologies, 4,* 227–233.

Guštin, S., & Marinšek-Logar, R. (2011). Effect of pH, temperature and air flow rate on the continuous ammonia stripping of the anaerobic digestion effluent. *Process Safety and Environmental Protection, 89,* 61–66.

Han, S. F., Jin, W. B., Tu, R. J., Abomohra, A., & Wang, Z. H. (2016). Optimization of aeration for biodiesel production by *Scenedesmus obliquus* grown in municipal wastewater. *Bioprocess and Biosystems Engineering, 39,* 1073–1079.

Harun, R., Singh, M., Forde, G. M., & Danquah, M. K. (2010). Bioprocess engineering of microalgae to produce a variety of consumer products. *Renewable and Sustainable Energy Reviews, 14,* 1037–1047.

Hernandez, D., Riano, B., Coca, M., Solana, M., Bertucco, A., & Garcia-Gonzalez, M. C. (2016). Microalgae cultivation in high rate algal ponds using slaughterhouse wastewater for biofuel applications. *Chemical Engineering Journal, 285,* 449–458.

Hoffmann, J. P. (1998). Wastewater treatment with suspended and nonsuspended algae. *Journal of Phycology, 34,* 757–763.

Jia, H., Yuan, Q., & Rein, A. (2016). Removal of nitrogen from wastewater using microalgae and microalgae—Bacteria consortia. *Cogent Environmental Science, 2,* 1275089.

Kochen, L. H. (2010). *Caracterização de fotobioreator air-lift para cultivo de microalgas.* Universidade Federal do Rio Grande do Sul.

Kong, Q. X., Li, L., Martinez, B., Chen, P., & Ruan, R. (2010). Culture of microalgae chlamydomonas reinhardtii in wastewater for biomass feedstock production. *Applied Biochemistry and Biotechnology, 160,* 9–18.

Lam, M. K., Yusoff, M. I., Uemura, Y., Lim, J. W., Khoo, C. G., Lee, K. T., et al. (2017). Cultivation of *Chlorella vulgaris* using nutrients source from domestic wastewater for biodiesel production: Growth condition and kinetic studies. *Renewable Energy, 103,* 197–207.

Larsdotter, K. (2006). Wastewater treatment with microalgae—A literature review. *Vatten, 62,* 31–38.

Lau, P. S., Tam, N. F. Y., & Wong, Y. S. (1995). Effect of algal density on nutrient removal from primary settled wastewater. *Environmental Pollution, 89,* 59–66.

Lavoie, A., & De La Noüe, J. (1983). Harvesting microalgae with chitosan. *Journal of the World Mariculture Society, 14,* 685–694.

Lee, K., & Lee, C.-G. (2001). Effect of light/dark cycles on wastewater treatments by microalgae. *Biotechnology and Bioprocess Engineering, 6,* 194–199.

Lee, R. E. (2008). *Phycology*. United States of America: Cambridge University Press.

Li, Y., Chen, Y.-F., Chen, P., Min, M., Zhou, W., Martinez, B., et al. (2011). Characterization of a microalga *Chlorella* sp. well adapted to highly concentrated municipal wastewater for nutrient removal and biodiesel production. *Bioresource Technology, 102,* 5138–5144.

Li, Y., Hosman, M., Wu, N., Lan, C. Q., & Dubois-Calero, N. (2008). Biofuels from microalgae. *Biotechnology Progress, 24*(4), 815–820.

Lower, S. K. (1999). *Carbonate equilibria in natural waters* [Online]. Available: http://www.chem1.com/acad/pdf/c3carb.pdf. Accessed April 8, 2017.

Lutzu, G. A., Zhang, W., & Liu, T. Z. (2016). Feasibility of using brewery wastewater for biodiesel production and nutrient removal by *Scenedesmus dimorphus*. *Environmental Technology, 37,* 1568–1581.

Martin, J. H., Knauer, G. A., Karl, D. M., & Broenkow, W. W. (1987). VERTEX: Carbon cycling in the northeast Pacific. *Deep Sea Research Part A. Oceanographic Research Papers, 34,* 267–285.

Mccarty, P. L., Bae, J., & Kim, J. (2011). Domestic wastewater treatment as a net energy producer-can this be achieved? *Environmental Science & Technology, 45,* 7100–7106.

Minster, J.-F., & Boulahdid, M. (1987). Redfield ratios along isopycnal surfaces—A complementary study. *Deep Sea Research Part A. Oceanographic Research Papers, 34,* 1981–2003.

Molina Grima, E., Belarbi, E. H., Acién Fernández, F. G., Robles Medina, A., & Chisti, Y. (2003). Recovery of microalgal biomass and metabolites: Process options and economics. *Biotechnology Advances, 20,* 491–515.

Molina Grima, E., Fernández, F. G. A., García Camacho, F., & Chisti, Y. (1999). Photobioreactors: Light regime, mass transfer, and scaleup. *Journal of Biotechnology, 70,* 231–247.

Mostafa, S. S. M. (2012). Microalgal biotechnology: Prospects and applications. In N. K. Dhal & S. C. Sahu (Eds.), *Plant Science*. InTech: Rijeka.

Mulbry, W., Kondrad, S., Pizarro, C., & Kebede-Westhead, E. (2008). Treatment of dairy manure effluent using freshwater algae: Algal productivity and recovery of manure nutrients using pilot-scale algal turf scrubbers. *Bioresource Technology, 99,* 8137–8142.

Munir, N., Imtiaz, A., Sharif, N., & Naz, S. (2015). Optimization of growth conditions of different algal strains and determination of their lipid contents. *The Journal of Animal & Plant Sciences, 25*(2), 546–553.

Muñoz, R., & Guieysse, B. (2006). Algal–bacterial processes for the treatment of hazardous contaminants: A review. *Water Research, 40,* 2799–2815.

Nascimento, I. A., Cabanelas, I. T. D., Santos, J. N. D., Nascimento, M. A., Sousa, L., & Sansone, G. (2015). Biodiesel yields and fuel quality as criteria for algal-feedstock selection: Effects of CO_2-supplementation and nutrient levels in cultures. *Algal Research, 8,* 53–60.

Novoveská, L., Zapata, A. K. M., Zabolotney, J. B., Atwood, M. C., & Sundstrom, E. R. (2016). Optimizing microalgae cultivation and wastewater treatment in large-scale offshore photobioreactors. *Algal Research, 18,* 86–94.

Olguín, E. J. (2003). Phycoremediation: Key issues for cost-effective nutrient removal processes. *Biotechnology Advances, 22,* 81–91.

Oswald, W., & Gotass, H. (1957). Photosynthesis in sewage treatment. Transactions of the *American Society* of *Civil Engineers,* 73 (United States), (Medium: X; Size).

Oswald, W. J. (2003). My sixty years in applied algology. *Journal of Applied Phycology, 15,* 99–106.

Park, J. B., Craggs, R. J., & Shilton, A. N. (2011). Wastewater treatment high rate algal ponds for biofuel production. *Bioresource Technology, 102,* 35–42.

Pires, J., Alvim-Ferraz, M., Martins, F., & Simoes, M. (2013). Wastewater treatment to enhance the economic viability of microalgae culture. *Environmental Science and Pollution Research, 20,* 5096–5105.
Pires, J. C. M. (2017). COP21: The algae opportunity? *Renewable and Sustainable Energy Reviews, 79,* 867–877.
Pizarro, C., Mulbry, W., Blersch, D., & Kangas, P. (2006). An economic assessment of algal turf scrubber technology for treatment of dairy manure effluent. *Ecological Engineering, 26,* 321–327.
Prandini, J. M., Da Silva, M. L. B., Mezzari, M. P., Pirolli, M., Michelon, W., & Soares, H. M. (2016). Enhancement of nutrient removal from swine wastewater digestate coupled to biogas purification by microalgae Scenedesmus spp. *Bioresource Technology, 202,* 67–75.
Pulz, O. (2001). Photobioreactors: Production systems for phototrophic microorganisms. *Applied Microbiology and Biotechnology, 57,* 287–293.
Pushparaj, B., Pelosi, E., Tredici, M. R., Pinzani, E., & Materassi, R. (1997). As integrated culture system for outdoor production of microalgae and cyanobacteria. *Journal of Applied Phycology, 9,* 113–119.
Rai, L. C., Gaur, J. P., & Kumar, H. D. (1981). Phycology and heavy-metal pollution. *Biological Reviews, 56,* 99–151.
Rawat, I., Ranjith Kumar, R., Mutanda, T., & Bux, F. (2011). Dual role of microalgae: Phycoremediation of domestic wastewater and biomass production for sustainable biofuels production. *Applied Energy, 88,* 3411–3424.
Richmond, A. (2004). *Handbook of microalgal culture—Biotechnology and applied phycology.* USA: Blackwell Science.
Rocha, L. G. (2012). Dossiê técnico: Cultivo de Microalgas.
Safonova, E., Kvitko, K. V., Iankevitch, M. I., Surgko, L. F., Afti, I. A., & Reisser, W. (2004). Biotreatment of industrial wastewater by selected algal-bacterial consortia. *Engineering in Life Sciences, 4,* 347–353.
Santos, L., Calazans, N., Marinho, Y., Santos, A., Nascimento, R., Vasconcelos, R., Dantas, D. & Gálvez, A. (2009). *Influência do fotoperíodo no crescimento da Chlorella vulgaris (Chlorophyceae) visando produção de biodiesel.* http://www.eventosufrpe.com.br/jepex2009/cd/resumos/R0358-1.pdf.
Shaffer, G., Bendtsen, J., & Ulloa, O. (1999). Fractionation during remineralization of organic matter in the ocean. *Deep Sea Research Part I: Oceanographic Research Papers, 46,* 185–204.
Sheehan, J., Dunahay, T., Benemann, J., & Roessler, P. (1998). *Look back at the U.S. department of energy's aquatic species program: Biodiesel from Algae; Close-out report.* Golden, CO. (US): National Renewable Energy Lab.
Sicko-Goad, L., & Andresen, N. A. (1991). Effect of growth and light/dark cycles on diatom lipid content and composition. *Journal of Phycology, 27,* 710–718.
Silva-Benavides, A. M., & Torzillo, G. (2011). Nitrogen and phosphorus removal through laboratory batch cultures of microalga *Chlorella vulgaris* and cyanobacterium *Planktothrix isothrix* grown as monoalgal and as co-cultures. *Journal of Applied Phycology, 24,* 267–276.
Silva, N. F. P., Gonçalves, A. L., Moreira, F. C., Silva, T. F. C. V., Martins, F. G., Alvim-Ferraz, M. C. M., et al. (2015). Towards sustainable microalgal biomass production by phycoremediation of a synthetic wastewater: A kinetic study. *Algal Research, 11,* 350–358.
Singh, D., & Yadav, K. (2015). Biofixation of carbon dioxide using mixed culture of microalgae. *Indian Journal of Biotechnology, 14,* 228–232.
Spolaore, P., Joannis-Cassan, C., Duran, E., & Isambert, A. (2006). Commercial applications of microalgae. *Journal of Bioscience and Bioengineering, 101,* 87–96.
Sydney, E. B., Da Silva, T. E., Tokarski, A., Novak, A. C., De Carvalho, J. C., Woiciecohwski, A. L., et al. (2011). Screening of microalgae with potential for biodiesel production and nutrient removal from treated domestic sewage. *Applied Energy, 88,* 3291–3294.
Takahashi, T., Broecker, W. S., & Langer, S. (1985). Redfield ratio based on chemical data from isopycnal surfaces. *Journal of Geophysical Research: Oceans, 90,* 6907–6924.

Tam, N. F. Y., & Wong, Y. S. (1989). Wastewater nutrient removal by *Chlorella pyrenoidosa* and Scenedesmus sp. *Environmental Pollution, 58,* 19–34.

Toro, J. E. (1989). The growth rate of two species of microalgae used in shellfish hatcheries cultured under two light regimes. *Aquaculture Research, 20,* 249–254.

Tredici, M. R. (2002). *Bioreactors, photo. Encyclopedia of Bioprocess Technology*. USA: Wiley.

Tripathi, R., Singh, J., & Thakur, I. S. (2015). Characterization of microalga *Scenedesmus* sp. ISTGA1 for potential CO_2 sequestration and biodiesel production. *Renewable Energy, 74,* 774–781.

Uduman, N., Qi, Y., Danquah, M. K., Forde, G. M., & Hoadley, A. (2010). Dewatering of microalgal cultures: A major bottleneck to algae-based fuels. *Journal of Renewable and Sustainable Energy, 2,* 012701.

USEPA. (2011). *Principles of design and operations of wastewater treatment pond systems for plant operators, engineers, and managers.*

Walter, A. (2011). *Estudo do processo biotecnológico para obtenção de ficocianina a partir da microalga Spirulina platensis sob diferentes condições de cultivo.* Pós-graduação, Universidade Federal do Paraná.

Wang, B., & Lan, C. Q. (2011). Biomass production and nitrogen and phosphorus removal by the green alga *Neochloris oleoabundans* in simulated wastewater and secondary municipal wastewater effluent. *Bioresource Technology, 102,* 5639–5644.

Wang, L., Min, M., Li, Y., Chen, P., Chen, Y., Liu, Y., et al. (2010). Cultivation of green algae *Chlorella* sp. in different wastewaters from municipal wastewater treatment plant. *Applied Biochemistry and Biotechnology, 162,* 1174–1186.

Wang, L., & Nancollas, G. H. (2008). Calcium orthophosphates: Crystallization and dissolution. *Chemical Reviews, 108,* 4628–4669.

Yaakob, Z., & Fakir, K. (2011). *An overview of microalgae as a wastewater treatment.* Jordan International Energy Conference, Amman.

Yau, C. C. (2016). *Tecnologias dos processos de lamas ativadas. University of Porto - Faculty of Engineering.*

Chapter 4
Process Intensification of Biofuel Production from Microalgae

Saurabh Joshi and Parag Gogate

Abstract A tremendous increase in population has also led to a significant increase in the demand for energy leading to search for alternatives which can match up with the current requirement quantitatively and also qualitatively as a green energy carrier. Fuels derived from algal biomass can be one of the potential alternatives, as microalgae possess higher nutrients, required lipids and CO_2 uptake capacity and can be grown quickly on nonarable land throughout the year without their interference in food supply chain. The quantum of biodiesel produced from microalgae can be about 10–20 times higher than that obtained from terrestrial plants. Microalgae also help in reducing global warming by capturing CO_2. The cost of production of biofuels from microalgae is the current setback which can be overcome by taking into consideration a biorefinery approach which can give multiple products with same expenditure as well as using some process intensification approaches. Process intensification plays a major role in reducing the cost and also can lead to use of less quantum of materials and lower operating temperatures. The present chapter will focus on analyzing the process intensification aspects applied to biofuels production from microalgae. The initial sections will cover the details of the types of microalgae and their harvesting techniques, followed by the discussion on the different approaches used to extract bio-oil from microalgae, and then the production of different biofuels. Intensification can be applied to both the extraction and the actual reaction for production of biofuels. The chapter will also focus on the mechanism of intensification using different approaches such as ultrasound, microwave, ultraviolet, and oscillatory baffled reactors. An overview of the literature will be presented so as to give guidelines about the possible reactor designs and operating parameters also highlighting the process intensification benefits that can be obtained. Overall, the work is expected to bring out critical analysis of the different approaches and the expected benefits due to the use of process intensification also enabling understanding of the reactor designs and operating parameters.

S. Joshi · P. Gogate (✉)
Chemical Engineering Department, Institute of Chemical Technology,
Matunga, Mumbai 40019, India
e-mail: pr.gogate@ictmumbai.edu.in

E. Jacob-Lopes et al. (eds.), *Energy from Microalgae*, Green Energy and Technology, https://doi.org/10.1007/978-3-319-69093-3_4

Keywords Microalgae · Process intensification · Biofuels · Biorefinery
Reactor designs

1 Introduction

1.1 Need of Biofuels

Greenhouse gases (GHG) are mainly produced by the transportation and energy-producing sectors. Along with GHG, other pollutants like SO_x, NO_x, CO, volatile compounds, and particulate matter are also released into the atmosphere. Day by day, the global energy consumption is increasing, also resulting in an increase in pollution which further raises the concern of global warming. To cope up with the energy requirements and at the same time reduce the pollution, development of sustainable alternative energy sources has become the major goal. Many countries are working on utilizing different alternatives like solar energy, geothermal, wind, hydroelectric, thermal or photovoltaic, and biofuels. Every alternative generally comes with its own pros and cons, and the development of optimum and feasible alternative with time is the desired solution. Among the biofuels, Second-generation biofuels (biodiesel, bioethanol, and biogas) offer important alternatives and can be produced from sustainable resources available, with lesser or practically no emissions on their combustion. Biodiesel can be produced from non-edible oils, waste cooking oil, waste grease, or animal fats, whereas bioethanol and biogas can be produced from agricultural waste (wheat straw, corn cobs, etc.) and other sustainable materials. The availability of these materials, expensive processing, and production cost cannot fulfill the current supply and demand of energy requirements in a most efficient manner. Biofuel production from microalgae which comes under third-generation biofuels has now become a significant research area. Advantages like easy cultivation, non-competitiveness with food supply chain, higher lipid content, and less processing are obtained based on the use of microalgae which help in overall reduction of biofuel production cost.

1.2 Microalgae

Microalgae are unicellular microscopic organisms found in both marine and freshwater environment. They perform photosynthesis with efficiency higher than that of crops and consist of various components which can be utilized for many commercial purposes such as in the food, cosmetic, and high-value specialty molecules industry. Capability to naturally produce many unusual and different fats, bioactive compounds, sugars, etc., comes from their diversified genetic group which also comes with different physiological and biological characteristics. Microalgae mainly consist of proteins, carbohydrates, fats, and nucleic acids which directs the utilization of microalgae in different ways (Fig. 1). These components vary according to the species observed in different areas depending on the surrounding conditions like temperature, nutrients, pH, and light intensity.

Microalgae production offers advantages like high rates of production, and less doubling time as compared to plants and other biomass feedstocks and can help in utilizing the non-arable land with possible cultivation using the saline or waste water. It has the ability to sustain in environments having nutrient limitations and varying pH. Actually, under specific stress conditions, it produces high levels of lipids which can be further converted to biofuels efficiently. Currently, the cost of cultivating and harvesting microalgae is a setback which requires a greater investment as compared to other options available. Study on microalgae production approaches is required on higher scale as they may consist of untapped information which can be utilized for further good of mankind, though this is not the focus of the current chapter.

Depending on the metabolism, microalgae can be classified into four groups, that is, photoautotrophic, heterotrophic, photoheterotrophic, and mixotrophic. Microalgae can also be differentiated based on the source of cultivation such as freshwater or marine water. Freshwater algae are found to be grown on rocks under water and in mud of streams and river but the growth observed is more in still water than in flowing water. *Chlorophyta* (green algae), *Rhodophyta* (red algae), and *Bacillariophyta* (diatoms) are the examples of freshwater algae. The main problem is, however, the contamination of freshwater caused due to algae growth. Marine algae cultivation can help in boosting the economics involved in biomass

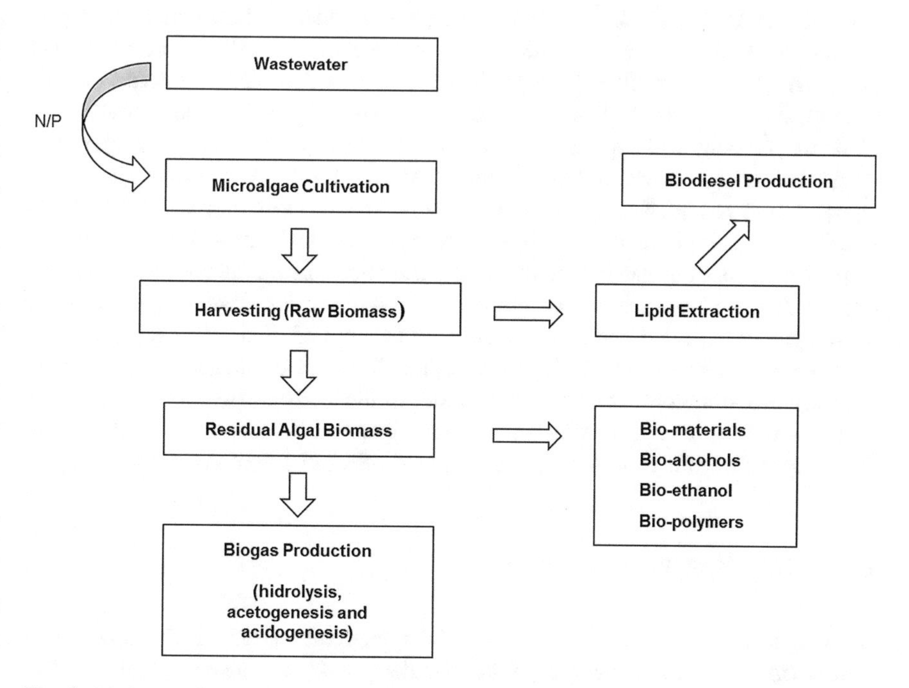

Fig. 1 Biofuels and other products which can be obtained by processing of microalgae

cultivation as they can be grown in brackish water, near coastal areas, floating on sea water, or in salt marshes. This also comes with some problems like effects on natural marine flora; premature rupture of microalgae cells due to high salinity of water and requirement of pretreatment of marine water, which adversely affects the economic feasibility. Overall, there is a need of deeper insight on the cultivation source and production approaches to be used for microalgae including the possible use of wastewater for microalgae cultivation which leads to solving of both the issues, that is, treatment of wastewater and cultivation source.

1.2.1 Lipid Content in Microalgae

Higher level of lipid content is an important parameter for utilization of microalgae. Few microalgae like *Botryococcus braunii* and *Chlorella emersonii* are naturally capable to produce up to 75% of lipid content (g lipids/dry weight). *Chlorella vulgaris and Dunaliella* sp. can reach up to 50% productivity under normal conditions. Lipid content in most of the microalgae species is generally between 20 and 50%. Profile of fatty acids also has a positive impact on biodiesel production (Priyadarshani and Rath 2012). Lipid profile is typically species-specific. Process improvement approaches can be efficiently applied to maintain desired specific conditions for microalgal growth (Patel et al. 2016). Growth parameters like nutrient availability, environmental factors, and cultivation type have a significant effect on microalgae lipid content. It has been reported that lipid production can be induced by nutrient-specific stress, for example, nitrogen starvation causes higher lipid production (Rodolfi et al. 2009). Similarly, phosphate content also has an effect on lipid productivity though it gives stronger increase in biomass content instead of lipid content (Xin et al. 2010). Salt stress can also have an impact on the production of lipids in microalgae as reported by Takagi et al. (2006). The microalgae grown in water with higher concentration of salts, that is, >1 M NaCl concentration were reported to have high lipid productivity as compared to those grown in 0.5 M NaCl solution. To get higher content of lipid is the main target which can be achieved with help of process optimization of required parameters. In the above cases, process improvement approaches can be helpful in identifying the desired conditions for microalgae growth and can increase the overall yield of lipids. The application of ultrasound as a process intensification approach can also enhance the growth of microalgae and increase the lipid production. In the study carried out by Han et al. (2016), it was reported that exposing the microalgae to different powers of ultrasound increased the overall yield and lipid content by 1.86 and 1.46 times, respectively.

2 Cultivation of Microalgae

Microalgae are cultivated using two main approaches based on the open pond system (raceway ponds, natural ponds, circular ponds, and inclined systems) and closed system (PBR-photobioreactor). Since 1950s, the open pond system has been used to cultivate the microalgae with the usage of natural water bodies like lakes, ponds or lagoons, and artificial water supply systems. Use of closed system prevents the contamination by other microbial species. Currently, many designs have been used for closed systems based on PBR such as flat plate, column or tubular and are classified on the basis of mode of operation and shape. Tubular and flat plate PBR are the most commonly used closed system PBR. The closed system comes with advantages but requires further detailed study on scale up, parameter control and cost, and currently, it is not considered economically feasible at large scale.

Process intensification can be applied to PBR considering aspects like carbon supply decoupling and mixing which will help significantly in ensuring proper supply of carbon dioxide and removal of oxygen. It has been reported that using a hollow fiber membrane can solve the problem of inefficient transfer up to an extent (Carvalho et al. 2006). Obtaining the desired increase in internal surface area and application of data on measurement, modeling, and control can also be a good process improvement approach. In one of the studies, concentrated microalgae cultivation in continuous mode was performed using resonant ultrasound field (RUF) which helped to enhance medium replacement process also resulting into process intensification benefits. The optimized process parameters reported were 1 MHz frequency and output intensity of 8 W/cm^2 with a circulating velocity of 2 mL/min leading to 93% collection of microalgae in 2 h (Lee and Li 2016). Pfaffinger et al. (2016) investigated the use of flat plate gas lift photobioreactor with continuous illumination using LEDs. The study showed an increase in algal productivity by 113% and lipid productivity by 59%. The design of the oscillatory baffled reactor was also utilized in developing the PBR which gave increased gas transfer and reported to improve overall economics of microalgae production (Abbott et al. 2015).

Other reactor configurations which have been also studied are rotating disk biofilm reactor and biofilm reactor which were reported to give a yield of 3.2 and 3.64 g/m^2/day, respectively, and also reported to help in overcoming the issues of suspended cultures (Sebestyen et al. 2016; Choudhary et al. 2017).

2.1 Cultivation of Microalgae from Wastewater

High content of nitrogen and phosphorous in wastewater makes it one of the best cultivation systems for microalgae. Total organic carbon of the wastewater can be utilized by some of the microalgal species as food source (Wang et al. 2010). Considering higher costs involved in microalgae production and wastewater treatment, it can be a boon if microalgae can be produced using wastewater as the cultivation medium. Algal ponds can be used for cultivation of microalgae using the municipal, industrial, and agricultural wastewaters. Secondary-treated wastewater contains nitrogen and phosphates in the range of 20–40 and 1–10 mg/L, respectively, which can help most microalgae strains to achieve high productivities (Olguín 2012). Microalgae release oxygen which in turn can be used by other microorganisms increasing the overall efficiency of aerobic degradation that can further decrease the BOD and COD of the wastewater, achieving the desired objective of wastewater treatment as well.

Several studies have been reported for reduction in nitrogen and phosphorous containing compounds coupled with biomass growth. Removal of nitrogen and phosphorus was reported using *C. vulgaris* with a removal efficiency of 72 and 28%, respectively (Aslan and Kapdan 2006). *Chlamydomonas polypyrenoideum* was used in a study of dairy wastewater treatment, and it was reported that nitrate level could be reduced by 90%, ammonia by 90%, phosphorus by 70%, and COD by 60% in 10 days (Lu et al. 2015). *Chlorella sorokiniana* when used for treatment of alcohol distillery wastewater in a 50 L PBR could decrease the nitrate content by 95%, phosphate by 77%, and sulfate by 35% in a time period of 3 days (Solovchenko et al. 2014). In a study performed by Li et al. (2011a), it has been reported that using bench scale continuous cultures, 0.92 g/L/d productivity of Chlorella strain was achieved using wastewater rich in ammonium, phosphorus, and organic matter with a COD of 1300 mg/L. Emerging contaminants (EC) can also be treated by microalgae to some extent as compared to other commonly available biological treatment. Microalgae can treat emerging contaminants in sequence of pharmaceuticals > PCPs (personal care products) > EDCs (endocrine disruption chemicals) > pesticides (Ahmed et al. 2017). Microalgae can also be used in the removal of heavy metals and can be employed based on the detoxification and biosorption techniques (Suresh Kumar et al. 2015). It can be clearly concluded from the studies mentioned above that cultivation of microalgae from wastewater not only reduces the pollution caused but also provides a rich sustainable feedstock in the form of algal growth which can be further utilized in biofuel production.

3 Harvesting

Process of harvesting consists of separation of biomass from the medium used for cultivation of microalgae. It is basically a separation process which separates microalgae biomass from cultivation medium. It is important that the process is a cost-effective one as it makes to about 20–30% of the total cost required for the whole process. Filtration, centrifugation, flocculation and floatation, gravity sedimentation, etc., are the techniques mostly used for this operation. The exact method is selected based on the cell size, cell density, and total quantity of the product to be separated. New techniques of harvesting and application of process intensification have also been reported based on techniques like flocculation assisted by the use of magnetic microparticles (Vergini et al. 2016), magnetic membrane filtration (Bilad et al. 2013), sedimentation assisted by the use of polymers (Zheng et al. 2015), electrical methods like electro-coagulation-filtration (ECF) (Gao et al. 2010) and electrochemical harvesting (ECH) (Misra et al. 2015). Low-frequency ultrasound can also be applied to the grown microalgal cells which results in decrease in the buoyancy and increases the sedimentation of the cells resulting in 90–92% as the harvesting efficiency (Kim et al. 2013).

4 Recovery of lipids and other products

Generally, the process of drying followed by disruption and solvent extraction is used for the recovery of desired products including the lipids from microalgae. Drying can be performed based on sun drying, spray drying, drum drying, and freeze drying. Sun drying is the most affordable option and can be employed effectively in biofuels production while spray drying hampers the overall economics of process when used for biofuels or protein production. Drum drying and freeze drying are also not the most viable options considering the application of biofuel production. After drying process, the dried biomass is subjected to disruption depending on the nature of desired product to be recovered or cell wall strength of microalgae, which affects the recovery. Disruption is carried out by mechanical processes (bead mills, autoclave, cell homogenizer, spray drying, ultrasound, etc.) and non-mechanical processes (using organic solvents, freezing, acid and osmotic shock, alkali, and enzyme treatment).

Microwave and ultrasound are emerging potential technologies which can be employed in the cell disruption process, also giving process intensification benefits. Application of Process intensification approaches helps this process to be performed with the achievement of economic feasibility. It has been reported that microwave and ultrasound are the technologies which result in higher amount of disruption as compared to other available technologies (Prabakaran and Ravindran 2011). Table 1 illustrates a few examples in which the ultrasound has been employed as a effective process for disruption of biomass.

Table 1 Ultrasound use for disruption of microalgae

Specie	Biomass concentration (g/L)	Frequency (kHz)	Time (min)	Yield of lipid	Reference
Chlorella sp.	5	50	15	156.6 mg/L	Prabakaran and Ravindran (2011)
C. vulgaris	5	10	5	6.1–8.8 mg/L	Lee et al. (2010)
C. vulgaris	2.5	–	17.5	2.9 times increase in lipid content	Zheng et al. (2011)
N. oculata	5	20	30	Increase in oil recovery to 0.24% from 0.15%	Adam et al. (2012)

Solvents such as ethanol, hexane, or a mixture of hexane-ethanol are generally used for extraction of lipids to be used in biodiesel production. Sometimes methanol can be also used which can serve both the purpose of extraction and as a reactant in the subsequent transesterification reaction. The extraction process is limited by mass transfer and hence the use of process intensification approaches can be very beneficial. Patil et al. (2011a, b) performed microwave-assisted direct transesterification of the microalgae using methanol and reported yield of up to 77% with process optimization. Wahlen et al. (2011) studied the comparison of the use of wet and dry algae biomass in direct transesterification and also the effect of water content on FAME yield. It was reported that 30 mg FAME can be produced from 100 mg of wet algae sample via this process of direct transesterification as compared to the conventional process which gave only 27 mg FAME from 100 mg. Study also concluded that the wet biomass of algae can be effectively utilized for biodiesel production based on nullifying the effects of water content by the addition of higher amount of methanol, also giving advantages of elimination of processing step.

Super critical extraction is also one of the techniques efficiently used for intensified extraction of lipids from microalgae. Many process intensification benefits have been reported with the use of super critical conditions as mentioned in Table 2. Similarly, ultrasound has also been applied in few studies to give intensified extraction. Ferreira et al. (2016) performed a study using low-frequency ultrasound with pure solvent (n-hexane and ethanol) and binary mixture of solvents (chloroform:ethanol; chloroform:isopropanol; chloroform:methanol; n-hexane: methanol; nhexane: isopropanol; n-hexane:ethanol, and n-hexane:2-butanol). It was reported that the frequency of 50/60 kHz with binary solvent of nhexane: isopropanol in 2:1 ratio was the most effective. It was also reported that the energy requirements were lesser as compared to conventional Soxhlet extraction and super critical extraction (SRE).

Table 2 Different super critical extraction (SRE) processes for microalgae processing with process parameters and yields (Lee et al. 2014)

Specie	Solvent/co-solvent	Temperature (°C)/ pressure (MPa)	Time (min)	Yield (%)
Chlorella vulgaris	Ethanol (6.6 ethanol/solids mass ratio) H_2O (10.1 wt%)	325	120	100
Chlorella vulgaris	Methanol (4 mL/g) H_2O (80 wt%)	175/2.2	240	89.7
Nannochloropsis (CCMP1776)	Methanol (9.0 mL/g) H_2O (ratio not mentioned)	255/8.27	25	84.1
Nannochloropsis salina	Ethanol (9 mL/g) H_2O (60 wt%)	265/8.27–9.30	20	67
Nannochloropsis salina	Ethanol (9 mL/g) H_2O (ratio not mentioned)	260/8.0	25	30.9
Chlorella protothecoids	Ethanol (20:1 ethanol/fatty acid molar ratio)	275/20.0	180	89

5 Process Intensification Strategies for Biofuel Production from Microalgae

Depletion of current fossil fuel reserves and pollution due to emissions from the usage of these fossil fuels have created a situation where there is a need to focus on fuels which can be produced in an easy manner and cause lesser emissions on use. Biofuels like biodiesel, bioethanol, and biogas are the alternatives which can replace the conventional fossil fuels. These biofuels can be produced from microalgae which are a rich source of biomass and can be utilized in a sustainable manner as it can be grown without the competition to food chain and has the energy content higher than that of the other biomass sources available. Biodiesel is produced via transesterification reaction of oil with a methanol to yield fatty acid methyl esters (FAME). Bioethanol is mostly produced by anaerobic fermentation of sources which are rich in sugars and starch using yeast as the microbial culture. Biogas is also obtained via anaerobic fermentation based on the use of methanogenic culture to utilize the biodegradable content available in feedstock. Due to high lipid content, microalgae can be effectively utilized for biodiesel production though their ability to accumulate starch and cellulose also make them suitable to be used for bioethanol and biogas production (Gendy and El-Temtamy 2013). Biodiesel is indeed the most common biofuel produced from microalgae as observed in open literature though some work has also been carried out to produce bioethanol and biogas from microalgae.

5.1 *Biodiesel*

Biodiesel is mostly produced from virgin vegetable oils, waste cooking oil, animal fats, and non-edible oils. The main advantage of biodiesel is that the physical properties are same as that of diesel obtained from crude oil, and hence, it can be used directly in diesel engine. The raw material which is selected for the production contributes a major proportion in the overall production cost as it depends on different factors like ease of availability, actual cost, and characteristics of oil. Thus, selection of the material plays a major role in process economics. The reaction involved in biodiesel production is the reaction of oil (triglycerides) with methanol in the presence of catalyst, and the product produced is fatty acid methyl ester (FAME) which is commonly called as the biodiesel.

Biodiesel can be produced by catalytic, non-catalytic, and in situ transesterification reactions. Catalytic process involves the use of homogenous, heterogeneous, and enzymatic catalyst. Non-catalytic process involves the use of methanol at critical temperature with possible simultaneous extraction and transesterification process. In situ transesterification is a process similar to the non-catalytic process performed at higher temperature and pressure and offers advantages as minimal usage of solvents, easy separation of products, and lesser reaction time. We now present an overview of important production approaches as catalytic (both homogeneous and heterogeneous) and in-situ transesterification.

5.1.1 Catalytic Homogenous Transesterification

Base Catalyst

In transesterification reaction, base catalysts are mostly used as they are cheaply available and allow the usage of moderate reaction temperature and pressure which helps in carrying out the process with favorable conditions. Base catalysts also give higher yield in shorter period of time as compared to other catalysts (Schuchardt et al. 1998). Bases such as KOH, CH_3ONa, NaOH, and others are reported to catalyze the reaction via deprotonating the alcohol to produce active RO^- species which further react with the carbonyl group and get converted into final transesterified product ($RCOOR^I$). The presence of free fatty acids in the feedstock is a hindrance for this process as it leads to soap formation due to the reaction of hydroxide groups of alkali catalyst and free fatty acid groups. Many studies of two step processing have been reported where the acid value of the oil has been reduced by esterification step initially and then the processed oil further utilized in transesterification step (Joshi et al. 2017). The requirement of two steps makes the overall cost of production much higher. Also it is rather difficult to develop a commercial process which will effectively separate the glycerol from FAME produced especially in the presence of soap, which can be formed based on the free fatty acid content. Handling of chemical waste generated from neutralization of base catalyst is also a major problem.

Acid Catalyst

AAcid catalysts find less application as compared with the base catalysts due to their slower reaction rates. They are used mostly with feedstocks which have a high free fatty acid content as they catalyze the reaction of esterification and transesterification simultaneously as well as does not give processing problems in terms of soap formation. Study was reported with mixotrophic approach first to increase the lipid content in the microalgae (*C. protothecoides*) and further sulfuric acid was used as a catalyst in acidic transesterification reaction performed with methanol in excess at 56:1 molar ratio (Miao and Wu 2006). Study related to the comparison of the use of H_2SO_4 with HCl in transesterification reaction established that HCl gave 10% higher yield as compared to $_2SO_4$ (Kim et al. 2015). Commercial application of the use of acid catalyst is not economically feasible as it leads to generation of waste and higher temperature and pressure are required for the reaction and also the slower reaction rate, which leads to higher energy consumption. The longer reaction time with high temperature may also lead to corrosion of reactor due to the prolonged use of acidic conditions.

Enzymatic Catalyst

Enzyme-based transesterification process is an attractive alternative to the chemical catalysts. Enzymes can work under mild reaction conditions with low temperature and pressure requirement and can also tolerate the FFA and water in reaction mixture. Typically the important operating parameters as the pH of the reaction, concentration of enzymes, and substrates, and the interactive distance between substrate and enzyme plays crucial role in deciding the rates of reactions carried out using enzymes. Enzymes can be denatured and destabilized by excess methanol and glycerol present in the reaction mixture. Also the prices of enzymes are higher which hamper the process economics (Suali and Sarbatly 2012). To overcome the above-mentioned issues, the enzymes can be used after they are immobilized transforming the system into heterogeneous. Immobilization is carried out via adsorption, encapsulation, entrapment, and cross-linking. Adsorption is the oldest and most commonly used method as it is less expensive as compared to other available methods. Many studies have been reported for the use of immobilized lipase in transesterification reaction for production of biodiesel (Subhedar et al. 2015). The advantage of reusability also makes immobilized form a more feasible option as compared to the free form, though the mass transfer limitations need to be looked at due to the heterogeneous nature of system.

5.1.2 Catalytic Heterogeneous Transesterification

The heterogeneous solid catalysts are environmentally friendly as they are easily separable from reaction mixture and hence reusable. Easy separation of catalyst from reaction mixture with simple filtration helps in improving the process economics. Solid catalysts are further classified as solid acid catalyst and solid base catalyst. Solid acid catalyst includes resins, polyaniline sulfate, zeolite tungstated and sulfated zirconia, sulfated tin oxide, heteropolyacid, metal complexes, and acidic ionic liquids. Solid base catalyst includes calcium oxide, hydrotalcite (also called layered double hydroxide), zeolites, and alumina. Yield of 97.5% was reported for the biodiesel production from lipids extracted from *Nannochloropsis oculata* when Al_2O_3-supported CaO and MgO were used as catalysts under processing conditions of excess of methanol (1:30) and catalyst loading (80% w/w) required for the completion of reaction (Umdu et al. 2009). Use of Mg–Zr was proposed in one of the study performed where in situ and two-step processes were compared and single-step approach was found to be more efficient. Reaction was performed with a mixture of methanol and methylene dichloride in ratio of 3:1 with 10% w/v catalyst at 65 °C for 4 h and a yield of 28% of methyl esters was reported (Li et al. 2011b). A study on utilization of hierarchal zeolites in transesterification was performed to establish the specific form of zeolite that can yield highest conversion rate. From the study, H-beta zeolite was established to give higher conversion rates as compared to other zeolites (Carrero et al. 2011). Study was performed using KOH/La–Ba–Al_2O_3 as the heterogeneous catalyst for conversion of microalgal lipids at 60 °C for 3 h with catalyst loading of 25% and the yield of biodiesel reported was 97.7% (Zhang et al. 2012). Syazwani et al. (2015) performed a study using CaO catalyst synthesized from angel wing shells in transesterification of N. oculata lipids. Yield of 84.1% was reported with 9% catalyst concentration and 1:150 molar ratio of lipids to methanol in a period of 1 h. Leaching of the heterogeneous catalyst into the final biodiesel product can be one of the concerns related to the use of these catalysts. There are many heterogeneous catalysts used for transesterification of edible oil and non-edible oil, but more thorough research needs to be performed for their application on microalgal lipids, as only limited information was observed in the literature for the algal lipids.

5.1.3 In situ Transesterification

In situ transesterification is the process where the extraction and transesterification reaction are carried out simultaneously. It has an advantage over conventional process as only a single step is required instead of two seperate steps of extraction and reaction. This approach of combination leads to intensification as it requires minimal amount of solvent, lesser reaction time, and easy separation of the products. The state of biomass is crucial in this approach as more amount of biodiesel is produced from dry biomass as compared to wet dry biomass.

Mechanically Catalyzed In situ Transesterification

The mechanically catalyzed in situ transesterification involves the use of mechanical processes based on the use of microwave (MW), ultrasound (US), and autoclave. These processes help in improving the surface area and local temperature of mixture leading to increased penetration of solvents to cells which further helps in enhanced extraction of lipids from microalgae. Microwave-assisted direct transesterification study was performed with dried *Nannochloropsis* and yield of 80.1% was reported under processing conditions of 1:12 (w/v) ratio of algae to methanol, 2% by weight KOH loading, and reaction time of 2–4 min at 60–65 °C (Patil et al. 2011a). Another study reported that with use of ultrasound, 91–96% yield was obtained in 20 min–2 h time with 1:105 to 1:315 algae to methanol molar ratio. The reaction time required for US is typically more as compared to MW (Ehimen et al. 2012), though the scale up prospects for MW need to be carefully evaluated.

Chemically Catalyzed In situ Transesterification

The chemically catalyzed reaction involves no use of mechanical energy. An important precondition of the chemical catalyst-based approach is that the process requires the use of dried biomass. Feedstock containing water more than 31.7% exhibit inhibition to transesterification reaction (Ehimen et al. 2010). These reactions are mostly performed using co-solvent and ionic liquids. Co-solvents increase the efficiency of lipid extraction and increase the overall yield. In one of the study, hexane was used as the co-solvent and was supplemented with sulfuric acid and methanol. It was reported that the yield of biodiesel increased from 16.6 to 94.5% with an increase in hexane supplementation from 2 to 10 mL (Sangaletti-Gerhard et al. 2015). Comparison of chloroform and hexane as co-solvent established that chloroform increases the yield of biodiesel more than hexane (Kim et al. 2015). The use of co-solvent not only increases the yield but also reduces the requirement of methanol, facilitating the downstream processing. It has been reported in a study that the use of diethyl ether as a co-solvent reduced the requirement of methanol from 105:1 to 79:1 (Ehimen et al. 2012). Study to evaluate the transesterification reaction with different co-solvents (petroleum ether, chloroform, n-hexane, ethyl ether, carbon tetrachloride, n-butanol, and acetone) established that the highest ester yields were obtained with use of petroleum ether, chloroform, and n-hexane. The yield increased from 48.3% to above 90% when a co-solvent was used with ethanol clearly confirming the role of co-solvent (Zhang et al. 2015). Research has been also focused toward development of green solvents which will eliminate the harmful effects of conventional solvents (Jeevan Kumar et al. 2017). In recent years, the ionic liquids (salts in liquid form) have been utilized in biodiesel production. Ionic liquids come with advantages like high solubility, inherent basicity or acidity, negligible vapor pressure, and are recyclable. They also possess the ability to immobilize the catalysts (acid/basic), and this makes them easily separable and recyclable. Cost of the ionic liquids is the major drawback currently restricting their application in biodiesel production process especially considering the requirement at large scale. There are very few studies reported on the use of ionic liquids, and more research needs to be carried out for their application in biodiesel production from microalgae with a focus on reducing the requirement and maximizing the reuse during the processing.

5.2 *Bioethanol*

The biofuel which accounts for a significant fraction of the total production is bioethanol. Majority of it is produced from sugarcane and the remaining comes from other crops. Bioethanol from biomass is produced via fermentation or gasification process, and the availability of the feedstock depends upon the season and geographical conditions. Microalgae can be one of the potential feedstock for bioethanol production as they are able to produce starch and cellulose and also do not compete with the food crops for land and water. The production of bioethanol from microalgae can be feasible on industrial scale when the method applied for hydrolysis is easy to handle, cost-effective, energy efficient, and maximum yield of reducing sugars is obtained. The absence of lignin makes the saccharification process easier and reduces the overall cost. Starch is stored by the microalgae inside the cells, and these cells can be separated periodically from photobioreactors and raceway ponds. Biomass harvested can be further disrupted, and starch extraction can be carried out via water or an organic solvent. Acids (concentrated and diluted) are mostly used for the disruption of the biomass. Zhou et al. (2011) reported that addition of 2.5% $MgCl_2$ in 2% HCl resulted in effective disruption and subsequent hydrolysis of the algal biomass, and 83% of the total sugars consisting of xylose, glucose, and arabinose were recovered via this process. Starch can also be saccharified using enzymes such as alpha amylase and gluco-amylase. Large amount of starch and glycogen have been reported to be present in microalgae like *Chlorella, Chlamydomonas, Dunaliella, Spirulina,* and *Scenedesmus* which can be processed for bioethanol production. The starch can be converted into ethanol with the step of anaerobic fermentation and pretreatment can be typically used to maximize the formation of sugars in first step and then ethanol in the second step. Study was performed using microalgal strains *M. afer* and *S. abundans* for bioethanol production, and it was reported from the study that dilute acid and cellulase-treated *S. abundans* was better feedstock yielding 0.103 g of ethanol per g of dry weight of microalgae. The process was optimized for sulfuric acid pretreatment, and 52% higher yield of ethanol was obtained with 10 mg/L microalgae using 3% v/v sulfuric acid treatment at 160 °C for 15 min (Guo et al. 2013).

The utilization of residual lipid extracted algae (LEA) for bioethanol production is also gaining attention in recent years. *Chlorococum* sp. was analyzed as a feedstock in a study to produce bioethanol. The lipid extraction was performed via supercritical method, and LEA was dried and further subjected to ethanol production giving a yield of 3.83 mg/L from 10 mg/L LEA (Harun et al. 2010). In another study, *C. vulgaris* FSP-E was reported to be used as biomass for bioethanol production with improvement based on pretreatment. Biomass was subjected to pretreatment using diluted acid and enzymes. It was reported that pretreatment with enzyme mixture of amylase/cellulase and dilute sulfuric acid were both effective techniques. The biomass was subjected to fermentation via SSF (simultaneous hydrolysis and fermentation) and SHF (separate hydrolysis and fermentation) processes. SHF process gave a higher ethanol yield of 11.66 mg/L as compared to SSF (Ho et al. 2013). El-Dalatony et al. (2016) performed a study on use of immobilized yeast and combination of sonication with enzymatic hydrolysis step. It was reported that sonication combined with hydrolysis gave higher yield of 445 mg/mg of total reducing sugars. Also it was reported that SSF gave higher ethanol yield compared to SHF, and energy recovery of the process was improved due to use of immobilized yeast cells. Regenerated beads exhibited fermentation efficiency of 79.8% for four cycles. The treatment of algal biomass with CaO before the process of hydrolysis can also help in giving overall increase in reducing sugar yield (Khan et al. 2017). In one of the studies, utilization of mixed microalgae culture has been reported for bioethanol production. The effects of different pre-treatment strategies (acidic, alkaline, and enzymatic) were also studied, and it was reported that dilute sulfuric acid with $MgSO_4$ gave higher yield of reducing sugars as compared to only dilute sulfuric acid. Among all the processes employed, enzymatic process was reported to give the highest yield of reducing sugars (Shokrkar et al. 2017). The analysis of literature reveals that many approaches are available to optimize the process and maximize the ethanol yield with better utilization of the resources. A well designed approach with optimization studies for specific system need to be developed to facilitate the commercial scale application.

There is also a possible solution of genetic modification in the microalgae which can induce the direct production of ethanol from lifecycle of microalgae. The functional genetic diversity of microalgae is very large and can be utilized in developing specialized strains to directly produce bioethanol. The activity of pyruvate decarboxylase (PDC) and alcohol dehydrogenase (ADH) enzymes in the microalgae needs to be increased which will convert the fixed carbon into bioethanol. To modify the microalgae genetically, it will require more focused research and time. Currently, genetic modifications have made possible to increase the carbohydrate accumulation in microalgae (Silva and Bertucco 2016), and hence it definitely offers as a possibility even for direct ethanol production.

5.3 *Biogas*

Biogas production is an anaerobic process in which a gas is generated by decomposition of organic materials with the help of specialized organisms. Biogas mainly consists of methane (55–75%) and carbon dioxide (25–45%) with other constituents like H_2, N_2, water vapor, and H_2S in minor fractions. The production process consists of stages like hydrolysis, acidogenesis, acetogenesis, and methanogenesis. Microalgae can also be a potential feedstock for biogas production, more promising than the utilization in other forms of biofuels due to the energy efficiency of the process for biogas. There is no requirement of lipid extraction process and the product, that is, biogas obtained in gaseous form does not require any separation. All the macromolecules present in the microalgae are typically utilized for the biogas fermentation process. The raw microalgae as well as the residuals from the other biofuels production process can be used in biogas production process. The factors affecting biogas production consist of retention time, organic loading, pH, temperature, quality of the substrates (characteristic of cell wall), pretreatment of substrate, and the presence of methanogenesis inhibitors (Jankowska et al. 2017). The digestibility of cell wall can be improved with the help of pretreatment which further increase the biogas yield and help in intensification of the process. The different pretreatment processes include mechanical (ultrasound, high pressure homogenization, and microwave), thermal, chemical (use of alkali, acids, and ionic liquids), and biological (enzymes). Ultrasound pretreatment can increase the methane yield by up to 91% (Park et al. 2013). Microwave irradiation also has an effect on the cell wall protein which results in the disruption of the cells leading to easy access to the cellular material. Irradiation of microalgae with MW has been reported to increase the production of biogas up to 79% (Passos et al. 2013). Microwave irradiation can be a efficient technique for pretreatment as the pretreatment time required is less but high energy requirements might be an issue when employed on large scale. Thermal pretreatment of microalgae cells is typically performed using autoclaves, heat chambers, or water bath. González-Fernández et al. (2012a) reported an increase in biogas yield by 123% with help of thermal pretreatment. Disadvantage of this method is it consumes large amount energy but the energy available after heating can be employed to maintain the temperature of reactor during anaerobic fermentation and hence some heat integration approaches can be thought of. Enzymes (mostly cellulase) can also be employed in biological pretreatment of microalgae as they are rich in cellulose. The lipid extraction efficiency can be increased by up to 56% with help of enzymes (Fu et al. 2010). Cost of enzymes is the major hindrance in the use of enzymes in pretreatment process. Acid and alkali pretreatment come under the category of chemical pretreatment which mostly uses sulfuric acid as the acid and sodium hydroxide as the alkali. With the chemical pretreatment, the biogas yield can be increased by threefold to fourfold (Jankowska et al. 2017). The summary of effects of different pretreatment processes on biogas production has been reported in Table 3. Combinations of pretreatment process like employing dilute acid pretreatment with microwave or ultrasound-assisted approach can further result in significant increase in biogas yield. Such combined processes also help in reduction of process cost and overcome the disadvantages of individual methods.

Table 3 Different methods employed for pretreatment of microalgae

Strain	Processing conditions	Method of pretreatment	Operating conditions	Biogas production		Reference
				Before pretreatment	After pretreatment	
Nannochloropsis salina	40 days (38 °C)	Freezing overnight	−15 °C, overnight	0.347 L biogas/g VS	0.233 L biogas/g VS	Schwede et al. (2011)
		Thermal	100 °C, 8 h		0.549 L biogas/g VS	
		Microwave	5 × until boiling at 600 W, 2450 MHz		0.487 L biogas/g VS	
		Ultrasound	3 × 45 s at 200 W, 30 kHz		0.274 L biogas/g VS	
		French Press	2 × 10 MPa		0.460 L biogas/g VS	
Spirulina maxima	60 days (35 °C)	Ultrasound	10 min (polytron generator)	0.19 L CH_4/g VS	0.17 L CH_4/g VS	Samson and Leduy (1983)
		Thermo-chemical	50 °C (1 h in water bath)		0.21 L CH_4/g VS	
			100 °C (1 h water bath)		0.22 L CH_4/g VS	
			150 °C (1 h in steam sterilizer)		0.24 L CH_4/g VS	
Scenedesmus sp.	34 days (35 °C)	Thermal	70 °C, 25 min	0.082 L CH_4/g COD	0.089 L CH_4/g COD	González-Fernández et al. (2012b)
			80 °C, 25 min		0.129 L CH_4/g COD	
		Ultrasound	130 MJ/kg, 30 min		0.154 L CH_4/g COD	

(continued)

Table 3 (continued)

Strain	Processing conditions	Method of pretreatment	Operating conditions	Biogas production		Reference
				Before pretreatment	After pretreatment	
Rhizoclonium	28 days (53 °C)	Size reduction (cutting)	5 cm, 1 cm	0.23–0.24 L CH_4/g TS	0.093–0.100 L CH_4/g TS	Ehimen et al. (2013)
		Size reduction (grinding)	<0.1 mm		0.100–0.113 L CH_4/g TS	
		Sonication	20 kHz, 10 min		0.113–0.127 L CH_4/g TS	
		Enzymes	Combination of enzymes (lipase, xylanase, α-amylase, protease, cellulase)		0.143 L CH_4/g TS	
Microalgal biomass	HRT 15 days (35 °C)	Microwave	900 W, t = 3 min, 110 °C, 200 kJ/kg VS	0.13 L CH_4/g VS	0.17 L CH_4/g VS	Passos et al. (2014)
	20 days (35 °C)			0.17 L CH_4/g VS	0.27 L CH_4/g VS	
Algae biomass	28 days (38 °C)	Thermal	100 °C, 8 h	0.34 g MSGP g/L	0.45 MSGP g/L	Chen and Oswald (1998)
Chlorella vulgaris	25 days (35 °C)	Ultrasound	200 J/mL	0.230 L CH_4/g VS	0.440 L CH_4/g VS	Park et al. (2013)
Scenedesmus obliquus, Chlorella vulgaris	46 days (35 °C)	Microwave	21,800 kJ/kg TS	0.172 L biogas/g VS	0.2120 L biogas/g VS	Passos et al. (2013)
			43,600 kJ/kg TS		0.245 L biogas/g VS	
			65,400 kJ/kg TS		0.307 L biogas/g VS	

(continued)

Table 3 (continued)

Strain	Processing conditions	Method of pretreatment	Operating conditions	Biogas production		Reference
				Before pretreatment	After pretreatment	
Isochrysis galbana	15 days (30 °C)	Mechanical	Stirring with 1 g of glass beads, 1 min	22 mL of biogas	12.7 mL of biogas	Santos et al. (2014)
		Chemical	40 °C, 0.2% v/v acid, t = 16 h		26 mL of biogas	
		Thermal	60 °C, 16 h		3.7 mL of biogas	
		Thermal	40 °C, 16 h		3.0 mL of biogas	
Scenedesmus	35 days (38 °C)	Lipid extraction	In hexane, Soxhlet apparatus, 6 h,	0.18 L CH_4/g VS	0.33 L CH_4/g VS	Keymer et al. (2013)
		High pressure thermal hydrolysis	170 °C, 800 kPa, 30 min		0.24 L CH_4/g VS	
		LE and HPTH			0.38 L CH_4/g VS	

MSGP methane specific gas production, *VS* volatile solids, *LE* Lipid extracted, *HPTH* high pressure thermal hydrolysis

6 Analysis of Reactor Configurations for Process Intensification

Biodiesel production from microalgae has been quite successful under laboratory scales and can be under serious consideration for commercialization. The biodiesel production can also be significantly improved based on the concept of process intensification that focuses on achieving shorter reaction time and high conversion with lower molar ratio of alcohol to oil and low catalyst concentration, also possibly giving lower operating cost and energy consumption for biodiesel purification with recovery of glycerol, catalyst, and excess alcohol. Reactor configurations which can be utilized for intensification of biodiesel production process are now discussed.

6.1 Cavitational Reactors

Application of cavitation in the field of biodiesel production has gained interest lately. Cavitation helps the reaction by providing mechanical energy for mixing and enhanced surface area for the transesterification reaction resulting in reduced reaction time and increased yield (Gogate and Pandit 2004). The main effects which are generated due to cavitation consist of (1) chemical effect which is produced due to generation of radicals (H^+, OH^-, and HO_2^+) from transient implosive collapse of the bubbles though this is not dominating in the case of biofuel production, (2) homogenization of the mixture, which is caused by micro-turbulence generated due to the collapse of bubbles. Due to the formation of fine emulsion, the interfacial region is increased between oil and alcohol which leads to increased reaction rate and high yield. There are mainly two types of cavitational reactors, ultrasonic (US) and hydrodynamic (HC). The ultrasonic reactors are operated in the frequency range of 20–40 kHz with lower range of power (120–220 W) (Gupta and Verma 2015), giving dominant physical effects controlling biodiesel production. Utilization of 40 kHz frequency has been reported to reduce the time required for reaction drastically (Stavarache et al. 2005). Ultrasound-assisted transesterification reactions are generally performed with reaction parameters as: molar ratio (1:6–1:10), catalyst loading (0.5–2 wt% of oil), and reaction time (15–20 min) with temperature over the range of 30 to 60 °C as observed in the literature (Gole and Gogate 2012). It is important to understand that most of the applications have been based on the use of ultrasonic horn and bath at the laboratory scale but application of ultrasound on continuous mode has not been reported. More research needs to be performed to utilize ultrasound effectively at commercial scale especially using continuous operation. Hydrodynamic cavitation (HC) produces similar effects to that of ultrasonic cavitation; only difference is in the method of generation of cavities. Cavity generation is due to sudden pressure drop with help of constriction introduced in the flow of the liquid. These reactors are generally more energy efficient and can work with large quantity reaction batch as compared to US and with similar reaction parameters as that of US but the reaction time required may be higher (45–60 min) than US (Ghayal et al. 2013). It is important to understand that the cavitational yield (amount of product per unit energy) of HC is also typically higher than US.

6.2 Microreactors

Miniature reaction systems have proven to provide sustainable and innovative solutions and have been utilized at both laboratory level and industrial level with good degree of intensification benefits. Intensified heat and mass transfer are achieved with these reactors as they have a small characteristic dimension and high surface-to-volume ratio offering proper temperature control. Immiscible liquid–liquid reactions can be carried out with higher efficiency with this reactor as it provides very high interfacial area between phases which further improves the rate of mass transfer (Kashid and Kiwi-Minsker 2009). Transesterification reaction consists of two immiscible reactants, that is, triglycerides and methanol. There are reports in which homogenous transesterification reaction have been performed using microreactor with significant reduction in reaction time (Mazubert et al. 2013; Wen et al. 2009). Typical operating conditions consist of methanol-to-oil molar ratio at 1:4–1:9 and catalyst loading in the range of 1–4.5% w/w of oil with a flow rate varied from 8 to 15 mL/h which can give yield of biodiesel up to 99% in very less reaction time of 1–6 min. More work is required to be performed to establish the design and scale up strategies for application of micro-reactors at the commercial scale of operation for the specific application of biofuel production from microalgae.

6.3 Microwave Reactor

Microwave reactors work on the principle of intensification based on the effects of dipolar polarization and ionic conduction. The dipolar polarization occurs when the alignment of the dipoles occurs in the direction of electric field imposed with help of microwave irradiation. Oscillation of the charged dissolved particles due to microwave results in the ionic conduction. Transesterification reaction performed using these reactors shows significant increase in reaction rate. Also it has been reported that intensification in the transesterification reaction is more sensitive for the use of methanol than ethanol due to low gyration radius and molecular inertia (Terigar et al. 2010). Intensification trends have been reported with different heterogeneous and homogenous transesterification reactions for the use of microwave reactors (Mazubert et al. 2013). It can be seen from the studies reported in the literature that the important reaction parameters are molar ratio (1:6–1:12), catalyst loading (0.15– 5 wt% of oil), temperature (40–60 °C), and power (300–1600 W) with required reaction time varying from 0.5 to 20 min.

6.4 Oscillatory Baffled Reactor

Oscillatory baffled reactor (OBR) consists of equally spaced orifice plate baffles arranged in a tube operating with an oscillatory or pulsed flow with generation of re-circulating flow pattern. This reactor provides enhanced mixing and inter-phase contact within sufficiently long residence time suitable for the reaction when employed in transesterification reaction. It can be established from the data available on transesterification reaction with OBR that the molar ratio in the range of 1:6–1:9 with flow rate of 0.12–3.12 L/h and residence time ranging from 20 to 40 min is able to yield 99% conversion in almost all the studies (Harvey et al. 2003; Zheng et al. 2007). OBR can also work with heterogeneous system and their scale up for effective operation at commercial scale is possible. It is important to understand that no direct work has been still reported for utilization of algal oil in biodiesel production with OBR, though the reported trends for other similar systems do induce a confidence for possible success.

6.5 Reactive Distillation

Reactive distillation (RD) combines the chemical reaction and product separation in a single unit. RD column boosts the conversion with improvement in selectivity by breaking the reaction equilibrium conditions (Estrada-Villagrana et al. 2006). The application of this reactor in biodiesel production can be very helpful as demonstrated by He et al. (2006). In the study, the canola oil and methanol feed were made to enter through an in-line static mixer into the RD column. Downward flow of reactant from the top of the RD column across the plate ensured efficient contact with vapors of methanol (produced in reboiler from product mixture) providing uniform mixing at each plate. Virtually, a series of "mini- reactors" were created in the reaction zone of RD column. Methanol from the distillate could be recycled and was combined with the feed methanol and then refluxed back to the RD column. This made the reactor to give 94.4% of yield with methanol: oil molar ratio of 4:1. From this study, it was also established that there is drastic decrease in quantity of methanol required and reaction time as compared to the conventional reactors. The requirement of extra unit operation required for recovery of solvent is also not present in RD giving lower capital costs. It is important to understand that not much work could be seen for the use of algal oil in the reactive distillation approach. More investigation needs to be carried out to employ this reactor for the algal oil with a detailed study on parameter optimization and also establish scale up strategies so that the commercial-scale biodiesel production from algal oil is a possibility.

6.6 *Centrifugal Contact Separator*

Centrifugal contact separator (CCS) employs the chemical reaction, and centrifugal separation in a single apparatus. Preheated oil is fed into the reactor, and the reaction is started by adding the methanol and the catalyst. The dispersion of the immiscible liquids takes place in the annular gap between the static housing and the rotating centrifuge. Further this mixture is transferred to the hollow centrifuge through a hole at bottom and separation into heavy and light layers take place via centrifugation. The optimum conditions reported by Kraai et al. (2009) for this system, though not for algal oil, were rotational frequency of 30 Hz, oil flow rate at 12.6 mL/min, sodium methoxide catalyst concentration of 1% w/w of oil, and reaction temperature of 75 °C. The FAME yield of 96% was reported in time period of 30 min. It was also reported that further increase in temperature and catalyst beyond optimum leads to excessive evaporation of methanol and soap formation which affects the overall reaction rates. The higher flow rates of oil also were reported to have a negative effect on the mean residence time of mixture lowering the yield of FAME. Again similar to the reactive distillation, more investigations are needed for the application of centrifugal contact separators for the specific feed stock of algal oil before firm conclusion can be made.

6.7 *Membrane Reactor*

In order to overcome the limitations of conventional biodiesel production processes, the development of membrane reactor can be a potential solution. Reaction and separation occur in a single chamber, and this ensures that the reversible reaction proceeds in the forward path with efficient removal of desired products from reaction mixture which leads to increase in yield (Cao et al. 2008, 2007; Dube et al. 2007). Membrane reactor works on the principle of utilizing the immiscibility of methanol with oil and miscibility of products (FAME and glycerol) in methanol. During the transesterification process, oil exists in the form of emulsion in methanol and reaction occurs at the surface of oil droplets. FAME produced via transesterification is soluble in methanol and is able to pass through the membrane with the by-product glycerol. The oil droplets being larger in size cannot pass through the membrane and remain in the reactor vessel. The simultaneous removal of product from a reversible reaction helps in the improvement of the reaction rates, and the permeate obtained is in pure form which requires less processing. Cao et al. (2008) have investigated the transesterification reaction using different feedstocks (soybean oil, canola oil, a hydrogenated palm oil/palm oil blend, yellow grease, and brown grease) having varied FFA presence. With efficient purification and separation process, membrane reactor was demonstrated to give high efficacy for different feedstocks making it a energy efficient, and environmental friendly reactor.

From the above-mentioned reactor configurations for process intensification, it can be said that the ultrasound and microwave reactors have been already employed for biodiesel production from microalgae as the feedstock and remaining need to be utilized via process intensification study to make process economically feasible. As mentioned in Sect. 5.2 and 5.3, microwave- and ultrasound-assisted processes have also been employed in enhanced bioethanol and biogas production from microalgae. It can be concluded that the study on US and MW is quite progressive as compared to other technologies available and more research needs to be performed for establishment of other available technologies in biodiesel production from microalgae as they might have a potential to overcome the disadvantages of US and MW, especially at large scale of operation.

7 Conclusions

Biofuels produced from microalgae can be considered as an effective alternative to petrochemical fuels but there are limited technologies available currently which can be commercially applied. Application of process intensification approaches at different stages of processing can give an energy efficient process with scope for commercialization as demonstrated in the current chapter. The techniques involved in harvesting as the very first stage of processing and subsequent lipid extraction need to be developed into efficient techniques based on process intensification to achieve economic feasibility of process. Innovative solutions are also required to build strategies for the subsequent reactions and separations which will give a possible solution maintaining the positive aspects of current methods and remove the undesired ones which will make the process costeffective and give positive energy gains. The development of microalgae biorefinery can be a feasible solution as high-value products which can be beneficial to the cosmetics, pharmaceutical, and nutritional industries remain largely unexplored, and this will essentially shift the current focus from only biofuels production to diversification of the other products with biofuels. The processes developed must be applicable to the microalgal species which are available commonly and should be easily transformed into continuous mode which can be applicable on commercial scale. Process intensification can help to improve the working of current processes making them efficient in aspects of time and energy. It has been established from the research articles available that biodiesel from microalgae is more feasible as compared to bioethanol and biogas. Biogas and bioethanol production from microalgae can also be improved via process intensification techniques like ultrasound and microwave with benefits as lower times, lower requirement of reactants, and lower temperature. Overall, it can be concluded that microalgae can be a potential feedstock for production of biofuels (biodiesel, bioethanol, and biogas) at commercial scale and process intensification aspects can be integrated to give production at lower cost and energy.

References

Abbott, M. S. R., Brain, C. M., Harvey, A. P., Morrison, M. I., & Valente, G. (2015). Liquid culture of microalgae in a photobioreactor (PBR) based on oscillatory baffled reactor (OBR) technology—A feasibility study. *Chemical Engineering Science, 138,* 315–323.

Adam, F., Abert-Vian, M., Peltier, G., & Chemat, F. (2012). "Solvent-free" ultrasound-assisted extraction of lipids from fresh microalgae cells: A green, clean and scalable process. *Bioresource Technology, 114,* 457–465.

Ahmed, M. B., Zhou, J. L., Ngo, H. H., Guo, W., Thomaidis, N. S., & Xu, J. (2017). Progress in the biological and chemical treatment technologies for emerging contaminant removal from wastewater: A critical review. *Journal of Hazardous Materials, 323,* 274–298.

Aslan, S., & Kapdan, I. K. (2006). Batch kinetics of nitrogen and phosphorus removal from synthetic wastewater by algae. *Ecological Engineering, 28,* 64–70.

Bilad, M. R., Discart, V., Vandamme, D., Foubert, I., Muylaert, K., & Vankelecom, I. F. J. (2013). Harvesting microalgal biomass using a magnetically induced membrane vibration (MMV) system: Filtration performance and energy consumption. *Bioresource Technology, 138,* 329–338.

Cao, P., Dubé, M. A., & Tremblay, A. Y. (2008). Methanol recycling in the production of biodiesel in a membrane reactor. *Fuel, 87,* 825–833.

Cao, P., Tremblay, A. Y., Dubé, M. A., & Morse, K. (2007). Effect of membrane pore size on the performance of a membrane reactor for biodiesel production. *Industrial and Engineering Chemistry Research, 46,* 52–58.

Carrero, A., Vicente, G., Rodríguez, R., Linares, M., & Del Peso, G. L. (2011). Hierarchical zeolites as catalysts for biodiesel production from Nannochloropsis microalga oil. *Catalysis Today, 167,* 148–153.

Carvalho, A. P., Meireles, L. A., & Malcata, F. X. (2006). Microalgal reactors: A review of enclosed system designs and performances. *Biotechnology Progress, 22,* 1490–1506. https://doi.org/10.1021/bp060065r.

Chen, P. H., & Oswald, W. J. (1998). Thermochemical treatment for algal fermentation. *Environment International, 24,* 889–897.

Choudhary, P., Prajapati, S. K., Kumar, P., Malik, A., & Pant, K. K. (2017). Development and performance evaluation of an algal biofilm reactor for treatment of multiple wastewaters and characterization of biomass for diverse applications. *Bioresource Technology, 224,* 276–284.

Dube, M. A., Tremblay, A. Y., & Liu, J. (2007). Biodiesel production using a membrane reactor. *Bioresource Technology, 98,* 639–647.

Ehimen, E. A., Holm-Nielsen, J. B., Poulsen, M., & Boelsmand, J. E. (2013). Influence of different pre-treatment routes on the anaerobic digestion of a filamentous algae. *Renewable Energy, 50,* 476–480.

Ehimen, E. A., Sun, Z. F., & Carrington, C. G. (2010). Variables affecting the in situ transesterification of microalgae lipids. *Fuel, 89,* 677–684.

Ehimen, E. A., Sun, Z., & Carrington, G. C. (2012). Use of ultrasound and co-solvents to improve the in-situ transesterification of microalgae biomass. *Procedia Environmental Sciences, 15,* 47–55. https://doi.org/10.1016/j.proenv.2012.05.009.

El-Dalatony, M. M., Kurade, M. B., Abou-Shanab, R. A. I., Kim, H., Salama, E. S., & Jeon, B. H. (2016). Long-term production of bioethanol in repeated-batch fermentation of microalgal biomass using immobilized *Saccharomyces cerevisiae*. *Bioresource Technology, 219,* 98–105.

Estrada-Villagrana, A. D., Quiroz-Sosa, G. B., Jiménez-Alarcón, M. L., Alemán-Vázquez, L. O., & Cano-Domínguez, J. L. (2006). Comparison between a conventional process and reactive distillation for naphtha hydrodesulfurization. *Chemical Engineering and Processing: Process Intensification, 45,* 1036–1040.

Ferreira, A. F., Dias, A. P. S., Silva, C. M., & Costa, M. (2016). Effect of low frequency ultrasound on microalgae solvent extraction: Analysis of products, energy consumption and emissions. *Algal Research, 14,* 9–16.

Fu, C. C., Hung, T. C., Chen, J. Y., Su, C. H., & Wu, W. T. (2010). Hydrolysis of microalgae cell walls for production of reducing sugar and lipid extraction. *Bioresource Technology, 101,* 8750–8754.

Gao, S., Yang, J., Tian, J., Ma, F., Tu, G., & Du, M. (2010). Electro-coagulation-flotation process for algae removal. *Journal of Hazardous Materials, 177,* 336–343.

Gendy, T. S., & El-Temtamy, S. A. (2013). Commercialization potential aspects of microalgae for biofuel production: An overview. *Egyptian Journal of Petroleum, 22,* 43–51.

Ghayal, D., Pandit, A. B., & Rathod, V. K. (2013). Optimization of biodiesel production in a hydrodynamic cavitation reactor using used frying oil. *Ultrasonics Sonochemistry, 20,* 322–328.

Gogate, P. R., & Pandit, A. B. (2004). Sonochemical reactors: Scale up aspects. *Ultrasonics Sonochemistry, 11,* 105–117.

Gole, V. L., & Gogate, P. R. (2012). Intensification of synthesis of biodiesel from nonedible oils using sonochemical reactors. *Industrial and Engineering Chemistry Research, 51*(37), 11866–11874.

González-Fernández, C., Sialve, B., Bernet, N., & Steyer, J. P. (2012a). Thermal pretreatment to improve methane production of *Scenedesmus* biomass. *Biomass and Bioenergy, 40,* 105–111.

González-Fernández, C., Sialve, B., Bernet, N., & Steyer, J. P. (2012b). Comparison of ultrasound and thermal pretreatment of *Scenedesmus* biomass on methane production. *Bioresource Technology, 110,* 610–616.

Guo, H., Daroch, M., Liu, L., Qiu, G., Geng, S., & Wang, G. (2013). Biochemical features and bioethanol production of microalgae from coastal waters of Pearl River Delta. *Bioresource Technology, 127,* 422–428.

Gupta, A., & Verma, J. P. (2015). Sustainable bio-ethanol production from agro-residues: A review. *Renewable and Sustainable Energy Reviews, 41,* 550–567.

Han, F., Pei, H., Hu, W., Jiang, L., Cheng, J., & Zhang, L. (2016). Beneficial changes in biomass and lipid of microalgae *Anabaena variabilis* facing the ultrasonic stress environment. *Bioresource Technology, 209,* 16–22.

Harun, R., Danquah, M. K., & Forde, G. M. (2010). Microalgal biomass as a fermentation feedstock for bioethanol production. *Journal of Chemical Technology and Biotechnology, 85,* 199–203.

Harvey, A. P., Mackley, M. R., & Seliger, T. (2003). Process intensification of biodiesel production using a continuous oscillatory flow reactor. *Journal of Chemical Technology and Biotechnology, 78,* 338–341.

He, B. B., Singh, A. P., & Thompson, J. C. (2006). A novel continuous-flow reactor using reactive distillation for biodiesel production. *Transactions of the ASABE, 49,* 107–112.

Ho, S. H., Huang, S. W., Chen, C. Y., Hasunuma, T., Kondo, A., & Chang, J. S. (2013). Bioethanol production using carbohydrate-rich microalgae biomass as feedstock. *Bioresource Technology, 135,* 191–198.

Jankowska, E., Sahu, A. K., & Oleskowicz-Popiel, P. (2017). Biogas from microalgae: Review on microalgae's cultivation, harvesting and pretreatment for anaerobic digestion. *Renewable and Sustainable Energy Reviews, 75,* 692–709.

Jeevan Kumar, S. P., Vijay Kumar, G., Dash, A., Scholz, P., & Banerjee, R. (2017). Sustainable green solvents and techniques for lipid extraction from microalgae: A review. *Algal Research, 21,* 138–147.

Joshi, S., Gogate, P. R., Moreira, P. F., & Giudici, R. (2017). Intensification of biodiesel production from soybean oil and waste cooking oil in the presence of heterogeneous catalyst using high speed homogenizer. *Ultrasonics Sonochemistry, 39,* 645–653.

Kashid, M. N., & Kiwi-Minsker, L. (2009). Microstructured reactors for multiphase reactions: State of the art. *Industrial and Engineering Chemistry Research, 48,* 6465–6485.

Keymer, P., Ruffell, I., Pratt, S., & Lant, P. (2013). High pressure thermal hydrolysis as pre-treatment to increase the methane yield during anaerobic digestion of microalgae. *Bioresource Technology, 131,* 128–133.

Khan, M. I., Lee, M. G., Shin, J. H., & Kim, J. D. (2017). Pretreatment optimization of the biomass of *Microcystis aeruginosa* for efficient bioethanol production. *AMB Express, 7,* 19.

Kim, B., Im, H., & Lee, J. W. (2015). In situ transesterification of highly wet microalgae using hydrochloric acid. *Bioresource Technology, 185,* 421–425.

Kim, J., Yoo, G., Lee, H., Lim, J., Kim, K., Kim, C. W., et al. (2013). Methods of downstream processing for the production of biodiesel from microalgae. *Biotechnology Advances, 31,* 862–876.

Kraai, G. N., Schuur, B., van Zwol, F., van de Bovenkamp, H. H., Heeres, H. J. (2009). Novel highly integrated biodiesel production technology in a centrifugal contactor separator device. *Chemical Engineering Journal, 154*, 384–389. https://doi.org/10.1016/j.cej.2009.04.047.

Lee, Y., & Li, P. (2016). Using resonant ultrasound field-incorporated dynamic photobioreactor system to enhance medium replacement process for concentrated microalgae cultivation in continuous mode. *Chemical Engineering Research and Design, 118,* 112–120.

Lee, K. T., Lim, S., Pang, Y. L., Ong, H. C., & Chong, W. T. (2014). Integration of reactive extraction with supercritical fluids for process intensification of biodiesel production: Prospects and recent advances. *Progress in Energy and Combustion Science, 45,* 54–78.

Lee, J. Y., Yoo, C., Jun, S. Y., Ahn, C. Y., & Oh, H. M. (2010). Comparison of several methods for effective lipid extraction from microalgae. *Bioresource Technology, 101,* S75–S77.

Li, Y., Chen, Y. F., Chen, P., Min, M., Zhou, W., Martinez, B., et al. (2011a). Characterization of a microalga *Chlorella* sp. well adapted to highly concentrated municipal wastewater for nutrient removal and biodiesel production. *Bioresource Technology, 102,* 5138–5144.

Li, Y., Lian, S., Tong, D., Song, R., Yang, W., Fan, Y., et al. (2011b). One-step production of biodiesel from *Nannochloropsis* sp. on solid base Mg–Zr catalyst. *Applied Energy, 88,* 3313–3317.

Lu, W., Wang, Z., Wang, X., & Yuan, Z. (2015). Cultivation of *Chlorella* sp. using raw diary wastewater for nutrient removal and biodiesel production: Characteristics comparison of indoor bench-scale and outdoor pilot-scale cultures. *Bioresource Technology, 192,* 382–388.

Mazubert, A., Poux, M., & Aubin, J. (2013). Intensified processes for FAME production from waste cooking oil: A technological review. *Chemical Engineering Journal, 233,* 201–223.

Miao, X., & Wu, Q. (2006). Biodiesel production from heterotrophic microalgal oil. *Bioresource Technology, 97,* 841–846.

Misra, R., Guldhe, A., Singh, P., Rawat, I., Stenstrom, T. A., & Bux, F. (2015). Evaluation of operating conditions for sustainable harvesting of microalgal biomass applying electrochemical method using non sacrificial electrodes. *Bioresource Technology, 176,* 1–7.

Olguín, E. J. (2012). Dual purpose microalgae-bacteria-based systems that treat wastewater and produce biodiesel and chemical products within a biorefinery. *Biotechnology Advances, 30,* 1031–1046.

Park, K. Y., Kweon, J., Chantrasakdakul, P., Lee, K., & Cha, H. Y. (2013). Anaerobic digestion of microalgal biomass with ultrasonic disintegration. *International Biodeterioration & Biodegradation, 85,* 598–602.

Passos, F., Hernández-Mariné, M., García, J., Ferrer, I. (2014). Long-term anaerobic digestion of microalgae grown in HRAP for wastewater treatment. Effect of microwave pretreatment. *Water Research, 49*.

Passos, F., Sole, M., Garcia, J., & Ferrer, I. (2013). Biogas production from microalgae grown in wastewater: Effect of microwave pretreatment. *Applied Energy, 108,* 168–175.

Patel, A., Gami, B., Patel, P., Patel, B. (2016). Microalgae: Antiquity to era of integrated technology. *Renewable and Sustainable Energy Reviews*. https://doi.org/10.1016/j.rser.2016.12.081.

Patil, P. D., Gude, V. G., Mannarswamy, A., Cooke, P., Munson-McGee, S., Nirmalakhandan, N., et al. (2011a). Optimization of microwave-assisted transesterification of dry algal biomass using response surface methodology. *Bioresource Technology, 102,* 1399–1405.

Patil, P. D., Gude, V. G., Mannarswamy, A., Deng, S., Cooke, P., Munson-McGee, S., et al. (2011b). Optimization of direct conversion of wet algae to biodiesel under supercritical methanol conditions. *Bioresource Technology, 102,* 118–122.

Pfaffinger, C. E., Schöne, D., Trunz, S., Löwe, H., & Weuster-Botz, D. (2016). Model-based optimization of microalgae areal productivity in flat-plate gas-lift photobioreactors. *Algal Research, 20,* 153–163.

Prabakaran, P., & Ravindran, A. D. (2011). A comparative study on effective cell disruption methods for lipid extraction from microalgae. *Letters in Applied Microbiology, 53,* 150–154.

Priyadarshani, I., & Rath, B. (2012). Commercial and industrial applications of micro algae—A review. *Journal of Algal Biomass Utilization, 3,* 89–100.

Rodolfi, L., Zittelli, G. C., Bassi, N., Padovani, G., Biondi, N., Bonini, G., et al. (2009). Microalgae for oil: Strain selection, induction of lipid synthesis and outdoor mass cultivation in a low-cost photobioreactor. *Biotechnology and Bioengineering, 102,* 100–112.

Samson, R., & Leduy, A. (1983). Influence of mechanical and thermochemical pretreatments on anaerobic digestion of *Spirulina maxima* algal biomass. *Biotechnology Letters, 5,* 671–676.

Sangaletti-Gerhard, N., Cea, M., Risco, V., & Navia, R. (2015). In situ biodiesel production from greasy sewage sludge using acid and enzymatic catalysts. *Bioresource Technology, 179,* 63–70.

Santos, N. O., Oliveira, S. M., Alves, L. C., & Cammarota, M. C. (2014). Methane production from marine microalgae *Isochrysis galbana. Bioresource Technology, 157,* 60–67.

Schuchardt, U., Sercheli, R., & Matheus, R. (1998). Transesterification of vegetable oils: A review general aspects of transesterification of vegetable oils acid-catalyzed processes base-catalyzed processes. *Journal of the Brazilian Chemical Society, 9,* 199–210.

Schwede, S., Kowalczyk, A., Gerber, M., Span, R. (2011). Influence of different cell disruption techniques on mono digestion of algal biomass. In *Bioenergy Technology—World Renewable Energy Congress* (pp. 41–47). https://doi.org/10.3384/ecp1105741.

Sebestyen, P., Blanken, W., Bacsa, I., Toth, G., Martin, A., Bhaiji, T., et al. (2016). Upscale of a laboratory rotating disk biofilm reactor and evaluation of its performance over a half-year operation period in outdoor conditions. *Algal Research, 18,* 266–272.

Shokrkar, H., Ebrahimi, S., & Zamani, M. (2017). Bioethanol production from acidic and enzymatic hydrolysates of mixed microalgae culture. *Fuel, 200,* 380–386.

Silva, C. E. F., & Bertucco, A. (2016). Bioethanol from microalgae and cyanobacteria: A review and technological outlook. *Process Biochemistry, 51,* 1833–1842.

Solovchenko, A., Pogosyan, S., Chivkunova, O., Selyakh, I., Semenova, L., Voronova, E., et al. (2014). Phycoremediation of alcohol distillery wastewater with a novel *Chlorella sorokiniana* strain cultivated in a photobioreactor monitored on-line via chlorophyll fluorescence. *Algal Research, 6,* 234–241.

Stavarache, C., Vinatoru, M., Nishimura, R., & Maeda, Y. (2005). Fatty acids methyl esters from vegetable oil by means of ultrasonic energy. *Ultrasonics Sonochemistry, 12,* 367–372.

Suali, E., & Sarbatly, R. (2012). Conversion of microalgae to biofuel. *Renewable and Sustainable Energy Reviews, 16,* 4316–4342.

Subhedar, P. B., Botelho, C., Ribeiro, A., Castro, R., Pereira, M. A., Gogate, P. R., et al. (2015). Ultrasound intensification suppresses the need of methanol excess during the biodiesel production with lipozyme TL-IM. *Ultrasonics Sonochemistry, 27,* 530–535.

Suresh Kumar, K., Dahms, H. U., Won, E. J., Lee, J. S., & Shin, K. H. (2015). Microalgae—A promising tool for heavy metal remediation. *Ecotoxicology and Environmental Safety, 113,* 329–352.

Syazwani, O., Rashid, U., & Taufiq Yap, Y. H. (2015). Low-cost solid catalyst derived from waste *Cyrtopleura costata* (Angel Wing Shell) for biodiesel production using microalgae oil. *Energy Conversion and Management, 101,* 749–756.

Takagi, M., Karseno, Yoshida, T. (2006). Effect of salt concentration on intracellular accumulation of lipids and triacylglyceride in marine microalgae Dunaliella cells. *Journal of Bioscience and Bioengineering, 101*, 223–226.

Terigar, B. G., Balasubramanian, S., Lima, M., & Boldor, D. (2010). Transesterification of soybean and rice bran oil with ethanol in a continuous-flow microwave-assisted system: Yields, quality, and reaction kinetics. *Energy & Fuels, 24,* 6609–6615.

Umdu, E. S., Tuncer, M., & Seker, E. (2009). Transesterification of *Nannochloropsis oculata* microalga's lipid to biodiesel on Al_2O_3 supported CaO and MgO catalysts. *Bioresource Technology, 100,* 2828–2831.

Vergini, S., Aravantinou, A. F., & Manariotis, I. D. (2016). Harvesting of freshwater and marine microalgae by common flocculants and magnetic microparticles. *Journal of Applied Phycology, 28,* 1041–1049.

Wahlen, B. D., Willis, R. M., & Seefeldt, L. C. (2011). Biodiesel production by simultaneous extraction and conversion of total lipids from microalgae, cyanobacteria, and wild mixed-cultures. *Bioresource Technology, 102,* 2724–2730.

Wang, L., Min, M., Li, Y., Chen, P., Chen, Y., Liu, Y., et al. (2010). Cultivation of green algae *Chlorella* sp. in different wastewaters from municipal wastewater treatment plant. *Applied Biochemistry and Biotechnology, 162,* 1174–1186.

Wen, Z., Yu, X., Tu, S. T., Yan, J., & Dahlquist, E. (2009). Intensification of biodiesel synthesis using zigzag micro-channel reactors. *Bioresource Technology, 100,* 3054–3060.

Xin, L., Hong-ying, H., Ke, G., & Ying-xue, S. (2010). Effects of different nitrogen and phosphorus concentrations on the growth, nutrient uptake, and lipid accumulation of a freshwater microalga *Scenedesmus* sp. *Bioresource Technology, 101,* 5494–5500.

Zhang, Y., Li, Y., Zhang, X., & Tan, T. (2015). Biodiesel production by direct transesterification of microalgal biomass with co-solvent. *Bioresource Technology, 196,* 712–715.

Zhang, X., Ma, Q., Cheng, B., Wang, J., Li, J., & Nie, F. (2012). Research on KOH/La-Ba-Al_2O_3 catalysts for biodiesel production via transesterification from microalgae oil. *Journal of Natural Gas Chemistry, 21,* 774–779.

Zheng, Y., Roberts, M., Kelly, J., Zhang, N., & Walker, T. (2015). Harvesting microalgae using the temperature-activated phase transition of thermoresponsive polymers. *Algal Research, 11,* 90–94.

Zheng, M., Skelton, R. L., & Mackley, M. R. (2007). Biodiesel reaction screening using oscillatory flow meso reactors. *Process Safety and Environment Protection, 85,* 365–371.

Zheng, H., Yin, J., Gao, Z., Huang, H., Ji, X., & Dou, C. (2011). Disruption of *Chlorella vulgaris* cells for the release of biodiesel-producing lipids: A comparison of grinding, ultrasonication, bead milling, enzymatic lysis, and microwaves. *Applied Biochemistry and Biotechnology, 164,* 1215–1224.

Zhou, N., Zhang, Y., Wu, X., Gong, X., & Wang, Q. (2011). Hydrolysis of Chlorella biomass for fermentable sugars in the presence of HCl and $MgCl_2$. *Bioresource Technology, 102,* 10158–10161.

Chapter 5
Microalgae Biorefineries for Energy and Coproduct Production

Pierre-Louis Gorry, León Sánchez and Marcia Morales

Abstract The 2015 Conference of the Parties (COP21) marked a turning point for global actions to mitigate atmospheric greenhouse gases, reduce the carbon dioxide emissions from fossil fuel combustion, and stabilize the global climate. On the other hand, the increase in energy demand asks for renewable sources and robust systems to supply energy and obtain product diversity like that obtained from a petroleum refinery. A biorefinery is the sustainable processing of biomass into a spectrum of profitable products and energy. Microalgal biomass is considered one of the most promising biorefinery feedstock providing alternatives for different areas, such as food, feed, cosmetics and health industries, fertilizers, plastics, and biofuels including biodiesel, methane, hydrogen, ethanol. Furthermore, microalgae can also be used for the treatment of wastewater and CO_2 capture. However, microalgal biofuels are not currently cost competitive at large scale and to develop a sustainable and economically feasible process, most of the biomass components should be valorized. High-value coproducts from microalgae include pigments, proteins, lipids, carbohydrates, vitamins, and antioxidants, and they can improve the process economics in the biorefinery concept. Therefore, mild and energy-efficient downstream processing techniques need to be chosen to maintain product properties and value. In this chapter, the existing products and microalgae biorefinery strategies will be presented, followed by new developments, sustainability assessments, and techno-economic evaluations. Finally, perspectives and challenges of microalgal biorefineries will be explored.

Keywords Biorefinery · Downstream processing · Biofuels · Microalgae products LCA

P.-L. Gorry · M. Morales (✉)
Processes and Technology Department and Doctoral Program in Natural Sciences and Engineering, Metropolitan Autonomous University Campus Cuajimalpa, Vasco de Quiroga, 4871, Col. Santa Fe, Cuajimalpa, 05300 Mexico City, Mexico
e-mail: mmorales@correo.cua.uam.mx

L. Sánchez
Doctoral Program in Biotechnology, Metropolitan Autonomous University Campus Iztapalapa, San Rafael Atlixco 186, Col.Vicentina, 09340 Iztapalapa, Mexico City, Mexico

E. Jacob-Lopes et al. (eds.), *Energy from Microalgae*, Green Energy and Technology, https://doi.org/10.1007/978-3-319-69093-3_5

1 Introduction

The World Meteorological Organization confirmed that 2016 was the hottest year on Earth; the global temperature rise is almost 1.1 °C higher than the value in the pre-industrial period (World Meteorological Organization 2017). In 2016, 195 countries ratified the Paris climate agreement including a commitment to keep the global warming below 2 °C before 2100. The global climate change is the result of a rise in Earth's temperature due to the presence of greenhouse gases from human activities. The limited availability of fossil resources, such as petroleum, and the strong dependence on the production of fuels and other chemicals provoke environmental, social, political, and economic concerns. Similar to the oil refinery, where the crude oil is processed and refined into different products of high and low values (liquefied petroleum gas, gasoline, naphtha for olefins and aromatics, kerosene, heating fuel, diesel, heavy fuel oil, and bitumen), the biorefinery is one of the most promising alternatives to obtain biofuels and chemicals from renewable sources (Chew et al. 2017). Biorefinery involves transformation of raw materials obtained from agriculture, silviculture, organic wastes, or any biomass through various unit processes to convert them in a wide range of products (Postma et al. 2016). Some definitions of biorefinery include the one provided by the National Renewable Energy Laboratory (NREL) "A biorefinery is a facility that integrates biomass conversion processes and equipment to produce fuels, power and (organic) chemicals from biomass" and by the International Energy Agency (IEA) "A biorefinery is the suitable processing of biomass into a spectrum of marketable products (food, feed, materials, and chemicals) and energy (fuels, power, heat)" (de Jong et al. 2012; Budzianowski 2017). Biorefineries are classified according to the biomass feedstock generation (Saai-Anuggraha et al. 2016; Hossain et al. 2017): The first-generation biorefineries use sugarcane, corn, or soybeans to produce value-added products for feed, food applications, fuels, and specialty chemicals. Almost all current biofuels (mainly ethanol, butanol, and biodiesel) and bio-based chemicals (lactic acid, itaconic acid, 1,3-propanediol, etc.) are produced in this type of biorefinery. The second-generation biorefineries are based on lignocellulosic materials, and they are composed of three main sections to convert lignocellulose into biofuels. The main product is the cellulosic ethanol; however, the biomass conversion by thermochemical platform involves the gasification of biomass to produce syngas (CO, CO_2, H_2, and CH_4), which can then be converted into various chemicals, such as ethanol, methanol, and butanol. The most advanced is the third-generation biorefinery that can use a mixture of biomass to produce a multitude of products using a combination of technologies. Microalgae biomass is considered as the most promising feedstock for the third-generation biorefineries. The microalgae might contribute to reduce the oil dependency and the rise in Earth's temperature. They have the ability to transform solar energy into chemicals by capturing CO_2 and releasing O_2. It is known that microalgae are one of the best technologies for carbon dioxide sequestration (Wiesberg et al. 2017) and their use as renewable energy source was long ago proposed by scientists. The patents and research papers indicate a strong interest in microalgae biorefinery looking for industrial-scale applications (Konur 2011; Mohan et al. 2016a, b; Xu and Boeing 2013; Zhu et al. 2016). The microalgae cultivation has high areal productivity, the possibility to grow in nonarable land, or wastewater used as nutrient source. From the biomass obtained, a spectrum of marketable products can be obtained, such as pigments, proteins, lipids, carbohydrates, vitamins, and antioxidants for applications like feed, food, polymers, pharmaceuticals, cosmetics, and biofuels (Borowitzka 2013; Budzianowski 2017; Suganya et al. 2016). Although microalgae biofuels are technically feasible, they remain strongly dependent on government subsidies and oil price, which make them economically nonviable for now (Wijffels and Barbosa 2010); therefore, primary strategies for bioenergy production from algae will need to rely on a multiproduct biorefinery approach (Laurens et al. 2017a). The CO_2 capture and the use of wastewater as nutrient source for microalgae growth combined with the production of high-value-added products and bioenergy make microalgae biorefinery potentially profitable.

2 Products Portfolio from Microalgae and Applications

The main goal of the biorefinery is to integrate the production of bioenergy (commodities: low-value high-volume products) and other chemicals (high-value low-volume products) to optimize the use of biomass resources by reducing wastes while maximizing profitability and benefits (Demibras 2009). Budzianowski (2017) categorized the high-value low-volume bioproducts from biorefineries into six groups: biopharmaceuticals, biocosmetics, bionutrients, biochemicals, biofertilizers, and biomaterials. All of them and biofuels can be obtained from microalgae (Chew et al. 2017; Milledge 2011). The microalgae products are reviewed in this section and summarized in Table 1 and Fig. 1.

Biopharmaceuticals: Microalgae are a source of many potential new drugs and bioactive molecules for health industry (Abd El Baky and El-Baroty 2013; Borowitzka 1995; Deniz et al. 2017; Mimouni et al. 2012). According to the number of patent publications, currently, biopharmaceutics is one of the most important innovation areas under development (Chilton et al. 2016). The bioactive molecules include applications, such as antioxidant, anti-inflammatory, antitumor, anticancer, antimicrobial, antiviral, and antiallergic agents along with other pharmaceutical properties (Deniz et al. 2017). Pigments, such as carotenoids (β-carotene and astaxanthin), phycobiliproteins (phycocyanin), and some polysaccharides or phenolic derivatives exhibit antioxidant and anti-inflammatory activities. Phycobilins have anti-inflammatory, antiallergic, antioxidant, and anticancer activities (Kim et al. 2016). Polyunsaturated fatty acids (PUFAs) are also of interest for human welfare, and there is a recent market of 11.5 billion dollars (Béligon et al. 2016). Molecules used for anticancer or antitumor effects include polysaccharides (carrageenan and fucoidan), PUFAs (eicosapentaenoic acid, EPA; or

Table 1 Microalgae portfolio adapted from Hamed (2016)

Application	Microalgae	Examples of functions and/or molecules	Market volume or price	References
Food and nutrition	*Spirulina platensis*, *Spirulina maxima*, *Chlorella*, *Muriellopsis* sp., *Crypthecodinium*, *Schizochytrium*	Food supplement. High content in proteins, iron, main unsaturated fatty acids, vitamin B12. Protection against hypertension and renal problems, support growth of lactobacillus (intestinal bacteria), immunostimulants, reduce free radical, effect on lipids levels in blood, can prevent gastric ulcers, wounds, and constipation. High levels of certain carotenoids to acting as antioxidant (like lutein) and used against degenerative diseases. PUFAs and EPA for infant formula	USD 12–50 millons per year for phycobiliproteins	Beheshtipour et al. (2013), Borowitzka (2013), Hamed (2016), Mata et al. (2010), Spolaore et al. (2006)
Feed	*Chlorella*, *Tetraselmis*, *Isochrysis*, *Pavlova*, *Phaeodactylum*, *Chaetoceros*, *Spirulina*, *Dunaliella*, *Skeletonema*, *Thalassiosira*	Used for their high nutritional value for zooplankton culture, fish farming, chicken breeding. As side effects, microalgae pigments can color flesh of salmon or shrimp exoskeleton and skin and other species. Gives additional value to aquaculture. It has positive impact on physiology for pets and farm animals and therefore on their appearance	USD 32 billions/year for *Chlorella* sp. (β-1,3-glucan, immune-stimulant in fish)	Brennan and Owende (2010), Das et al. (2012), Gouveia et al. (2008), Hemaiswarya et al. (2011), Mata et al. (2010), Priyadarshani and Sahu (2012), Sirakov et al. (2015), Spolaore et al. (2006)

(continued)

Table 1 (continued)

Application	Microalgae	Examples of functions and/or molecules	Market volume or price	References
Fertilizer and biostimulants	*Anabaena*, *Nostoc* *Spirulina*, *Haematococcus*, *Chlorella* sp., *Nannochloropsis* sp.	Upgrade soil quality and fertility by acting as water retainer, releasing essential nutrients (phosphate, nitrogen, and trace elements). Require less chemical fertilizer and act as promoters for plant growth and preventing plant infection by synthetizing biochemical components with antiviral and antibacterial activities	Biostimulant represents 27% of the USD 3.4 billons with prices of €1300–1500/ton for microalgal biostimulant	Abd El Baky and El-Baroty (2013), Hannon et al. (2010), van der Voort et al. (2015)
Cosmetic	*Spirulina*, *Chlorella*, *Haematococcus*	Proteins, vitamins, minerals, and pigments which have effects on skin. Skin-care and hair-care products already for sale. Help tissue to regenerate, collagen synthesis is improved	Astaxanthin produced by *Haematococcus pluvialis* has market prices between 1900 and 7000 USD/kg	Adarme-Vega et al. (2012), Hariskos et al. (2014), Shah et al. (2016), Spolaore et al. (2006), Yaakob et al. (2014)
Pharmaceuticals	*Nannochloropsis oculata*, *Tolypothrix byssoidea*, *Chlamydomonas*	Zeaxanthin pigment reduces tyrosinase activity by inhibition. Used in whitening creams. Tubercidin has activities in lymphocytic leukemia. Growth of lymphosarcoma (blood tumor cells) in mice has been stopped by L-asparaginase	Until today no microalgal-derived pharmaceuticals have entered the market yet. But total global pharmaceutical industry had a value of USD 955 billion in 2011 (LEK 2013), and the market of active ingredients for pharmaceuticals had a size of USD 101 billions	Ahmad et al. (2012), Paul (1982), van der Voort et al. (2015)

(continued)

Table 1 (continued)

Application	Microalgae	Examples of functions and/or molecules	Market volume or price	References
Fuels and energy	*Scenedesmus almeriensis*, *Chlorella vulgaris*, *Chlamydomonas*, *Dunaliella*	Fatty acids for either biodiesel or other hydrocarbon fuels. Biogas from anaerobic digestion. Ethanol produced through fermentation	209,000,000 ton for fatty acids at 920 USD/ton	Chew et al. (2017), Chilton et al. (2016), Laurens et al. (2017a)
Materials	*Chlorella infusionum*, *Dunaliella*, *Schizochytrium*	Starch and proteins for a wide range of bioplastics. Phytol for and sterols for surfactants. Amino acids or peptides for either polyurethane and plasticizers	1.5 millons ton for thermoplastics from protein at 1900 USD/ton	Laurens et al. (2017a)
Chemical	*Dunaliella*, *Schizochytrium*	Lactic acid, biomethanol, glutamic acid, sorbitol, glycerol for chemical industries	36,000–2,300,000 ton for glycerol at 1550–3400 USD/ton	Bozell and Petersen(2010, Budzianowski (2017), Laurens et al. (2017a)

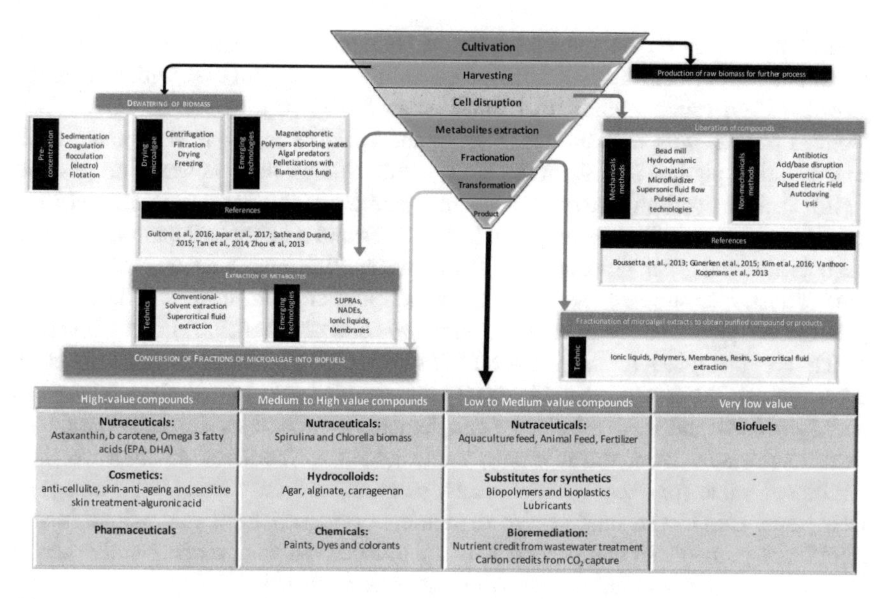

Fig. 1 Downstream processing in a biorefinery

docosahexaenoic acid, DHA) and phycobiliproteins (phycocyanin). Antimicrobial and antiviral applications are related to fatty acids, pigments, peptides, and polyphenols.

Biocosmetics: Microalgae components are often used in cosmetics as water-binding agents, texturizing agents, antioxidants providing dermal protection as well as skin-whitening agents (Jahan et al. 2017; Stolz and Obermayer 2005; Wang et al. 2015). Compounds such as sporopollenin, mycosporine-like amino acids, and scytonemin protect against UV-A or UV-B radiation. Carotenoids are also used as stabilizers in cosmetics and solar protection products. Phycocyanin, a natural blue pigment, is extensively used in cosmetics, including lipsticks, eyeliners, nail polishes, and eye shadows. Red-phycoerythrin is used as an alternative for synthetic pigments in creams or other cosmetics. Collagen-like proteins are included in creams and gels with high moisturizing action, but their other activities are also known, including antiaging and antiwrinkling. Some polysaccharides, such as chitin or fucoidan, have protective and moisturizing properties. Other polysaccharides, such as agar, carrageenan, and alginates, are used as stabilizers, thickeners, and emulsifiers. Skin-whitening and antimicrobial agents are related mostly to compounds synthetized by macroalgae; however, microalgae extracts have exhibited antimicrobial effects too (Martins et al. 2008). Based on patent landscape, this sector is dominated by European countries and, specifically, by France (Chilton et al. 2016).

Bionutrients: Microalgal biomass is a promising source of nutrients. Indeed, both food and feed sectors have been quickly increasing in recent years to replace animal protein. Nourishment is the fourth area of applications of microalgae feedstock, and based on patent landscape, human nutrition is the second most important sector (Chilton et al. 2016). Aquaculture is a special case with an average growth (35%) much higher than other areas (20%) (Chilton et al. 2016). Dry biomass powder with high nutrient content and valuable compounds included, such as fatty acids, pigments, and antioxidants, is the main product presentation (Hamed 2016). Proteins, carbohydrates, lipids, and vitamins are of great interest for nutrition as well as pigments like: yellow-orange carotenes and xanthophylls, the red or blue phycobilins, and green chlorophylls. They have applications as natural colorants for food industry or supplements for both human and animal nutrition. High-quality proteins are produced by microalgae like *Spirulina* (Becker 2007) and *Chlorella*, which are identified as "super food" (Milledge 2011) and commercialized as nutraceuticals. Microalgae food is mainly commercialized as dried algae (*Chlorella* and *Spirulina*) and sold as dietary supplements or found as specialty products, extracted/isolated from the microalgae and added to food/feed to improve their nutritional value (pigments, antioxidants, proteins, and fatty acids, e.g., omega-3, DHA, and EPA). The market size of nutrients obtained from microalgae is still significantly smaller in comparison with the one derived from crops, but this sector has an impressive and unique growth (Vigani et al. 2015).

Biochemicals: Market projection predicts that 17–38% of total organic chemicals will be provided by biochemicals around 2050 (Budzianowski 2017). The US Department of Energy (DOE) registered ten biochemicals with high future potential for the market (Bozell and Petersen 2010): biohydrocarbons, succinic acid, furanic, glycerol and derivatives, lactic acid, levulinic acid, hydroxypropionic acid/aldehyde, xylitol, sorbitol, and ethanol. Numerous biochemicals, such as biomethanol, lactic acid, glutamic acid, sorbitol, glycerol, and 3-hydroxypicolinic acid (3-HPA), are already used in industries like BioMCN or Roquette Freres SA (Broeren et al. 2013). Further reduction of production costs will allow expanding their applications. Other products, such as alginates, xylose, or glucaric acid, are however unique, and their specific market does not exist yet (Budzianowski 2017). Microalgae produce various building blocks for biochemicals, and these are the largest class of high-value bioproducts that could be obtained in a biorefinery, such as pigments and PUFAs (Budzianowski 2017).

Biofertilizers: They have great potential to replace chemical fertilizers and avoid the aggressive use of chemicals that leads to soil erosion and degradation of local ecosystems through eutrophication when they run off into rivers or percolate into groundwater. Likewise, their use contributes indirectly toward greenhouse gas emissions as their production depends on fossil fuels. Biofertilizers include the nitrogen-fixing, phosphate solubilizing, and plant growth-promoting microorganisms. Microalgae have important role in soil ecosystems (Pulz and Gross 2004). According to Chatterjee et al. (2017), microalgae contribute to soil fertilization through: (1) enhancement of soil porosity because of the filamentous structure and production of adhesive substances of certain cyanobacteria; (2) release of growth promoters, such as amino acids, hormones (auxins, gibberellins, cytokinins), and vitamins (Pulz and Gross 2004; Singh et al. 2016); (3) increase in water retention capacity through their thickened structure (Hamed 2016); (4) soil enrichment with organic matter and minerals after death and decomposition of microalgae biomass (Saadatnia and Riahi 2009); (5) reduction in soil salinity (Al-sherif et al. 2015); (6) prevention of weed growth and production of antiviral and antibacterial substances to protect plants (Abd El Baky and El-Baroty 2013; Dahms et al. 2006; Hannon et al. 2010); and (7) increase in soil phosphate by excretion of organic acids (Singh et al. 2016). Some nitrogen-fixing species, such as *Anabaena* and *Nostoc*, can be directly used as fertilizers for agricultural purposes (Hamed 2016) through direct inoculation in soils, or green algae can be applied as dry powder with high percentage of macronutrients, considerable amounts of micronutrients, and amino acids (Faheed and Abd-El Fattah 2008; Garcia-Gonzalez and Sommerfeld 2016).

Biomaterials: Biomaterials use complex structures of biomass for application in plastics, coatings and surface treatment materials, packaging materials, fibers and textiles, elastomers, lubricants and fillers, surfactants, and functional materials (Budzianowski 2017). Biomaterials have a bright future in replacing materials from fossil resources. The biochemical composition of biomass defines the potential biomaterial that can be produced. Proteins are the main platform molecules to make thermoplastics, foams, adhesives, biocomposites, and flocculants, and bioplastics are made from starch (Laurens et al. 2017a). In the case of microalgae biomass, bioplastics can be derived from any of the three major component fractions (lipids, proteins, and carbohydrates) (IEA 2017). Some researchers have described the use of the whole algae as filler material for different types of plastics, such as polypropylene (Zhang et al. 2000a), polyvinyl chloride (Zhang et al. 2000b, polyethylene (Otsuki and Zhang 2004; Zeller et al. 2013), blends of algae and starch (Kipngetich and Hillary 2013), or proteins (Reddy et al. 2013; Shi et al. 2011). But microalgae can also produce high-quality biodegradable plastics, such as polyhydroxyalkanoates (PHA) (Balaji et al. 2013; Chaogang et al. 2010; Haase et al. 2011; Rahman and Miller 2017). Surfactants can also be produced from microalgal sterols and phytol and have a high market potential of around 8,436 billion dollars for a five-year period (IEA 2017; Laurens et al. 2017a). Furthermore, asphalts can be made from microalgae biomass as well (Chailleux et al. 2012).

Bioenergies: A wide range of biofuels for bioenergy can be produced from microalgae biomass and all petroleum fuels, such as hydrocarbons, asphalts, liquid (kerosene, gasoline, diesel), and gaseous fuels (methane, syngas); even more, the biocrude can be made from microalgae biomass (Bahadar and Bilal Khan 2013; Budzianowski 2017; Chew et al. 2017). Biofuels are the third sector in terms of patent applications due to decades of research. However, given the noneconomic viability, this area has experienced a slow growth although it has aroused a lot of interest (Chilton et al. 2016). Hydrogen can be produced directly by microalgae photolysis. Other biofuels, such as ethanol and biogas, can be obtained from transformation of carbohydrates (starch, sugars, or other polymers) by fermentation into bioethanol (Chng et al. 2015) and/or anaerobic digestion, respectively. Bioelectricity can be also generated by integration of microalgae into a microbial fuel cell using microalgae in the cathode compartment and bacteria in the anode (Gouveia et al. 2014; Lee et al. 2015). This integration becomes especially favorable when considering that phototrophic organisms act as in situ generators of oxygen facilitating the reaction in the cathode of the chamber. Bioelectricity is produced by bacteria in the anode, which oxidize organic matter and produce electrons. Those electrons are transferred to the cathode electrode with an external circuit and produce electricity. The bacteria can be used for biodegradable waste treatment, and with the help of microalgae, the organic and inorganic load of the water can be reduced.

From all the above-mentioned bioproducts, the role of microalgae in the human diet is well established, but other applications are currently under development: biofuel production of pharmaceutical compounds, bioremediation, cosmetic active ingredients. Furthermore, microalgae produce many environmental benefits, such as carbon fixation, oxygen release, heavy metal removal, and wastewater treatment that provide energy savings and supply oxygen to anaerobic bacteria (Uggetti and Puigagut 2016). However, market is clearly dependent on actual investigation of new technologies, and mainly on governmental policies such as subsidies and mandated use of biofuels (Gorry et al. 2017).

3 Microalgal Biomass Processing

In order to maximize the potential of microalgae biomass, the whole chain process development should be defined in an integrated way, starting from an adequate supply of nutrients and CO_2, good harvesting methods, dewatering, and downstream processing (Mata et al. 2010; Toledo-Cervantes and Morales 2014). For this, it is necessary to know not only the potential added value that can be obtained, the microalgae cell wall strength and the composition and localization of cellular components in order to break down the cell wall properly to avoid product loss (Gerardo et al. 2015; Pei et al. 2010; Roux et al. 2017), but also the available processing technologies and the sequence of separation; these latter ones are needed to maintain the integrity of the possible products maximizing the recovery and to produce biofuels. Each biorefinery stage for processing microalgal biomass would be linked to the characteristics of each specific strain and biochemical composition, and the route to obtain bioenergy must be defined too. Main downstream processing technologies are explained in the following sections (see Fig. 1), and the routes to obtain diverse biofuels are shown as well.

3.1 Downstream Processing

3.1.1 Harvesting Technologies

Harvesting accounts for 20–30% of microalgae biomass production cost that is associated with the recovery of microalgae biomass from diluted streams (Barros et al. 2015; Pei et al. 2010; Tan et al. 2014). Harvesting is an energy-intensive process; therefore, it is necessary to choose an effective procedure to concentrate the biomass with low energy to minimize separation costs. The main technologies and emerging options for microalgae recovery are shown in Fig. 1 and Table 2. Harvesting includes physical (centrifugation, sedimentation, and filtration), chemical (flocculation, flotation, auto-flocculation, and bioflocculation), and electric (electro-coagulation–filtration or electrochemical harvesting) alternatives. Japar et al. (2017) established that filtration, flocculation, bioflocculation, and electro-coagulation-filtration and further drying using solar heat to process the algal cake are the most feasible solutions to remove water due to their high harvesting efficiencies, moderate operational and logistic costs, no negative impacts on the environment, and the shortest harvesting time. However, in a biorefinery, the combination of separation processes is recommended: a first step where the biomass is concentrated with a mechanical or chemical process to obtain a final concentration around 2–7% of total suspended solids, and a second, dewatering step to produce a microalgal cake (Barros et al. 2015; Gerardo et al. 2015). The pre-concentration step reduces the energy necessary to separate biomass from water. However, the dewatering process must be established depending on the strain and the final product requirements, so more research is required to reduce the energy requirement and lower microalgal harvesting costs (Barros et al. 2015). New emerging technologies include ultrasound, magnetophoretic procedures, the use of polymers to absorb water, and co-culture with fungi to form flocs (Xia et al. 2011; Zhou et al. 2013) that favor the removal of solids; however, additional research is still required.

3.1.2 Cell Disruption

Microalgae store most of their valuable components inside the cell, behind a thick and resistant cell wall. Therefore, energy - or solvent- consuming steps are needed to alter this physical barrier and to efficiently extract the desired compounds. A mild cell disruption method is necessary to make cell components available without losses. Cell disruption technologies can be divided into two main categories: mechanical and nonmechanical methods (see Fig. 1). Mechanical methods include: bead milling, homogenization, sonication, microwaving, thermolysis, freezing, use of chemicals, electroporation, supersonic flow, among others. Detailed information about principles, advantages, and disadvantages can be found elsewhere (Halim et al. 2012b; Günerken et al. 2015; Postma et al. 2016; Toledo-Cervantes and Morales 2014). The cell disruption method depends on the cell wall characteristics (Eppink et al. 2017) and must be carefully selected because some cell components can be denatured (Günerken et al. 2015; Pei et al. 2010). The cell wall is a barrier that separates the cellular content from the surrounding aqueous medium. Its composition is strain-dependent, but is usually composed of polysaccharides (cellulose, hemicellulose, etc.), lipids, and membrane proteins, which can adopt different structures (Baudelet et al. 2017). For instance, Chlorella has two or three layers with different structures, such as a transparent microfibrillar layer and

Table 2 Harvesting technologies, adapted from Barros et al. (2015) and Gerardo et al. (2015)

	Technology	Function	Advantages	Difficulties
Primary step: pre-concentration	Sedimentation	Based on gravity for the cell to freely settle	No energy needed Very low cost	Slow process that can affect biomass, relatively high superficies required to treat high volume Final concentration still very low
	Coagulation–flocculation (electro)	Use of a chemical or bioflocculants that use the electrostatic properties of a microalgae cell to make it aggregate in larger flocs to enhance settling velocity	Simple, better, and faster sedimentation as microalgae forms aggregates No energy use if not electro-coagulation	Use of additives which can be expensive and toxic And/or high additional energy for electro-coagulation Recycling of culture medium is limited
	Flotation	Gas vesicles are used to help microalgae going upward and float near the surface; a surfactant can be aggregated	Low cost Low space required Short operation times	Use of additives to make flocculates Inapplicable in case of marine microalgae
Secondary step: biomass recovery or dewatering	Centrifugation	Based on centrifugal force	High efficiency Fast Almost any microalgae can be centrifuge	Expensive as it requires high quantity of energy Viable only for recovery of high-value products Possible cell damage
	Filtration	Filtration with membranes	High efficiency Free of any additive chemical hazardous	Fouling/clogging which increases operational cost Membranes must be cleaned regularly Relatively high cost due to pumping and membrane replacement

(continued)

Table 2 (continued)

	Technology	Function	Advantages	Difficulties
Emerging technologies	Ultrasound	Ultrasound is used to exploit dielectric properties of microalgae cells by forcing them to move to the nodules of the standing waves, flocculate, and sediment	High efficiency Potential for multipurpose use as: cell lysing with lower frequencies and higher pressures	High energy cost compared to other technologies including filtration and centrifugation
	Magnetophoretic (Japar et al. 2017)	Use of magnetophoretic properties of microalgae by adding a magnetic nanoparticle which will link microalgae cells into flocculates and be removed	Easily remove from broth High efficiency of recovering biomass High efficiency of recycling magnetic particle and can be reused	Energy intensive due to need for agitation and use of compressors and shear mills High efficiency can be maintained only for low flow rates (<0.6 L/h)
	Polymers absorbing water (Japar et al. 2017)	Super-absorbent polymers which will immobilize biomass	Fast High concentration factor	Separation of biomass for polymers still unclear
	Algal predators (bacteria) (Sathe and Durand 2015)	Flocculating agent	Potent flocculating agent Inexpensive Eco-friendly	More investigation necessary
	Attached microalgal growth	High velocity culture in PBR have a support inside the culture medium where microalgae can fix itself	Easy recovery	Depend on the microalgae ability to form film
	Pelletizations with filamentous fungi (Gultom et al. 2014; Tan et al. 2014, Xia et al. 2011)	Fungi will trap microalgae cell in their filamentous and make pellets of high weight which make them sediment	Low cost of culture (wastewater, sucrose) (Zhou et al. 2013)	2 days to sediment Need to be further filtrated

trilaminated structure (Yamada and Sakaguchi 1982); hence, the cell disruption of Chlorophyta is hard because they have rigid and thick cell walls (Baudelet et al. 2017). Additionally, culture conditions could alter the cell wall structure and composition (Eppink et al. 2017).

According to Parniakov et al. (2015), the pulse electric field (PEF) seems promising for a controlled cell wall disruption as pre-treatment or combined with other treatment processes, such as sonication or extraction with a green solvent (Postma et al. 2016). But this technology has some disadvantages; e.g., the solution must be free of ions, and energy consumption is strongly dependent on biomass concentration (Günerken et al. 2015). Nevertheless, more research is needed to improve the efficiency of the cell wall disruption for different microalgae biomass concentrations and the liberation of products needs to be increased. Electrical arc treatment is a relatively recent technique for extraction from biomass. This technique was applied for polyphenol extraction and resulted in lower energy consumption, 16 kJ/kg compared to 53–267 kJ/kg for PEF (Boussetta et al. 2013). This could be of high interest for microalgae biorefinery. Two other processes are promising: the subcritical water use and the high-pressure homogenization (Roux et al. 2017), which suggest a positive energy balance for cell disruption.

3.1.3 Metabolite Extraction

After cell disruption, the next step is the extraction of products. Extraction methods include the use of solvents, super- or subcritical fluids, polymers, ionic liquids, membranes, or resins (see Table 3). The main objective of extraction is to obtain all fractions with no loss either in quantity or in quality (avoiding alteration/loss in functions). Reviews about this topic have been done by Eppink et al. (2017), Gong et al. (2017), González-Delgado and Kafarov (2011), Michalak and Chojnacka (2014), Postma et al. (2016), Roux et al. (2017), and improvements have mainly focused on fuel/lipid extraction by either solvent or supercritical fluid extraction. Conventional extraction procedures for lipids are hydraulic pressing, expeller pressing, and solvent extraction (Cuellar-Bermudez et al. 2015; Ranjith Kumar et al. 2015). For solvent extraction, hexane, hexane-isopropanol, or chloroform-methanol are the main solvents used (Cuellar-Bermudez et al. 2015; Ranjith Kumar et al. 2015). The adequate solvent blend must be chosen depending on lipid polarity and solubility (Cuellar-Bermudez et al. 2015). The ideal solvent blend for lipid extraction from microalgae seems to be chloroform-methanol in a 1:1 (%v/v) proportion (Ryckebosch et al. 2012). A wet technology for lipid extraction was studied by the NAABB (National Alliance for Advanced Biofuels and Bioproducts) at laboratory-scale and showed good performance with selective separation of free fatty acids and tocopherol; this alternative offers energy savings because harvesting and drying operations are not necessary (Marrone et al. 2017). Recent advances have also been made in supercritical fluid extraction (SFE) (Nobre et al. 2013; Yen et al. 2015). One advantage of SFE is the application for extraction of both lipids and pigments. Nobre et al. (2013) achieved $33 g_{lipid}/100\ g_{dry\ biomass}$

Table 3 Extraction technics

Extraction techniques	Advantages	Disadvantage	References
Water-soluble techniques			
Autoclaving (subcritical water)	High efficiency, solvent eco-friendly, recyclable, nontoxic solvent, direct process from culture with concentration step, subcritical water	High energy consumption, not suitable for delicate compounds sensible to temperature	Gong et al. (2017)
Boiling	Quick, solvent eco-friendly, recyclable, nontoxic solvent, direct process from culture with concentration step, good to extract phenols	High energy consumption, not suitable for delicate compounds sensible to temperature, low efficiency	Godlewska et al. (2016)
Homogenization (high pressure)	High efficiency, co-extraction, can be eco-friendly and nontoxic (use of green solvent)	High energy consumption not adequate for compounds sensible to temperature and/or pressure	Mulchandani et al. (2015)
Novel extraction techniques			
Supercritical fluid extraction (SFE)	High efficiency, cheap, high removal rate, eco-friendly, with CO_2 for thermolabile molecules and other fragile compounds, recyclable	High inversion, supercritical CO_2 nonpolar	Cuellar-Bermudez et al. (2015), Grosso et al. (2015), Michalak and Chojnacka, (2014), Nobre et al. (2013), Ranjith Kumar et al. (2015), Taher et al. (2014), Yen et al. (2015)
Ultrasound-assisted extraction (UAE)	Reduce quantity of solvents used, high performance and process faster due to quicker kinetics, equipment cheaper than other novel techniques, scalable, combine to MAE and SFE possible, reaction and extraction can be joint	Increase in energy consumption	Michalak and Chojnacka (2014)

(continued)

Table 3 (continued)

Extraction techniques	Advantages	Disadvantage	References
Microwave-assisted extraction (MAE)	Reduce quantity of solvents used, high performance and process faster, reaction and extraction can be joint	Increase in energy consumption, a further step is needed to split solid residue from liquid phase	Michalak and Chojnacka (2014)
Pressurized liquid extraction (PLE)	Reduce quantity of solvents used, faster than other solvent extraction, shorter time of process, an extended variety of solvents can be occupied for PLE, more compliant for bioactive molecules than SFE	Cannot be used for bioactive compounds susceptible to temperature due to high temperature and pressure of the process	Michalak and Chojnacka (2014)
Enzyme-assisted extraction (EAE)	Bioproducts easily freed, eco-friendly, nontoxic, no increase in energy consumption, easy separation of molecules, scalable	High cost of enzymes, enzyme selectivity which will increase cost to make an efficient cocktail, yield depending on enzyme selectivity	Michalak and Chojnacka (2014)
NADESs	Eco-friendly, biodegradable, nontoxic, low cost compared to DESs Made of primary metabolites such as amino acids, organic acids, sugars, and choline derivatives Used for phenols and flavonoids extraction	Further study of lipids extraction with theses solvents needs to be done	Espino et al. (2016), Jeevan Kumar et al. (2017)
SUPRASs	Nanostructured amphiphile liquids, solvent improvability is high, appropriate for extraction as existence of various polarity areas	Efficiency needs to be demonstrated for extraction	Ballesteros-Gómez et al. (2010), Jeevan Kumar et al. (2017)

(continued)

Table 3 (continued)

Extraction techniques	Advantages	Disadvantage	References
Deep eutectic solvents (DESs)	Safe, cheap, multi eutectic fluid system (two or more solvents), biodegradable, nontoxic, production of such solvents is inexpensive, extensive polarity	Promising results where demonstrated, application area needs to be defined	Jeevan Kumar et al. (2017), Jeong et al. (2015), Paiva et al. (2014)
Fluorous solvents	Nontoxic, hydrophobic and lipophobic, inert in nature, employed for trace metals extraction and fractionation of oils, phase easily split up	Lipid extraction through fluorous solvents needs to be studied	Horváth (1998), Jeevan Kumar et al. (2017)
Acid and alkaline hydrolysis	High efficiency, fast	Use of high concentration of acids or alkaline	Dong et al. (2016), Jeevan Kumar et al. (2017), Roux et al. (2017)
Conventional solvent extraction			
Extraction in Soxhlet apparatus	High efficiency, easy to scale up thanks to its simple operating system, safety	Toxicity	Halim et al. (2012a), Michalak and Chojnacka (2014)
Solid–liquid extraction (SLE)	Easy to scale up	Toxicity	Michalak and Chojnacka (2014)
Liquid–liquid extraction (LLE)	Easy to scale up	Toxicity	Michalak and Chojnacka (2014)

from *Nannochloropsis* sp. biomass when supercritical CO_2 was used, and they observed an increase of 36% when ethanol was added as co-solvent; in this case, the global recovery was around 85% for lipids, and 70% of pigments. Ethanol allows faster extraction and is suitable for feed and nutraceutical applications. The main advantage of SFE with CO_2 is the use of a nontoxic, cheap, safe, and chemically inert solvent at adequate critical temperature and pressure. Water is another good candidate for SFE being nontoxic and cheap, but high pressure and temperature necessary to reach its critical point involve higher energy costs than SFE with CO_2.

On the other hand, innovative solvent-free methods such as osmotic pressure or isotonic (also called ionic) solvent have started to be investigated (Ranjith Kumar et al. 2015). They are eco-friendly, since they avoid the use of toxic solvents, have lower energy consumption, and are considered cheaper alternatives (Adam et al. 2012). Ionic liquids, nonaqueous salt solutions that comprise an organic cation and a polyatomic inorganic anion, are becoming popular and high extraction yields are expected due to their chemical nature. They are considered as green solvents (Eppink et al. 2017; Halim et al. 2012a; Kumar et al. 2016), because they reduce energy consumption, allow the use of alternative solvents and renewable natural products, and ensure a safe and high-quality extract/product" (Chemat et al. 2012).

Ionic liquids are nonvolatile, thermally stable, and also have the capacity to disturb cells and destabilize them (Park et al. 2015). Grosso et al. (2015) suggest the use of switchable solvents to improve the extraction, using an alcohol and an amine base in a nonionic state that after injection of CO_2 turn into an ionic liquid; finally, to recycle the solvent, N_2 is injected through the solvent turning it back to nonionic state. There is another kind of switchable solvents, such as hydrophilic solvents (Boyd et al. 2012; Jessop et al. 2012). Some interesting reviews about green extraction were made by Du et al. (2015) and Jeevan Kumar et al. (2017). Green techniques allow a lower use of solvent, improve product quality, do not affect other biocompounds, and, moreover, induce a decrease in energy consumption (Jeevan Kumar et al. 2017). Emerging green solvents include natural or deep eutectic solvents (NADESs) and supramolecular solvents (SUPRASs). NADESs play a role as alternative media to water in living organisms. The main reason to use this other medium is to help survival of any organism under harsh conditions, such as cold or dryness, and therefore, NADESs are mostly made up of sugars, urea, choline chloride, and organic acids (Jeevan Kumar et al. 2017). SUPRASs are nanostructured liquids that consist of assemblies of amphiphiles dispersed in a continuous phase.

Novel approaches consist in combining a disruption method with an extraction method to enhance the global process and make it greener as is the case with microwave-assisted extraction (MAE) or ultrasound-assisted extraction (UAE), enzyme-assisted extraction (EAE), pressurized liquid extraction (PLE) combined with solvent extraction or other techniques (Ibañez et al. 2012; Kadam et al. 2013).

Regarding protein extraction, fragile proteins are of economic interest and extraction of the protein fraction after cell lysis using mild technologies is incipient (Eppink et al. 2017). Proteins are mainly recovered with solvents by filtration (micro- and ultrafiltration) (Marrone et al. 2017) or through precipitation by pH shifting (Ursu et al. 2014). Extraction of proteins through tangential ultrafiltration and neutral pH has a relatively high yield without alteration of protein functionality (Ursu et al. 2014). Filtration requires little energy and is considered a green and mild process because it does not change protein state compared to extraction with solvents (Safi et al. 2017). However, precipitation is considered a better option to obtain protein powder and also to reduce the operating costs (Ursu et al. 2014). A recent interest has emerged for the use of polymers within the aqueous two-phase system (ATPS) looking for mild separation and extraction of proteins. This system is prepared using a polymer–polymer and a polymer–salt mixture in such a way that two water-rich phases are formed, thus providing the necessary gentle solvent for proteins that does not affect their functionality. Zhao et al. (2014) proposed a multiple stage ATPS extraction in order to increase purity of C-phycocyanin from *Spirulina platensis*. Phong et al. (2017a) combined UAE and ATPS for protein recovery of *Chlorella sorokiniana* finding that phases could be recycled at least five times.

On the other hand, high-value pigments are commonly extracted using solvents or supercritical fluids. Enhancement of pigment extraction has also been investigated through combination of solvent extraction with other methods, such as ultrasound or microwaves (Halim et al. 2010; Pasquet et al. 2011).

Lastly, classical solvent extraction can be used for carbohydrate recovery but improvements in recovery have been reported using fluidized bed extraction or ultrasonic-assisted extraction. However, this enhancement was associated with higher operating costs (Zhao et al. 2013). Wu et al. (2017) proposed a high-speed counter current chromatography (HSCCC) combined with ATPS extraction to recover high-purity polysaccharides in a single-step extraction process. Carbohydrates from microalgae have aroused recent interest in the biorefinery process (IEA 2017; Templeton et al. 2012).

Emergent technologies for protein recovery include the liquid biphasic flotation based on the combination of ATPS and solvent sublation. This technique allows the integration of concentration, separation, and extraction into one step, along with a higher concentration coefficient (Phong et al. 2017a, b).

3.1.4 Fractionation

Fractionation could be required in purification train after extraction and depending on the application of microalgae products. It focuses on the primary recovery and partial purification of products with no loss in products and functionality. The goals of fractionating microalgae biomass are either to separate lipids, proteins, and carbohydrates for further valorization of each fraction or to obtain a specific compound. Hence, the microalgal extracts from either hydrophobic and hydrophilic phase can be separated using common techniques based on density differences and further selective techniques (see Table 4), such as ionic exchange chromatography, charged membranes or protein precipitation (Schwenzfeier et al. 2011, 2014) allow isolation of proteins from a common hydrophilic phase where carbohydrates are also present. On the other hand, complex high-cost downstream processing is used when isolation of a specific compound, such as PUFAs, from lipid fraction (Dibenedetto et al. 2016) or high-grade protein is required (Halim et al. 2016). Therefore, developments in fractionation are still limited for high-value products due to its high cost and feasibility is only attainable in domains, such as food, health, and cosmetics. Indeed, this is an incipient area that needs development but thanks to biopharmaceutical field, mild extraction techniques are being adopted for microalgae specialty products.

Membrane technologies are commonly used for biomass harvesting (Drexler and Yeh 2014). They provide a thin barrier to restrict the interactions between the solvent and solute depending on their properties and membrane characteristics; however, finest filtration methods, such as microfiltration (MF), ultrafiltration (UF), nanofiltration (NF), and reverse osmosis (RO), allow selective product separation.

Table 4 Fractionation technics applications and advantages

	Methods	Applications and other advantages
Pre-fractioning		
Polymers (Cuellar-Bermudez et al. 2015; Grosso et al. 2015)	Aqueous two-phase system (ATPS) separation: mix of a polymer–polymer, polymer–salt beyond a critical concentration will form two phases	For protein extraction and purification Alternative to chromatography
Ionic liquids (Grosso et al. 2015; Suarez Ruiz et al. 2017)	Simple molten salts in forms of cations and anions, separate in hydrophilic and hydrophobic phase	Applications for organic synthesis, liquid-phase extraction, and catalysis for clean technology and separations Able for recycling with minimum pollution compared to organic solvents Separation of hydrophilic (e.g., carbohydrates, proteins) and hydrophobic (e.g., pigments, lipids) Novel approach: mix of organic solvents (e.g., ethyl acetate) and ionic liquids
Fractionation (Pei et al. 2010; Ranjith Kumar et al. 2015; Taher et al. 2014)		
Membranes (Demmer et al. 2005; Gerardo et al. 2014; Marcati et al. 2014; Safi et al. 2014b; Schwenzfeier et al. 2014, 2011; Van Reis and Zydney 2001)	Separation of carbohydrates and proteins and pigments traces	Tangential flow presents the advantages of fractionation with different filter sizes, from 1 to 1000 kDa, under mild conditions New developments of filter: dead-end filtration with a layer of specific ligands (charged, hydrophobic, hydrophilic) with the objective to extract a precise protein component; further fractionation is possible
Resins (Bermejo et al. 2006; Cuellar-Bermudez et al. 2015; Schwenzfeier et al. 2014)	Chromatographic separation (size exclusion or ionic exchange)	Protein mixture was fractionated with ionic exchange chromatography. Technology mainly used for high-value products in food/health/cosmetics
Extraction (Cuellar-Bermudez et al. 2015; Gilbert-López et al. 2015; Grosso et al. 2015; Taher et al. 2014)	Solvent extraction or supercritical fluids Nanofiltration	Recent research points out the supercritical fluids as a promising technology for scalable extraction of pigments and lipids

Residual pigments and carbohydrates are separated from the hydrophilic phase through dead-end or tangential flow membrane filtration (Gerardo et al. 2014; Lorente et al. 2017; Marcati et al. 2014; Safi et al. 2014a, b; Schwenzfeier et al. 2011, 2014; Van Reis and Zydney 2001). Besides, membrane technologies can be combined with other processes to increase selectivity by combining principles of other fields (Demmer et al. 2005) for isolation of specific proteins using adsorbent particles embedded in membrane pores or selective aqueous buffer systems for the next fractioning step of carbohydrates and/or proteins (Weaver et al. 2013).

High-resolution chromatography has also been used for fractioning product recovery. In a first step, Schwenzfeier et al. (2011) characterized *Tetraselmis* sp. fractioning with a mild process and ionic exchange chromatography was used to obtain protein. High purity can be also reached through ionic exchange chromatography and size exclusion chromatography for phycoerythrin from *Porphyridium cruentum* (Bermejo et al. 2006; Cuellar-Bermudez et al. 2015). Those highly purified proteins could be of interest for clinical and pharmacological research as they can present some properties interesting for health, such as antioxidant or anticancer activities.

3.1.5 Selective Extraction

Some techniques, such as ionic liquids, SFE, and ATPS, are applied to hydrophobic phase for further separation of their different compounds (PUFAs, glycolipids, phospholipids) from oily fraction. Solvent extraction or SFE is specifically used to split up lipids and pigments (Cuellar-Bermudez et al. 2015; Grosso et al. 2015). Innovative processes, such as direct transesterification during SFE, are commented by Ranjith Kumar et al. (2015) and Taher et al. (2014) in reviews, but they need deeper research for scaling up to industrial scale.

3.2 *Processing Biomass to Obtain Energy*

Processes involved in microalgae biomass transformation in the biofuel-driven biorefineries are classified into direct combustion, thermochemical or biochemical processing, and chemical transformation (see Fig. 2) involving the chemical transformation of lipids extracted from biomass to produce biodiesel through transesterification. All of them are explained in the following sections.

3.2.1 Direct Combustion

It is the most direct route to utilize microalgae biomass as fuel. Direct combustion is a thermochemical technique used to burn biomass in the presence of excess air. In theory, algae can be dried and burned. Combustion of algae for power generation

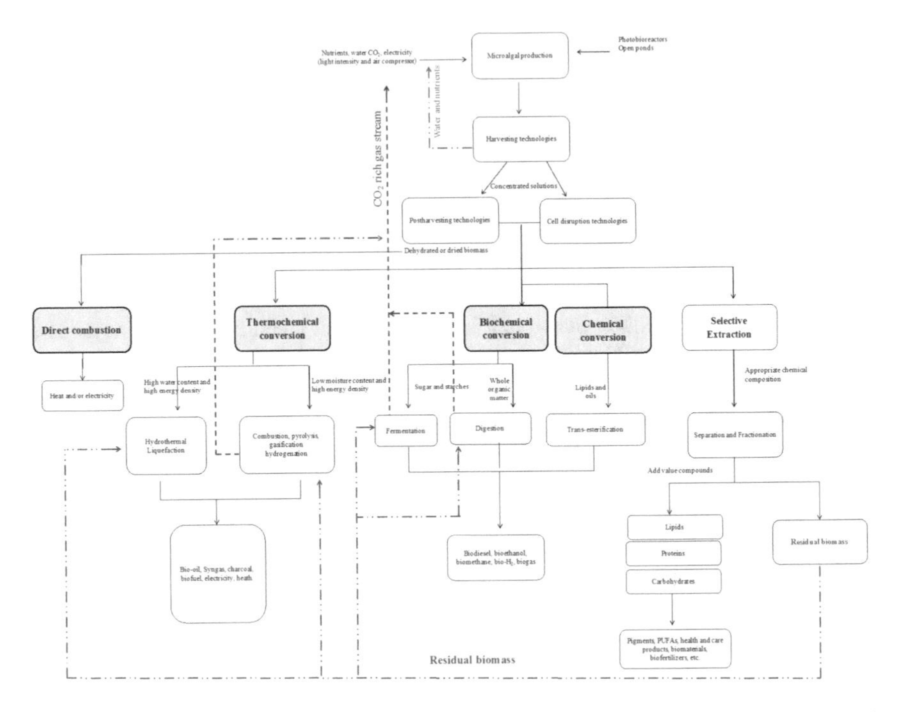

Fig. 2 Thermochemical and biochemical processing in a microalgae biorefinery for production of biofuels and value-added compounds (adapted from Toledo-Cervantes and Morales 2014). Brown dotted lines mean the incorporation of residual biomass to produce biofuels, blue dotted lines indicate water recycling, and red dotted lines are CO_2 gas stream generated during microalgae biofuel processing and reincorporated for biomass growth

has never been attempted on a large scale, in part because the large-scale cultivation operations have focused on other, more economical uses of algae. Drying is not difficult, and several methods have been standardized, including sun-drying, drum drying, vacuum drying, and freeze-drying. These methods are capable of reducing the moisture content to $\sim$2%. Because it is not possible to burn biomass directly in an internal combustion engine, the technology for power generation could be a Rankine engine using a forced convection industrial boiler. A system with this configuration would require an area of $\sim 20 \times 20$ m and a continuous power output of a kilowatt, at a cost of roughly USD 0.95 kWh, which is approximately four times the cost of current diesel generation in off-grid areas. Furthermore, the practical aspects of pumping algae from a separator to a dryer and handling and feeding algal solids into a combustor have not been standardized or automated for any commercial power generation scheme as of yet, so while this configuration appears quasi-attractive from a thermal efficiency, land footprint, and cost perspective, it would still require significant engineering input to be realized (Orosz and Forney 2008).

3.3 Thermochemical Processing

Thermochemical processing is the decomposition of organic materials from biomass for conversion at elevated temperatures and pressures into fuels. It comprises: pyrolysis, torrefaction, hydrothermal liquefaction, and gasification (Chen et al. 2015; Tan et al. 2014; Toledo-Cervantes and Morales 2014). Through these conversion technologies, solid, liquid, and gaseous biofuels are produced for heat and power generation. Pyrolysis is the combustion taking place at high temperatures (350–800 °C) in the absence of oxygen. It produces fuels with medium–low calorific power (Brennan and Owende 2010), such as charcoal, gas, and biocrude. Torrefaction is a mild pyrolysis at lower temperatures (200–300 °C) lasting minutes to hours, whose main product is a solid biofuel. Biocrude is also produced by hydrothermal liquefaction performed at 300–350 °C and pressures of 5–20 MPa to convert wet microalgal biomass into liquid fuel without using hot compressed or subcritical water (Chen et al. 2015). The refining of biocrude produces fuels and lubricants, and some of the byproducts form materials, such as plastics, detergents, solvents, elastomers, and fibers, such as nylon and polyesters, and asphalts (Chailleux et al. 2012). On the other hand, the main product of gasification is syngas (CO, CO_2, H_2), which is obtained when dry microalgae react with an oxidizer, such as air, oxygen, and water or steam, in a partial oxidation environment at a temperature ranging between 800 and 1000 °C, within a pressure range of 1–10 bar in an environment of insufficient oxidizer used for producing fuels and chemical intermediates. Comprehensive reviews of recent progresses and development of thermochemical processing are found in the latest literature (Toledo-Cervantes and Morales 2014; Chen et al. 2015; Chiaramonti et al. 2015).

3.3.1 Biochemical Processing

Biochemical conversion depends on the cell wall digestibility, which could be enhanced by physical, chemical, or biological pre-treatment of either whole or residual defatted biomass to reduce the processing time and increase the biomass. Pre-treatments are classified into physical, chemical, or biological. They include bead milling, ultrasound, alkaline, acidic or thermal hydrolysis, ionic liquid, pulsed electric field, microwave or enzymatic pre-treatment, among others (Eldalatony et al. 2016; Jankowska et al. 2017).

3.3.2 Anaerobic Digestion

Anaerobic digestion is the bacterial decomposition of organic biopolymers (i.e., carbohydrates, lipids, and proteins) into monomers in the absence of oxygen over a temperature range of about 30–65 °C. These monomers are easier to convert into a methane-rich gas via fermentation (typically 50–75% CH_4), and CO_2 is the second main component found in biogas (approximately 25–50%). Biogas can be upgraded up to >97% methane content and used as a substitute for natural gas (Toledo-Cervantes et al. 2017) to generate electricity.

The first mention of using microalgal biomass to produce biogas was long ago (Golueke et al. 1957), but the idea was taken to the modern times with the work of Sialve et al. (2009) and mainly due to the efforts to improve the economy and sustainability of biodiesel production from microalgae lipids (Harun et al. 2011) using the waste defatted biomass. Biogas production from anaerobic digestion of microalgae biomass is primarily affected by organic loads, temperatures, pH, and retention times in the reactor used. Besides, it was demonstrated that biogas potential is also strongly dependent on the microalgae species and biomass pre-treatment (Alzate et al. 2012; Jankowska et al. 2017; Harun et al. 2011; Mussgnug et al. 2010).

As was previously mentioned, biomethane from microalgae biomass can be used as gaseous fuel and to generate electricity, whereas the spent biomass can be used to make biofertilizers or a wide range of biofuels and chemicals in a thermochemical approach. Although microalgae biomass offers good potential for biogas production, industrial production has still not been fully implemented.

3.3.3 Fermentation

Bioethanol is usually obtained by alcoholic fermentation from carbohydrates, such as sugars, cellulose, or starch (Harun et al. 2014; Ho et al. 2013; Tan et al. 2014) or previously hydrolyzed lignocellulosic feedstocks. Microalgal bioethanol can be produced through two distinct processes: via dark fermentation or yeast fermentation.

The dark fermentation of microalgae consists of anaerobic bioethanol production by the microalgae themselves through the consumption of intracellular starch (Ueno et al. 1998). The yeast fermentation process of microalgal biomass is well known industrially, and to achieve higher yields, it is necessary to screen microalgal strains with high carbohydrate content or induce accumulation of intracellular starch. On the other hand, polysaccharides on the microalgal cell walls are not easily fermentable for bioethanol production by microorganisms. For fermentation, an acid pre-treatment has been proposed as the best option compared to other pre-treatment methods, namely in terms of cost-effectiveness and low energy consumption (Harun and Danquah 2011). During the bioethanol fermentation process, the pH is maintained in the range of 6–9, because a pH below 6 or above 9 could slow down bioethanol production. The fermentation process consumes less energy, and the process is much simpler in comparison with the biodiesel production system. In addition, the CO_2 produced as a by-product from the fermentation process can be recycled as carbon source for microalgae cultivation, thus reducing greenhouse gas emissions as well.

Hydrogen can be also produced by dark fermentation (DF) through the spore-forming bacteria, such as *Clostridium*. There are comprehensive reviews on hydrogen production from microalgae biomass (Buitrón et al. 2017; Sambusiti et al. 2015; Xia et al. 2015). Recent results show a clear potential of microalgae as feedstock for DF, achieving molar yields up to 3 mol H_2/mol sugar, which represents 75% of the maximum theoretical yield (Nayak et al. 2014). Such values are obtained only with other carbohydrate-rich substrates operated under thermophilic conditions or with a reduced hydrogen partial pressure. In DF, the highest yields are produced with the simplest carbohydrate molecules (Quemeneur et al. 2011); hence, carbohydrates must be released to be assimilated for hydrogen production when microalgae are used as substrate (Nguyen et al. 2010). Hydrolysis for cell wall disruption is a usual method to obtain fermentable sugars (Günerken et al. 2015).

Therefore, the major constraint to the use of microalgae for DF is related to the hydrolysate quality in terms of reducing sugar concentration and the pre-treatment efficiencies. Methane production is a frequent concern in DF systems because methanogenic microorganisms can be presented in the inoculum used. For instance, using wet untreated biomass, Kumar et al. (2016) produced methane rather than H_2 because of an inefficient inoculum heat pre-treatment.

As in other technologies for biofuel production using microalgae biomass, the suitable DF application depends on its insertion into an integrated scheme. The final by-product of DF is a mixture of volatile fatty acids and solvents, depending on the operational conditions and the microorganisms present.

3.3.4 Chemical transformation

Transesterification

Microalgae biodiesel is generally produced through the extraction and further transesterification of algal oil. Transesterification is the reaction of triglycerides (TAGs) with alcohol or methanol, in the presence of a catalyst that produces glycerol and fatty acid methyl esters (FAME or biodiesel) derived from TAGs. The complete biomass conversion depends on lipid profile, oil impurities, catalyst nature, temperature, and time. Transesterification can be catalyzed by acids, alkalis, or lipase enzymes (Chisti 2007). Recently, Lemões et al. (2016) have studied direct wet-transesterification using ethanol with yields similar to those obtained from extracted lipids. Furthermore, contributions to sustainability are claimed based on savings related to the unnecessary dewatering of the microalgae biomass, and the use of ethanol as renewable feedstock. Other innovation is transesterification in supercritical conditions, a catalyst-free chemical reaction that enables the full transformation of TAG (Ngamprasertsith and Sawangkeaw 2011), dramatically accelerated under supercritical conditions.

4 Biorefinery Strategies and Current Concepts

Fluctuations in fossil fuels prices, diminution of oil reserves, and COP21 agreements to reduce the GHG emissions encourage the biomass-biofuel industry and enhance the microalgae biofuel research (Pires 2017). Microalgae can play a dual role: They capture CO_2, and the resulting biomass can be used to produce a wide range of materials. It is important to mention that some chemicals derived from microalgae biomass cannot be synthesized from fossil fuels. Some of them are high-value products and can be used directly after separation or with slight structural adjustments. For these reasons, microalgae are of great interest for research and development of industrial processes to make viable their use in a biorefinery concept. As was mentioned previously, in the biorefinery concept, it is necessary to valorize most of the constituents of the microalgal biomass. Figure 3 shows options for valorization of the algae biomass that include its use as: (1) intact algae cells, (2) the disrupted whole cell content, or (3) fractionation of the biomass into different biochemical groups or specific compounds (Bastiaens et al. 2017). In the first alternative, the microalgae biomass is mostly commercialized as dry powder for feed, food, or aquaculture and the volume market is constantly increasing (IEA 2017; Vigani et al. 2015). Manufacturing commercial products derived from whole cells does not generate residual biomass, and extraction or fractioning of biomass is not necessary. On the contrary, the other two options allow the application of the biorefinery concept. The following sections present the main strategies to cultivate/process/valorize the microalgae biomass, and they are globally classified into four categories: (i) biofuels production only (low-value compounds), (ii) high-value-added products

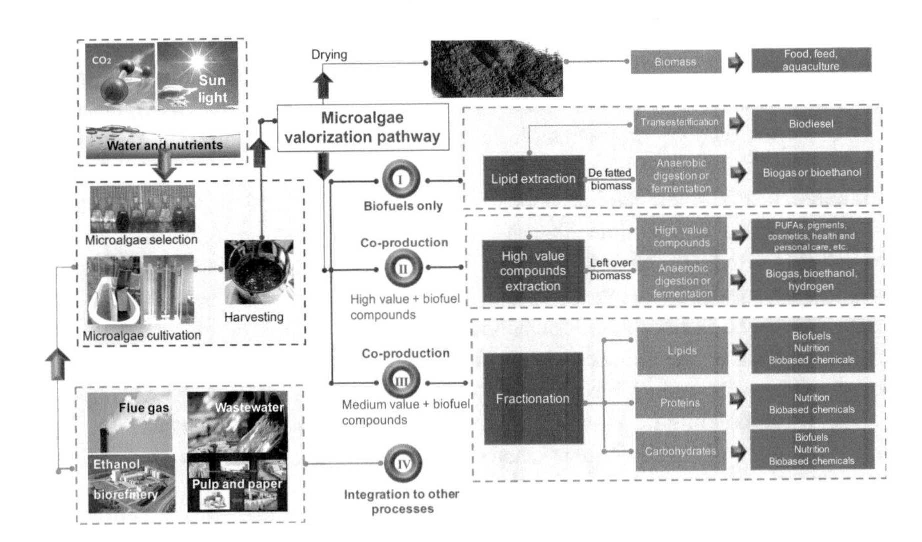

Fig. 3 Biorefinery strategies for microalgae biomass valorization

Table 5 Examples of biorefineries classified according the different strategies

Target compound	Substrate	System/conditions	Downstream processing	Production Scale	Market situation	Algae	References
Only biofuels							
Sugars (bioethanol) Residual biomass (biocrude)	BG-11 cultivation medium	Hybrid vertical flexible films, indoor and outdoor conditions	Vapor compression steam Stripper	Pilot scale	Marketable products distributed by Algenol Company	Cyanobacteria	Sterner (2013)
Lipids (biodiesel) Residual biomass (biomethane)	Post-transesterified microalgae residues (*Chlorella* sp.)	Anaerobic digester 2 L, continuous stirring at 300 rpm, semi-continuous fed	Anaerobic digestion	Laboratory	–	Digester bacteria	Ehimen et al. (2011)
Lipids (biodiesel)	Bold's Basal Medium	Raceway pond, 2000 L, greenhouse conditions, semi-continuous mode	–	Pilot scale	–	*Scenedesmus* sp. SCX2	Fernández-Linares et al. (2017)
Lipids (biodiesel) Residual biomass (methane)	Liophilized biomass	–	SCF extraction with CO_2, anaerobic digestion	Laboratory	–	*Scenedesmus almeriensis*	Hernández et al. (2014)
Lipids (biodiesel) Carbohydrates (bioethanol)	Codafol 14.6.5 plant fertilizer	Plastic sleeves, 300 L, outdoor conditions	Cell disruption by steam explosion and solvent extraction with hexane	Pilot plant	–	*Nannochloropsis gaditana*	Nurra et al. (2014)
Lipids (biodiesel)	Wastewater and gas emissions from refinery oil industry	Bubble column photobioreactor, 2 L	Lipid extraction by Bling and Dryer method	Laboratory	–	*Aphanothece microscopica Nägeli*	Jacob-Lopes and Franco (2013)

(continued)

Table 5 (continued)

Target compound	Substrate	System/conditions	Downstream processing	Production Scale	Market situation	Algae	References
Hydrogen residual biomass Biogas	TAP medium and ProF medium	–	Drying and anaerobic fermentation	Laboratory	–	*Chlamydomonas reinhardtii* and *Scenedesmus obliquus*	Mussgnug et al. (2010)
Biofuels and high-value compounds							
Carotenoid (pigments) Lipids (biodiesel)	Mineral medium culture	Bubble column 5L light intensity 25.7 μmol/m^2/s and temperature of 25 °C	Freeze-drying and milling and acetone, and Sohxlet extraction with hexane	Laboratory	–	*Chlorella protothecoides*	Campenni et al. (2013)
PHB (bioplastic) Lipids (biodiesel) Residual biomass (biogas)	Mineral medium culture	Tubular reactor	Solvent extraction with hexane, Glycerol purification	Model conceptual	–	*Haematococcus pluvialis*	García-Prieto et al. (2014)
Carotenoids (pigments) Lipids (biodiesel) Residual biomass (hydrogen)	GPM medium	Polyethylene bag, 10 L, light intensity 25.7 μmol/m^2/s and temperature of 25 °C	SCF extraction with CO_2/ethanol, dark fermentation	Laboratory	–	*Nannochloropsis* sp.	Nobre et al. (2013)
β-carotene, phytosterols fatty acids and biochar	Walne's medium	Vertical tubular photobioreactor, 400L, outdoor conditions, semi-continuous mode	Pyrolysis	Pilot scale	STAR facility center	*Dunaliella tertiolecta*	Francavilla et al. (2015)

(continued)

Table 5 (continued)

Target compound	Substrate	System/conditions	Downstream processing	Production Scale	Market situation	Algae	References
Biofuels and other medium products							
Lipids (biodiesel) Proteins (food) Reduced sugars (bioethanol)	BG-11 mineral medium	Raceway 300 m^3, outdoor conditions	Pre-treatment, cell disruption Lysis buffer solution for protein and H_2SO_4 and autoclaving for reduce sugars	Laboratory	–	*Scenedesmus obliquus*	Ansari et al. (2015)
Lipids (biodiesel) Protein (food or feed) Carbohydrates (polymers)	Domestic wastewater	Reactor 25L, with 16:8 h, Light-dark (80 $\mu mol/m^2/s$) cycle and temperature of 25 °C	Sequential biomass extraction (protein, lipids and carbohydrates)	Laboratory	–	*Scenedesmus obliquus*	Ansari et al. (2017)
Sugars (bioethanol) Lipids (biodiesel) Proteins (feed)	Mineral medium culture	Flat panel photobioreactor, 650 L, outdoor conditions	Acid pre-treatment, solid liquid separation, fermentation, solvent extraction	Pilot scale	High-value coproduct opportunities possess potential to significantly reduce the high cost	*Scenedesmus acutus*	Dong et al. (2016)

(continued)

Table 5 (continued)

Target compound	Substrate	System/conditions	Downstream processing	Production Scale	Market situation	Algae	References
Fucoxanthin carbohydrates and proteins with antioxidant activity	Mineral medium culture	Vertical photobioreactor, 400 L, outdoor conditions	(1) SCF extraction with CO_2, (2) SCF extraction with CO_2/ethanol, (3) pressurized liquid extraction (PLE) with ethanol, (4) PLE with pure water	Pilot scale	–	*Isochrysis galbana*	Gilbert-López et al. (2015)
Proteins	–	Horizontal tubular photobioreactor, 2000 L, outdoor conditions	Cell disruption by enzymatic treatment and ultrafiltration/ diafiltration	Laboratory	–	*Nannochloropsis gaditana*	Safi et al. (2017)
Biorefinery integration with other process							
Residual biomass organic acids and biogas Lipids (biodiesel) Sugars (bioethanol)	Mineral medium culture	Mathematical model developed in MATLAB Open pond, environmental conditions	Pre-treatment, cell disruption, supercritical fluid extraction (SCF) with CO_2 and ethanol as a co-solvent	Laboratory	Still not economic profitable	*Chlorella vulgaris*	Albarelli et al. (2017)
Lipids (biodiesel) Fatty acids (food) Proteins (feed)	Wastewater	Photobioreactors and open ponds, the LCA represents 100-ha facility	Membrane filtration, hexane extraction,	Demonstration facility	Cellna's Kona Demonstration Facility	*Nannochloropsis oceanica*	Barr and Landis (2017)

(continued)

Table 5 (continued)

Target compound	Substrate	System/conditions	Downstream processing	Production Scale	Market situation	Algae	References
Lipids (biodiesel) Residual biomass	Mixed municipal and industrial wastewater	Laboratory tubes	Drying, lipids extraction through Folch method	Laboratory-scale	–	*Scenedesmus* sp.	Gentili (2014)
Residual biomass (agricultural or food industry) Hydrogen	Residual biomass for dark fermentation and wastewater for algae culture	PBR and anaerobic digester for dark fermentation PBR (photo-fermentation)	Anaerobic digestion (dark fermentation)	Laboratory-scale	–	*Chlamydomonas reinhardtii* *Scenedesmus obliquus* *Chlorococcum littorale* *Platymonas subcordiformis* *Chlorella fusca*	Kapdan and Kargi (2006)
Organic acids	Microalgal biomass used as substrate	Incubator chamber	Alkaline pre-treatment, enzymatic saccharification and anaerobic digestion with Clostridium	Laboratory	–	*Chlorella* sp. and *Tetraselmis suecica*	Kassim and Meng (2017)
High-value microalgae products, fertilizer, and biogas	Secondary streams from pulp and paper industry	Conceptual configuration, liquid effluent from the mechanical dewatering of digested residues and flue gases from biomass boiler at the mill		Conceptual biorefinery CO_2 Wastewater	–	*Nannochloropsis*	Kouhia et al. (2015)

(continued)

Table 5 (continued)

Target compound	Substrate	System/conditions	Downstream processing	Production Scale	Market situation	Algae	References
Joint production of sugar, ethanol, and electricity from sugar cane biorefinery and biodiesel and glycerol from microalgae	Cultivation using CO_2-rich streams derived from fermentation and cogeneration system	Conceptual configuration and simulation of scenarios including economic evaluation		Conceptual biorefinery under the knowledge-based approach CO_2	–	*Chlorella* sp.	Moncada et al. (2014)
Lipids (biodiesel) Protein (food) Pigments (food) Carbohydrates (bioethanol)	Wastewater from fish processing industry	Bubble column photobioreactor, 5 L	Coagulation–flocculation–sedimentation and membrane microfiltration	Laboratory	–	*Aphanothece microscopica Nägeli*	Queiroz et al. (2013)
Lipids (biodiesel) Residual biomass (biosyngas and methanol)	CO_2-flue gas from a coal energy plant	–	Oil extraction, gasification of remaining biomass to obtain syngas conversion of biosyngas to methanol	Conceptual model	–	*Chlorella pyrenoidosa*	Wiesberg et al. (2017)

and biofuels, (iii) medium-value-added products (bulk compound) plus biofuels, and finally, (iv) coupled to other processes in the context of circular economy. Illustrative works about these categories are shown in Table 5 and Fig. 3.

4.1 Strategy I: Biofuel Production Only (Low-Value Compound)

Biofuels are required at low price and in large volume; however, among all products obtained from microalgae, biofuels have the lowest cost and their price is always compared with fossil alternatives; moreover, the energy balance of the production process must be positive. As was shown in Sect. 2 and Table 5, most of the biofuels can be conceptually obtained from microalgae, but biodiesel is one of the most interesting alternatives as liquid fuel for transportation (Amaro et al. 2011; Mallick et al. 2016; Mondal et al. 2017). Nonetheless, it has been demonstrated that currently microalgae biodiesel is not cost competitive when compared with fossil diesel (Chisti 2007). Therefore, the idea of using the residual biomass to obtain more energy was introduced looking for a better revenue, by minimizing wastes to complete biomass use in the biorefinery concept (Collet et al. 2011; Maurya et al. 2016). In the strategy of biofuel only, there are three scenarios to obtain more energy: (i) the wastes can be burned to generate heat and electricity, (ii) they can be converted to biogas using anaerobic digestion or to ethanol using fermentation from carbohydrate, and (iii) they can be thermochemically processed to obtain other biofuels (Brownbridge et al. 2014). But in the structure of the biorefinery with bioprocessing only, selection will depend on the stoichiometric composition of the cell; lipid content in the cell can vary from 15 to 80% depending on the strain, leaving over a huge amount of wastes. Residual deoiled biomass composition is rich in carbohydrates or proteins, and its composition defines the potential application. Residual biomass with high C/N ratio is beneficial for the production of biomethane, bioethanol, and biohydrogen, while a low C/N ratio means high protein content, which may be beneficial for its use as fertilizer, fermentation medium for microorganisms, or feed supplement for animals and fish (Maurya et al. 2016). Biogas production from residual deoiled biomass was suggested to produce heat or electricity and contribute to the energy balance (Collet et al. 2011; Ward 2015). Laurens et al. (2017b) established that fractionation approach to microalgal biomass can create three different potential fuel streams, which allow a 35% reduction in the overall minimum fuel selling price as compared to the biodiesel-only strategy. As can be seen from Table 5, the general strategy is focused on exploiting lipids for biodiesel production and there is a preference for biogas production first from residual biomass and then from bioethanol. One strategy attempts to diversify its production of biofuels with hydrogen. In this approach, ethanol is obtained from fermentation of the released carbohydrates, biodiesel, or jet fuel from the lipid fraction through hydrotreating and isomerization and finally mixed alcohols (isobutanol, isopentanol, and others) from the protein fraction. However, it becomes clear that the use of residual biomass alone to produce energy is not as favorable as it looks at first glance.

4.2 Strategy II: Coproduction of High-Value-Added Products and Biofuels

Microalgae are important producers of many high-value nutraceutical compounds, such as polyunsaturated fatty acids or astaxanthin that can justify the high cost of microalgae cultivation and processing technologies (Liang et al. 2015; Shah et al. 2016). Under this scenario, the fixed CO_2 is valorized and biofuels are produced after extraction of high-value products. Some examples include astaxanthin produced by *Haematococcus pluvialis*, this high-value-added molecule is already commercialized (Lorenz and Cysewski 2000), and its market price is 7000 USD/kg (Hariskos and Posten 2014; Shah et al. 2016). This microalga is an excellent candidate for this strategy because astaxanthin accounts for approximately 5% of the total cell dry weight, representing only a small fraction of it. Astaxanthin is produced under nitrogen limitation and simultaneously with triglycerides that constitute up to 60% of dry weight (Solovchenko 2015) and can be utilized for biodiesel. In this way, after astaxanthin extraction, biodiesel can be produced from lipids and biogas from residual biomass (Shah et al. 2016). Another organism suitable for the biorefinery of high-value-added compound is *Nannochloropsis* (Chua and Schenk 2017) due to its rapid growth, high oil productivities, and omega-3 fatty acid content, specifically the EPA whose market price is up to USD 100 per liter. Furthermore, in the biorefinery configuration, the high edible protein content makes this alga feasible for food or feed. Other high-value molecules in microalgae biomass are: vitamins, pigments, etc., and other examples of studies exploring this strategy are shown in Table 5. As can be seen, pigment as high-value compound coupled to production of biodiesel from lipid fraction or other biofuels from the leftover biomass is the most commonly studied scenario.

One of the major challenges of this strategy is that despite the fact that the projected demand for high-value products from microalgae is increasing, these products are still produced in relatively small amounts; therefore, the biofuel supply cannot be guaranteed.

4.3 Strategy III: Coproduction of Medium-Value-Added Compounds and Biofuels

The approach of integration of medium-value products (carbohydrates, lipids, and proteins) into biofuels production was proposed by Wijffels et al. (2010).

They emphasized that a process for biodiesel from microalgae lipid production only is unlikely to be economically viable and that all biomass bulk components should be valorized in order to develop a feasible and sustainable process. This idea promotes diversification of market sectors, introducing microalgae products not only in the energy sector. According to Hariskos and Posten (2014), bulk chemicals constitute a market volume of >10,000 tons/year with prices from only a few USD/kg up to 100 USD/kg and represent 11% of crude oil destined to petrochemical synthesis. As was seen before, the main components of microalgal biomass depend on the strain and common contents are: lipids (30–50%), proteins (50–70%), carbohydrates (50%), and pigments (Chew et al. 2017). These biochemicals involve the use of protein for feed or food; carbohydrates for bioactive materials, cosmetics, nutritional, and pharmaceutical applications; and lipids, which depending on the length chain, have application as surfactants, cosmetics, solvents, lubricants, or biopolymers (Hariskos and Posten 2014; IEA 2017). Most of the studies focused mainly on lipids for biodiesel and proteins for food or feed. But others include bioethanol production (Table 5).

Under the coproduction strategies (II and III), the selection of mild and selective separation techniques is important to keep the properties of the most biomass components. Therefore, an adequate progression of harvesting followed by cell disruption (Lee et al. 2012) and a further suitable mild and selective extraction and purification sequence of metabolites of interest must be chosen.

Regarding the definition of the best order of extraction of metabolites, it depends on the strain and properties of the products to be recovered. Ansari et al. (2017) showed for *Scenedesmus obliquus* that the sequence of extraction: proteins–lipids–carbohydrates was the most convenient. However a different strategy was proposed by Dong et al. (2016), who suggested that a combined algal processing (CAP) is much better than parallel algal processing (PAP). Instead of extracting lipids from algal biomass prior to alcoholic fermentation as in PAP, lipids are extracted from the anaerobic digestion cake. CAP turns out to be highly efficient for sugar conversion, and lipid loss is negligible. CAP reduces the biofuel cost of microalgae by 9%. However, it is important to mention that this study did not evaluate the cost of bioproducts in a complete biorefinery scheme, using the whole biomass. Table 5 shows different studies using this strategy.

Under this strategy, the high efficiency of fractionation in a sequential process is one of the principal bottlenecks and represents a challenge to overcome. The main goal of this approach is to maximize biomass production and valorization in order to prioritize the use of biomass to obtain products of value by giving more importance to the production of materials, rather than its use for energy (Keegan et al. 2013).

4.4 Strategy IV: Integration of Microalgae into Other Processes and Circular Economy

In recent years, the importance of the new concepts for biorefineries and in particular for microalgae's has been highlighted. Biorefinery obeys the principles of a circular economy in the sense that all waste streams are valued (Mohan et al. 2016a) promoting the use/transformation of secondary or residual streams into value-added products (Yuan et al. 2015). It includes the use of: (i) wastewater as nutrient source (Barr and Landis 2017; Delrue et al. 2016; Gouveia, et al. 2016; Olguin 2012; Queiroz et al. 2013; Zhu 2015), (ii) digestate from a wastewater treatment of the pulp and paper industry to internally recycle the nutrients for microalgae cultivation; this is an example that leads to a notably lower cost of microalgae biomass production (Kouhia et al. 2015) or (iii) gaseous waste streams with high CO_2 content (Moncada et al. 2014; Wiesberg et al. 2017).

On the other hand, in all the above-mentioned studies, the potential of microalgae for producing different forms of bioenergy and chemicals has been presented as separated concepts that are not integrated into the first and second biorefineries in multiproduct portfolios. Recently, it has been recognized that the biorefinery concept plays an important role in the future development of a bio-based economy and integration of the first-, second-, and third-generation biorefineries has been recently proposed to develop a complete bioindustry (Moncada et al. 2014). That work analyzed the integration of microalgae into a second-generation sugarcane biorefinery, including the joint production of sugar, ethanol, and electricity, by introducing the cultivation of microalga *Chlorella* to use CO_2-rich streams derived from fermentation and cogeneration systems and subsequently produce biodiesel, glycerol, sugar, fuel ethanol, heat, and power. Table 5 shows the concept of incorporation of microalgae into other biorefinery or production schemes. These scenarios are as diverse as the existing types of biomass; some works include microalgae in pulp and paper industry or sugar cane biorefineries for treatment and valorization of effluents, using combustion gas or wastewater as presented in Table 5.

It is important to note that under this scenario, the possibility of biorefinery schemes is infinite. Despite the complexity it involves, the biorefinery needs to promote a circular economy to achieve viability (Mohan et al. 2016a, b). In the proposals, the concept of biorefining for one single product was abandoned.

4.5 Design Strategy and Current Status

In order to conceptualize a biorefinery, the sequence of decisions includes (Toledo-Cervantes and Morales 2014): (1) microalgae rich in target products. (2) cultivation conditions and operation strategies, (3) conversion processes of whole microalgae/defatted microalgal biomass to biofuels, (4) biomass harvesting, post-harvesting technology, (5) methods and sequence of extraction of coproducts/processing the whole biomass/the pre-extracted product to final products, (6) process integration of streams and recycling of materials, reducing wastes, and finally (7) a life cycle analysis. In the case of high-value products, Eppink et al. (2017) suggested specific recommendations about harvesting, cell disruption, and extraction methods. They indicated the need to know the composition and strength of the cell wall, and localization of the target compound in the cytoplasm, as well as to select moderated harvesting methods and mild cell disruption, removing first the cell wall and after, the organelle disruption. The next step is the selective separation of hydrophobic (lipids, pigments) and hydrophilic (proteins, carbohydrates) fractions while keeping full functionality and finally, fractioning the hydrophilic and hydrophobic component mixtures to recover target compounds for different market applications. The final step for the implementation of a biorefinery process is to connect different stages into the complete chain process. All efforts in designing multiproduct scenarios are looking to increase the economic feasibility of global production of biofuels, but, as can be seen, a biorefinery configuration involves many decisions to be made, including alternatives for cultivation, water consumption and nutrients supply, culture condition to trigger accumulation target products, and a complicated downstream processing to maximize the product recovery-exploitation. Therefore, in order to define the adequate processing pathway of the microalgae biomass, different scenarios have to be analyzed using optimization techniques, including techno-economic aspects and sustainability issues.

5 Economic and Sustainability Aspects

Detailed information and fundamentals about economic aspects and life cycle assessment (LCA) of microalgae biofuels will be presented in Chaps. 6 and 7 of this book, and this section is specific for the biorefinery strategy.

Most of the current research about microalgae biofuels is based on the potential of biodiesel from microalgae lipids (Chisti 2007); this optimistic study was performed in an extraordinary situation with extremely high oil prices, favoring the biofuel development. At that time, the initial challenge to decrease its production cost to 0.48 USD/L was established. Afterward, Wijffels et al. (2010) established that this biodiesel price was achievable, when considering a microalgae biorefinery, based on the valorization of different biochemical fractions and high-value-added products. Up to date, several studies have been published on the economic analysis of microalgae-based processes (Acién et al. 2017; Davis et al. 2016; Douskova et al. 2009; Hoffman et al. 2017; Norsker et al. 2011; Slade and Bauen 2013; Tredici et al. 2015). Recently, reviews about techno-economic evaluations of microalgae biofuels, from a biorefinery point of view, were performed by analyzing a great variety of scenarios (de Boer and Bahri 2015; Laurens et al. 2017b). Those works report prices ranging from 0.88 USD/L up to 24.60 USD/L, concluding, on the basis of available techno-economic studies and current technologies, that microalgal biofuel production is 4–5 times more expensive than current fossil fuels; and actions to reduce that cost involve: (1) productivities of at least 30 $g/m^2/day$ and minimum lipid content of 30%, (2) lowering the capital cost, discarding the use of photobioreactors and centrifuges, reducing costs of dewatering, and finally identifying opportunities for lower-cost carbon and nutrient sources.

In addition to economic evaluation, biofuels from microalgae must also meet favorable life cycle goals on energy return, and carbon and water footprint to provide quantitative improvements to current fuels.

However, there is no common conclusion on sustainability of microalgae biofuels (Gnansounou and Raman 2017; Quinn and Davis 2015). The significant variance in the studies could be due to diverse choices regarding technical (microalgae species, production units, downstream processing, and technology for energy production, coproducts) and methodological alternatives (functional units, boundaries, coproduct allocation methods) (Collet et al. 2015; Thomassen et al. 2017). But there is a general agreement that producing only biodiesel from algae is not favorable and, in order to reduce the overall cost, the following have been suggested: (i) process integration (CO_2 capture, wastewater treatment, and biofuel production); (ii) optimization of photobioreactor design and conditions to improve biofuel yield; and (iii) extraction of valuable products from algal biomass (biorefinery concept). Therefore, a multiproduct strategy in a biorefinery is indicated as the future trend. Nevertheless, the absence of facilities for microalgae biofuels production at industrial scale with accurate/reliable information entails theoretical assumptions or extrapolation of laboratory information to make predictions, whereas the design problem is mathematically formulated to describe the production systems and its performance. Recent theoretical studies about economic aspects or LCA in a multiprocessing-downstream processing-multiproduct strategy have been published (Gutiérrez-Arriaga et al. 2014; Martinez-Hernandez et al. 2013; Menetrez 2012; Posada et al. 2016). Also, multiobjective optimization approaches to trade off different criteria simultaneously have been performed (Andiappan et al. 2014; Brunet et al. 2015; Rizwan et al. 2015; Santibañez-Aguilar et al. 2014) by applying mixed integer nonlinear programming (MINLP) models or Monte Carlo simulations to maximize incomes or production yields, determine economic viability, and minimize the environmental impact to find the optimal processing pathway for the production of biodiesel from microalgal biomass and treating wastes. Although computational tools are developed, no scenario has reached 0.48 USD/L necessary to compete with the fossil alternative.

In general, most LCA studies concluded that bioenergy from algae has lower greenhouse gas (GHG) emissions than fossil fuels and that energetically viable process must use raceway ponds, process wet biomass (avoid drying), minimize energy required for cell disruption, and minimize solvent use (de Boer and Bahri 2015).

6 Remarks and Conclusion

The lineal system of our current economy (extraction, manufacture, use, and disposal) has reached its limits, entailing depletion of a number of natural resources and fossil fuels (Mohan et al. 2016a). In a bio-based economy, biorefinery strategy is a key factor to close the loop in a circular economy with a restorative and regenerative production model that values waste and minimizes negative environmental impacts through a transition to renewable energy sources. Microalgae biomass is one of the best alternatives for a biorefinery due to the diverse products that can be obtained. However, current applications of microalgae biomass are mainly for food and feed, and biofuels are not produced at industrial scale. At present, the high production cost of biomass and its subsequent fractionation make it economically nonviable, particularly when the production focuses on a single product, such as fuel. Advances in genetic modification of strains, production systems, and downstream processing, besides valorization and acceptance of a broad range of products, could contribute to assess sustainability and profitability.

Acknowledgements This work was supported by CONACYT (Mexican Council for Science and Technology) through project numbers 247402 and 247006.

References

Abd El Baky, H. H., & El-Baroty, G. S. (2013). Healthy benefit of microalgal bioactive substances. *Journal of Aquatic Science, 1*(1), 11–22.

Acién, F. G., Molina, E., Fernández-Sevilla, J. M., Barbosa, M., Gouveia, L., Sepúlveda C., et al. (2017). Economics of microalgae production. In R. Muñoz, C. González (Eds.), *Microalgae-based biofuels and bioproducts* (pp. 485–503). Amsterdam: Elsevier Ltd. ISBN: 9780081010235.

Adam, F., Abert-vian, M., Peltier, G., & Chemat, F. (2012). "'Solvent-free'" ultrasound-assisted extraction of lipids from fresh microalgae cells: A green, clean and scalable process. *Bioresource Technology, 114,* 457–465.

Adarme-Vega, T., Lim, D. K. Y., Timmins, M., Vernen, F., Li, Y., & Schenk, P. M. (2012). Microalgal biofactories: A promising approach towards sustainable omega-3 fatty acid production. *Microbial Cell Factories, 11*(1), 96.

Ahmad, N., Pandit, N., & Maheshwari, S. (2012). L-asparaginase gene-a therapeutic approach towards drugs for cancer cell. *International Journal of of Biosciences,* 2(4), 1–11. Retrieved from http://sanjiv08.6te.net/ijb1.pdf.

Albarelli, J. Q., Santos, D. T., Ensinas, A. V., Marechal, F., Cocero, M. J., & Meireles, M. A. A. (2017). Product diversification in the sugarcane biorefinery through algae growth and supercritical CO_2 extraction: Thermal and economic analysis. *Renewable Energy*, 1–10.

Al-Sherif, E. A., Ab El-Hameed, M. S., Mahmoud, M. A., & Ahmed, H. S. (2015). Use of cyanobacteria and organic fertilizer mixture as soil bioremediation. *American-Eurasian Journal of Agricultural and Environmental Science, 15,* 794–799.

Alzate, M. E., Muñoz, R., Rogalla, F., Fdz-Polanco, F., & Pérez-Elvira, S. I. (2012). Biochemical methane potential of microalgae: Influence of substrate to inoculum ratio, biomass concentration and pretreatment. *Bioresource Technology, 123,* 488–494.

Amaro, H. M., Guedes, A. C., & Malcata, F. X. (2011). Advances and perspectives in using microalgae to produce biodiesel. *Applied Energy, 88,* 3402–3410.

Andiappan, V., Ko, A. S. Y., Lau, V. S. S., Ng, L. Y., Ng, R. T. L., Chemmangattuvalappil, N. G., et al. (2014). Synthesis of sustainable integrated biorefinery via reaction pathway synthesis: Economic, incremental enviromental burden and energy assessment with multiobjective optimization. *AIChE Journal, 61*(1), 132–146.

Ansari, F. A., Shriwastav, A., Gupta, S. K., Rawat, I., & Bux, F. (2017). Exploration of microalgae biorefinery by optimizing sequential extraction of major metabolites from *Scenedesmus obliquus. Industrial and Engineering Chemistry Research, 56*(12), 3407–3412.

Ansari, F. A., Shriwastav, A., Gupta, S. K., Rawat, I., Guldhe, A., & Bux, F. (2015). Lipid extracted algae as a source for protein and reduced sugar: A step closer to the biorefinery. *Bioresource Technology, 179,* 559–564.

Bahadar, A., & Bilal Khan, M. (2013). Progress in energy from microalgae: A review. *Renewable and Sustainable Energy Reviews, 27,* 128–148.

Balaji, S., Gopi, K., & Muthuvelan, B. (2013). A review on production of poly β hydroxybutyrates from cyanobacteria for the production of bio plastics. *Algal Research, 2,* 278–285.

Ballesteros-Gómez, A., Sicilia, M. D., & Rubio, S. (2010). Supramolecular solvents in the extraction of organic compounds. A review. *Analytica Chimica Acta, 677*(2), 108–130.

Barr, W. J., & Landis, A. E. (2017). Comparative life cycle assessment of a commercial algal multiproduct biorefinery and wild caught fishery for small pelagic fish. *The International Journal of Life Cycle Assessment.*

Barros, A. I., Gonçalves, A. L., Simões, M., Pires, J. C. M., Gonçalves, A. L., Simões, M., et al. (2015). Harvesting techniques applied to microalgae: A review. *Renewable and Sustainable Energy Reviews, 41,* 1489–1500.

Bastiaens, L., Van Roy, S., Thomassen, G, & Elst, K. (2017). Biorefinery of algae: Technical and economic considerations. In R. Muñoz & C. González (Eds.), *Microalgae-based biofuels and bioproducts* (pp. 327–345). Amsterdam: Elsevier.

Baudelet, P. H., Ricochon, G., Linder, M., & Muniglia, M. (2017). A new insight into cell walls of Chlorophyta. *Algal Research, 25,* 333–371.

Becker, E. W. (2007). Microalgae as a source of protein. *Biotechnol Advances, 25,* 207–210.

Beheshtipour, H., Mortazavian, A. M., Mohammadi, R., Sohrabvandi, S., & Khosravi-Darani, K. (2013). Supplementation of *Spirulina platensis* and chlorella vulgaris algae into probiotic fermented milks. *Comprehensive Reviews in Food Science and Food Safety, 12*(2), 144–154.

Béligon, V., Christophe, G., Fontanille, P., & Larroche, C. (2016). Microbial lipids as potential source to food supplements. *Current Opinion in Food Science, 7,* 35–42.

Bermejo, R., Felipe, M., Talavera, E. M., & Alvarez-Pez, J. M. (2006). Expanded bed adsorption chromatography for recovery of phycocyanins from the Microalga *Spirulina platensis. Chromatographia, 63*(1–2), 59–66.

Bocchiaro, P., & Zamperini, A. (2016). World's Largest Science, Technology & Medicine Open Access Book Publisher c. RFID Technol. Secur. Vulnerabilities, Countermeas.

Borowitzka, M. (1995). Microalgae as sources of pharmaceuticals and other biologically active compounds. *Journal of Applied Phycology, 7*(1), 3–15.

Borowitzka, M. A. (2013). High-value products from microalgae-their development and commercialisation. *Journal of Applied Phycology, 25*(3), 743–756.

Boussetta, N., Lesaint, O., & Vorobiev, E. (2013). A study of mechanisms involved during the extraction of polyphenols from grape seeds by pulsed electrical discharges. *Innovative Food Science and Emerging Technologies, 19,* 124–132.

Boyd, A. R., Champagne, P., McGinn, P. J., MacDougall, K. M., Melanson, J. E., & Jessop, P. G. (2012). Switchable hydrophilicity solvents for lipid extraction from microalgae for biofuel production. *Bioresource Technology, 118,* 628–632.

Bozell, J. J., & Petersen, G. R. (2010). Technology development for the production of biobased products from biorefinery carbohydrates—The US Department of Energy's "Top 10" revisited. *Green Chemistry, 12*(4), 539–554.

Brar, S. K., Sarma, S. J., & Pakshirajan K. (2016). *Platform chemical biorefinery. Future green chemistry* (pp. 438–450). The Netherlands: Elsevier. ISBN978-12-802980-0.

Brennan, L., & Owende, P. (2010). Biofuels from microalgae—A review of technologies for production, processing, and extractions of biofuels and co-products. *Renewable and Sustainable Energy Reviews, 14*(2), 557–577.

Broeren, M., Kempener, R., Simbolotti, G., & Tosato, G. (2013). Production of bio-methanol-technology brief. *Technology Brief, 108*(January), 1–24. Retrieved from https://iea-etsap.org/E-TechDS/HIGHLIGHTS%20PDF/I09IR_Bio-methanol_MB_Jan2013_final_GSOK%201.pdf. September 2017.

Brownbridge, G., Azadi, P., Smallbone, A., Bhave, A., Taylor, B., & Kraft, M. (2014). The future viability of algae-derived biodiesel under economic and technical uncertainties. *Bioresource Technology, 151,* 166–173.

Brunet, R., Boer, D., Guillén-Gosálbez, G., & Jiménez, L. (2015). Reducing the cost, environmental impact and energy consumption of biofuel processes through heat integration. *Chemical Engineering Research and Design, 93,* 203–212.

Budzianowski, W. M. (2017). High-value low-volume bioproducts coupled to bioenergies with potential to enhance business development of sustainable biorefineries. *Renewable and Sustainable Energy Reviews, 70,* 793–804.

Buitrón, G., Carrillo-Reyes, J., Morales, M., Faraloni, C., & Torzillo, G. (2017). Biohydrogen production from microalgae. In R. Muñoz & C. González (Eds.), *Microalgae-based biofuels and bioproducts* (pp. 210–234). Amsterdam: Elsevier. ISBN 9780081010235.

Campenni, L., Nobre, B. P., Santos, C. A., Oliveira, A. C., Aires-Barros, M. R., Palavra, A. M. F., et al. (2013). Carotenoid and lipid production by the autotrophic microalga *Chlorella protothecoides* under nutritional, salinity, and luminosity stress conditions. *Applied Microbiology Biotechnology, 97,* 1383–1393.

Chailleux, E., Audo, M., Bujoli, B., Queffiec, C., Lagrand, J., & Lepine, O. (2012). Alternative binder from microalgae algoroute project. *Alternative Binders for Sustainable Asphalt Pavements*, 23–36.

Chaogang, W., Zhangli, H., Anping, L., & Baohui, J. (2010). Biosynthesis of poly-3-hydroxybuturate (PHB) in the transgenic green alga *Chlamydomonas reinhardtii. Journal of Phycology, 46,* 396–402.

Chatterjee, A., Singh, S., Agrawal, C., Yadav, S., Rai, R., & Rai, L. C. (2017). Role of algae as a biofertilizer. In R. Prasad Rastogi, D. Madamwar, & A. Pandey (Eds.), *Algae green chemistry. Recent progress in biotechnology* (pp. 189–200). Amsterdam: Elsevier.

Chemat, F., Vian, M. A., & Cravotto, G. (2012). Green extraction of natural products: Concept and principles. *International Journal of Molecular Sciences, 13,* 8615–8627.

Chen, W. H., Lin, B. J, Huang, M.-Y., & Chang, J. S. (2015). Thermochemical conversion of microalgal biomass into biofuels: A review. *Bioresource Technology, 184*, 314–327, ISSN 0960-8524.

Chew, K. W., Yap, J. Y., Show, P. L., Suan, N. H., Juan, J. C., Ling, T. C., et al. (2017). Microalgae biorefinery: High value products perspectives. *Bioresource Technology, 229,* 53–62.

Chiaramonti, D., Prussi, M., Buffi, M., Casini, D., & Rizzo, A. M. (2015). Thermochemical conversion of microalgae: Challenges and opportunities. *Energy Procedia, 75,* 819–826.

Chilton, V., Mantrand, N., & Morel, B. (2016). *Patent landscape report: Microalgae-related technologies*. Patent Landscape Report. Retrieved from http://www.wipo.int/edocs/pubdocs/en/wipo_pub_947_5.pdf.

Chisti, Y. (2007). Biodiesel from microalgae. *Biotechnology Advances, 25*(3), 294–306.

Chisti, Y. (2008). Biodiesel from microalgae beats bioethanol. *Trends in Biotechnology, 26*(3), 126–131.

Chng, L. M., Chan, D. J. C., & Lee, K. T. (2015). Sustainable production of bioethanol using lipid-extracted biomass from *Scenedesmus dimorphus. Journal of Cleaner Production, 130,* 68–73.

Chua, E. T., Schenk, P. M. (2017). A biorefinery for Nannochloropsis: Induction, harvesting, and extraction of EPA-rich oil and high-value protein. *Bioresource Technology, 244*(2), 1416–1424.
Collet, P., Hélias, A., Lardon, L., Ras, M., Goy, R. A., & Steyer, J. P. (2011). Life-cycle assessment of microalgae culture coupled to biogas production. *Bioresource Technology, 102* (1), 207–214.
Collet, P., Hélias, A., Lardon, L., Steyer, J. P., & Bernard, O. (2015). Recommendations for Life Cycle Assessment of algal fuels. *Applied Energy, 154,* 1089–1102.
Cuellar-Bermudez, S. P., Garcia-Perez, J. S., Rittmann, B. E., & Parra-Saldivar, R. (2015). Photosynthetic bioenergy utilizing CO_2: An approach on flue gases utilization for third generation biofuels. *Journal of Cleaner Production, 98,* 53–65.
Dahms, H. U., Xu, Y., & Pfeiffer, C. (2006). Antifouling potential of cyanobacteria: a mini-review. *Biofouling, 22,* 317–327.
Das, P., Mandal, S. C., Bhagabati, S. K., Akhtar, M. S., & Singh, S. K. (2012). Important live food organisms and their role in aquaculture. In K. Sundaray, M. Sukham, R. K. Mohanty, & S. K. Otta (Eds.), *Frontiers in aquaculture* (1st ed., pp. 69–86). New Delhi: Narendra Publishing House.
Davis, R., Jennifer, M., Kinchin, C., Grundl, N., Tan, E. C. D., & Humbird, D. (2016). Process design and economics for the production of algal biomass: Algal biomass production in open pond systems and processing through dewatering for downstream conversion. Technical Report NREL/TP-5100-64772. Retrieved from https://www.nrel.gov/docs/fy16osti/64772.pdf. September 2017.
de Boer K., & Bahri P. A. (2015). Economic and energy analysis of large-scale microalgae production for biofuels. In N. Moheimani, M. McHenry, K. de Boer, & P. A. Bahri (Eds.), *Biomass and biofuels from microalgae. Biofuel and biorefinery technologies* (Vol. 2, pp. 347–365). Cham: Springer. ISBN: SBN 978-94-007-5479-9.
de Jong E., Higson A., Walsh P., & Wellisch M. (2012). IEA bioenergy—Task 42 biorefinery: Biobases chemicals-value added products from biorefineries. http://www.ieabioenergy.com/publications/bio-based-chemicals-value-added-products-from-biorefineries/ Consult done in August 2017.
Delrue, F., Álvarez-Díaz, P. D., Fon-Sing, S., Fleury, G., & Sassi, J. F. (2016). The environmental biorefinery: Using microalgae to remediate wastewater, a win-win paradigm. *Energies, 9*(3), 1–19.
Demibras, A. (2009). Biorefineries: Current activities and future developments. *Energy Conversion and Management, 50,* 2782–2801.
Demmer, W., Fischer-Fruehholz, S., Kocourek, A., Nusbaumer, D., & Wuenn, E. (2005, April 28). Adsorption membrane comprising microporous polymer membrane with adsorbent particles embedded in the pores, useful in analysis, for purification or concentration. Google Patents. Retrieved from http://www.google.com.pg/patents/DE10344820A1?cl=en.
Deniz, I., García-Vaquero, M., & Imamoglu, E. (2017). Trends in red biotechnology: Microalgae for pharmaceutical applications. In R. Muñoz & C. González (Eds.), *Microalgae-based biofuels and bioproducts* (pp. 420–440). Amsterdam: Elsevier. ISBN: 9780081010235.
Dexler, I. L. C., & Yeh, D. H. (2014). Membrane applications for microalgae cultivation and harvesting: A review. *Reviews inEnvironmental Science and Bio/technology, 13,* 487–504.
Dibenedetto, A., Colucci, A., & Aresta, M. (2016). The need to implement an efficient biomass fractionation and full utilization based on the concept of "biorefinery" for a viable economic utilization of microalgae. *Environmental Science and Pollution Research, 23*(22), 22274–22283.
Dierkes, H., Steinhagen, V., Bork, M., Lütge, C., & Knez, Z. (2012). Cell lysis of plant or animal starting materials by a combination of a spray method and decompression for the selective extraction and separation of valuable intracellular materials. Patent no. EP 2315825 A1. Retrieved from http://www.google.com/patents/EP2315825B1?cl=en.

Dong, T., Knoshaug, E. P., Davis, R., Laurens, L. M. L., Van Wychen, S., Pienkos, P. T., et al. (2016). Combined algal processing: A novel integrated biorefinery process to produce algal biofuels and bioproducts. *Algal Research, 19,* 316–323.

Douskova, I., Doucha, J., Livansky, K., MacHat, J., Novak, P., Umysova, D., et al. (2009). Simultaneous flue gas bioremediation and reduction of microalgal biomass production costs. *Applied Microbiology and Biotechnology, 82,* 179–185.

Du, Y., Schuur, B., Kersten, S. R. A., & Brilman, D. W. F. (2015). Opportunities for switchable solvents for lipid extraction from wet algal biomass: An energy evaluation. *Algal Research, 11,* 271–283.

Ehimen, E. A., Sun, Z. F., Carrington, C. G., Birch, E. J., & Eaton-Rye, J. J. (2011). Anaerobic digestion of microalgae residues resulting from the biodiesel production process. *Applied Energy, 88,* 3454–3463.

Eldalatony, M. M., Kabra, A. N., Hwang, J. H., Govindwar, S. P., Kim, K. H., Kim, H., et al. (2016). Pretreatment of microalgal biomass for enhanced recovery/extraction of reducing sugars and proteins. *Bioprocess and Biosystems Engineering, 39*(1), 95–103.

Eppink, M. H. M., Olivieri G., Reith, H., van den Berg, C., Barbosa M. J., Wijffels, R. H. (2017). From current algae products to future biorefinery practices: A review. In *Advances in biochemical engineering/biotechnology*. Berlin, Heidelberg: Springer.

Espino, M., de los Angeles Fernandez, M., Gomez, F. J. V, & Silva, M. F. (2016). Natural designer solvents for greening analytical chemistry. *TrAC—Trends in Analytical Chemistry, 76,* 126–136.

Faheed, F. A., & Abd-El Fattah Z. (2008). Effect of *Chlorella vulgaris* as bio-fertilizer on growth parameters and metabolic aspects of lettuce plant. *Journal of Agriculture and Social Sciences, 4,* 165–169. Retrieved from http://www.fspublishers.org/published_papers/71170_..pdf.

Fernandez-Linares, L. C., González-Falfán, K. A., & Ramirz-López, C. (2017). Microalgal biomass: A biorefinery approach. In J. S. Tumuluru (Ed.), *Biomass volume estimation and valorization for energy microalgal biomass—A biorefinery approach* (pp. 1–23). InTech. ISBN 978-953-51-2938-7.

Francavilla, M., Kamaterou, P., Intini, S., Monteleone, M., & Zabaniotou, A. (2015). Cascading microalgae biorefinery: Fast pyrolysis of *Dunaliella tertiolecta* lipid extracted-residue. *Algal Research, 11,* 184–193.

Garcia-Gonzalez, J., & Sommerfeld, M. (2016). Biofertilizer and biostimulant properties of the microalga *Acutodesmus dimorphus*. *Journal of Applied Phycology, 28,* 1051–1061.

García-Prieto, C. V. G., Ramos, F. D., Estrada, V., & Díaz, M. S. (2014). Optimal design of an integrated microalgae biorefinery for the production of biodiesel and PHBS. *Chemical Engineering Transactions, 37,* 319–324.

Gentili, F. G. (2014). Microalgal biomass and lipid production in mixed municipal, dairy, pulp and paper wastewater together with added flue gases. *Bioresource Technology, 169,* 27–32.

Gerardo, M. L., Oatley-Radcliffe, D. L., & Lovitt, R. W. (2014). Integration of membrane technology in microalgae biorefineries. *Journal of Membrane Science, 464,* 86–99.

Gerardo, M. L., Van Den Hende, S., Vervaeren, H., Coward, T., & Skill, S. C. (2015). Harvesting of microalgae within a biorefinery approach: A review of the developments and case studies from pilot-plants. *Algal Research, 11,* 248–262.

Gilbert-López, B., Mendiola, J. A., Fontecha, J., van den Broek, L. A. M., Sijtsma, L., Cifuentes, A., et al. (2015). Downstream processing of Isochrysis galbana: A step towards microalgal biorefinery. *Green Chemistry, 17*(9), 4599–4609.

Gnansounou, E., Raman, J.K. (2017). Life cycle assessment of algal biorefinery. In E. Gnansounou & A. Pandey (Eds.), *Life-cycle assessment of biorefineries* (pp. 199–219). Amsterdam: Elsevier. ISBN: 978-0-444-63585-3.

Godlewska, K., Michalak, I., Tuhy, L., & Chojnacka, K. (2016). Plant growth biostimulants based on different methods of seaweed extraction with water. *BioMed Research International*, 2016.

Golueke, C. G., Oswald, W. J., & Gotaas, H. B. (1957). Anaerobic Digestion of Algae. *Applied Microbiology, 5*(1), 47–55.

Gong, M., Hu, Y., Yedahalli, Sh, & Bassi, A. (2017). Oil extraction processes in microalgae. *Recent Advances in Renewable Energy, 1,* 377–411.
González-Delgado, Á.-D., & Kafarov, V. (2011). Microalgae based biorefinery: Issues to consider. *CT&F - Ciencia, Tecnología y Futuro, 4*(4), 5–22.
Gorry, P. L., Morales, M., Gorry, Ph. (2017). Science and technology indicators of microalgae-based biofuel research. *Proceedings of ISSI 2017 Wuhan: 16th International Society of Scientometrics and Informetrics Conference.* Retrieved from http://www.issi2017.org/media/2017ISSI%20Conference%20Proceedings.pdf.
Gouveia, L., Batista, A. P., Sousa, I., Raymundo, A., & Bandarra, N. M. (2008). Microalgae in novel food product. In K. Papadoupoulos (Ed.), *Food chemistry research developments* (pp. 75–112). Nova Science Publishers. ISBN 978-1-60456-262-0.
Gouveia, L., Graça, S., Sousa, C., Ambrosano, L., Ribeiro, B., Botrel, E. P., et al. (2016). Microalgae biomass production using wastewater: Treatment and costs. Scale-up considerations. *Algal Research, 16,* 167–176.
Gouveia, L., Neves, C., Sebastião, D., Nobre, B. P., & Matos, C. T. (2014). Effect of light on the production of bioelectricity and added-value microalgae biomass in a photosynthetic alga microbial fuel cell. *Bioresource Technology, 154,* 171–177.
Grosso, C., Valentão, P., Ferreres, F., & Andrade, P. B. (2015). Alternative and efficient extraction methods for marine-derived compounds. *Marine Drugs, 13*(5), 3182–3230.
Gultom, S. O., Zamalloa, C., & Hu, B. (2014). Microalgae harvest through fungal pelletization—Co-culture of *Chlorella vulgaris* and *Aspergillus niger*. *Energies, 7,* 4417–4429.
Günerken, E., D'Hondt, E., Eppink, M. H. M., Garcia-Gonzalez, L., Elst, K., & Wijffels, R. H. (2015). Cell disruption for microalgae biorefineries. *Biotechnology Advances, 33*(2), 243–260.
Gutiérrez-Arriaga, C. G., Serna-González, M., Ponce-Ortega, J. M., & El-Halwagi, M. M. (2014). Sustainable integration of algal biodiesel production with steam electric power plants for greenhouse gas mitigation. *ACS Sustainable Chemistry & Engineering, 2*(6), 1388–1403.
Haase, S. M., Huchzermeyer, B., & Rath, T. (2011). PHB accumulation in *Nostoc muscorum* under different carbon stress situations. *Journal of Applied Phycology, 24,* 157–162.
Halim, R., Danquah, M. K., & Webley, P. A. (2012a). Extraction of oil from microalgae for biodiesel production: A review. *Biotechnology Advances, 30*(3), 709–732.
Halim, R., Harun, R., Danquah, M. K., & Webley, P. A. (2012b). Microalgal cell disruption for biofuel development. *Applied Energy, 91,* 116–121.
Halim, R., Hosikian, A., Lim, S., & Danquah, M. K. (2010). Chlorophyll extraction from microalgae: A review on the process engineering aspects. *International Journal of Chemical Engineering*, 2010.
Halim, H., Webley, P. A., & Martin, G. J. O. (2016). The CIDES process: Fractionation of concentrated microalgal paste for co-production of biofuel, nutraceuticals, and high-grade protein feed. *Algal Research, 19,* 299–306.
Hamed, I. (2016). The evolution and versatility of microalgal biotechnology: A review. *Comprehensive Reviews in Food Science and Food Safety, 15,* 1104–1123.
Hannon, M., Gimpel, J., Tran, M., Rasala, B., & Mayfield, S. (2010). Biofuels from algae: Challenges and potential. *Biofuels, 1*(5), 763–784.
Hariskos, I., & Posten, C. (2014). Biorefinery of microalgae—Opportunities and constraints for different production scenarios. *Biotechnology Journal, 9,* 739–752.
Harun, R., & Danquah, M. K. (2011). Influence of acid pre-treatment on microalgal biomass for ethanol production. *Process Biochemistry, 46,* 304–309.
Harun, R., Davidson, M., Doyle, M., Gopiraj, R., Danquah, M., & Forde, G. (2011). Technoeconomic analysis of an integrated microalgae photobioreactor, biodiesel and biogas production facility. *Biomass and Bioenergy, 35*(1), 741–747.
Harun, R., Yip, J. W. S., Thiruvenkadam, S., Ghani, W. A. W. A. K., Cherrington, T., & Danquah, M. K. (2014). Algal biomass conversion to bioethanol—A step-by-step assessment. *Biotechnology Journal, 9,* 73–86.

Hemaiswarya, S., Raja, R., Kumar, R. R., Ganesan, V., & Anbazhagan, C. (2011). Microalgae: A sustainable feed source for aquaculture. *World Journal of Microbiology & Biotechnology, 27*(8), 1737–1746.

Hernández, D., Solana, M., Riaño, B., García-gonzález, M. C., & Bertucco, A. (2014). Biofuels from microalgae: Lipid extraction and methane production from the residual biomass in a biorefinery approach. *Bioresource Technology, 170,* 370–378.

Ho, Sh-H, Huang, Sh-W, Chen, Ch-Y, Hasunuma, T., Kondo, A., & Chang, J-Sh. (2013). Bioethanol production using carbohydrate-rich microalgae biomass as feedstock. *Bioresource Technology, 135,* 191–198.

Hoffman, J., Pate, R. C., Drennen, T., & Quinn, J. C. (2017). Techno-economic assessment of open microalgae production systems. *Algal Research, 23,* 51–57.

Horváth, I. T. (1998). Fluorous biphase chemistry. *Accounts of Chemical Research, 31*(10), 641–650.

Hossain, G. S., Liu, L., & Du, G. C. (2017). Industrial bioprocesses and the biorefinery concept. In C. Larroche, M. Ángeles Sanromán, G. Du, A. Pandey (Eds.), *Current developments in biotechnology and bioengineering. bioprocesses, bioreactors and controls* (pp. 3–27). Amsterdam: Elsevier.

Ibañez, E., Herrero, M., Mendiola, J. A., & Castro-Puyana, M. (2012). Extraction and characterization of bioactive compounds with health benefits from marine resources: Macro and micro algae, cyanobacteria, and invertebrates. In M. Hayes (Ed.), *marine bioactive compounds: Sources, characterization and applications* (pp. 55–98). Boston, MA: Springer.

IEA. (2017). State of technology review—Algae bioenergy an IEA Bioenergy inter-task strategic project. Report coordinated by Lieve M.L. Laurens, National Renewable Energy Laboratory, Published by IEA Bioenergy: Task 39: January 2017.

Jacob-Lopes, E., & Franco, T. T. (2013). From oil refinery to microalgal biorefinery. *Journal of CO_2 Utilization, 2,* 1–7.

Jahan, A., Ahmad, I. Z., Fatima, N., Ansari, V. A., & Akhtar, J. (2017). Algal bioactive compounds in the cosmeceutical industry: A review. *Phycologia, 56*(4), 410–422.

Jankowska, E., Sahu, A. K., & Oleskowicz-Popiel, P. (2017). Biogas from microalgae: Review on microalgae's cultivation, harvesting and pretreatment for anaerobic digestion. *Renewable and Sustainable Energy Reviews, 75,* 692–709.

Japar, A. S., Takriff, M. S., & Yasin, N. H. M. (2017). Harvesting microalgal biomass and lipid extraction for potential biofuel production: A review. *Journal of Environmental Chemical Engineering, 5*(1), 555–563.

Jeevan Kumar, S. P., Vijay Kumar, G., Dash, A., Scholz, P., & Banerjee, R. (2017). Sustainable green solvents and techniques for lipid extraction from microalgae: A review. *Algal Research, 21,* 138–147.

Jeong, K. M., Lee, M. S., Nam, M. W., Zhao, J., Jin, Y., Lee, D. K., et al. (2015). Tailoring and recycling of deep eutectic solvents as sustainable and efficient extraction media. *Journal of Chromatography A, 1424,* 10–17.

Jessop, P. G., Mercer, S. M., & Heldebrant, D. J. (2012). CO2-triggered switchable solvents, surfactants, and other materials. *Energy & Environmental Science, 5*(6), 7240.

Kadam, S. U., Tiwari, B. K., & O'Donnell, C. P. (2013). Application of novel extraction technologies for bioactives from marine algae. *Journal of Agricultural and Food Chemistry, 61*(20), 4667–4675.

Kapdan, I. K., & Kargi, F. (2006). Bio-hydrogen production from waste materials. *Enzyme and Microbial Technology, 38*(5), 569–582.

Kassim, M. A., & Meng, T. K. (2017). Carbon dioxide (CO_2) biofixation by microalgae and its potential for biorefinery and biofuel production. *Science of the Total Environment, 584–585,* 1121–1129.

Keegan, D., Kretschmer, B., Elbersen, B., & Panoutsou, C. (2013). Cascading use: A systematic approach to biomass beyond the energy sector. *Biofuels, Byproducts and Biorefining, 7,* 193–206.

Kim, D. Y., Vijayan, D., Praveenkumar, R., Han, J. I., Lee, K., Park, J. Y., et al. (2016). Cell-wall disruption and lipid/astaxanthin extraction from microalgae: Chlorella and Haematococcus. *Bioresource Technology, 199,* 300–310.

Kipngetich, T. E., & Hillary, M. (2013). A blend of green algae and sweet potato starch as a potential source of bioplastic production and its significance to the polymer industry. *International Journal of Emerging Technology and Advanced Engineering, 2,* 15–19.

Konur, O. (2011). The scientometric evaluation of the research on the algae and bio-energy. *Applied Energy, 88*(10), 3532–3540.

Kouhia, M., Holmberg, H., & Ahtila, P. (2015). Microalgae-utilizing biorefinery concept for pulp and paper industry: Converting secondary streams into value-added products. *Algal Research, 10,* 41–47.

Kumar, G., Zhen, G., Kobayashi, T., Sivagurunathan, P., Kim, S. H., & Xu, K. Q. (2016). Impact of pH control and heat pre-treatment of seed inoculum in dark H_2 fermentation: A feasibility report using mixed microalgae biomass as feedstock. *International J. Journal of Hydrogen Energy, 41,* 4382–4392.

Laurens, L. M. L., Chen-Glasser, M., & McMillan, J. D. (2017a). A perspective on renewable bioenergy from photosynthetic algae as feedstock for biofuels and bioproducts. *Algal Research, 24,* 261–264.

Laurens, L. M. L., Markham, J., Templeton, D. W., Christensen, E. D., Van Wychen, S., Vadelius, E. W., et al. (2017b). Development of algae biorefinery concepts for biofuels and bioproducts; A perspective on process-compatible products and their impact on cost-reduction. *Energy & Environmental Science, 10*(8), 1716–1738.

Lee, D., Chang, J Sh, & Lai, J. Y. (2015). Microalgae–microbial fuel cell: A mini review. *Bioresource Technology, 198,* 891–895.

Lee, A. K., Lewis, D. M., & Ashman, P. J. (2012). Disruption of microalgal cells for the extraction of lipids for biofuels: Processes and specific energy requirements. *Biomass and Bioenergy, 46,* 89–101.

Lemões, J. S., Rui C. M. Sobrinho, A., Farias, S. P., de Moura, R. R., Primel, E. G., et al. (2016). Sustainable production of biodiesel from microalgae by direct transesterification. *Sustainable Chemistry and Pharmacy, 3,* 33–38.

Liang, Y., Kashdan, T., Sterner, C., Dombrowski, L, Petrick, I., Kröger, M., et al. (2015). Algal biorefineries. In A. Pandey, R. Höfer, M. Taherzadeh, M. Nampoothiri, & C. Caroche (Eds.), *Industrial biorefineries and white biotechnology* (pp. 36–90). Amsterdam: Elsevier. ISBN: 978-0-444-63453-5.

Lorente, E., Hapońska, M., Clavero, E., Torras, C., & Salvadó, J. (2017). Microalgae fractionation using steam explosion, dynamic and tangential cross-flow membrane filtration. *Bioresource Technology, 37,* 3–10.

Lorenz, R. T., & Cysewski, G. R. (2000). Commercial potential for *Haematococcus microalgae* as a natural source of astaxanthin. *Trends in Biotechnology, 18,* 160–167.

Mallick, N., Bagchi, S. K., Koley, S., & Singh, A. K. (2016). Progress and challenges in microalgal biodiesel production. *Frontiers in Microbiology, 7*(1019), 1–11.

Marcati, A., Ursu, A. V., Laroche, C., Soanen, N., Marchal, L., Jubeau, S., et al. (2014). Extraction and fractionation of polysaccharides and B-phycoerythrin from the microalga *Porphyridium cruentum* by membrane technology. *Algal Research, 5*(1), 258–263.

Marrone, B. L., Lacey, R. E., Anderson, D. B., Bonner, J., Coons, J., Dale, T., et al. (2017). Review of the harvesting and extraction program within the National Alliance for Advanced Biofuels and Bioproducts. *Algal Research* (in press).

Martinez-Hernandez, E., Campbell, G., & Sadhukhan, J. (2013). Economic value and environmental impact (EVEI) analysis of biorefinery systems. *Chemical Engineering Research and Design, 91*(8), 1418–1426.

Martins, R. F., Ramos, M. F., Herfindal, L., Sousa, J. A., Skærven, K., & Vasconcelos, V. M. (2008). Antimicrobial and cytotoxic assessment of marine Cyanobacteria—Synechocystis and Synechococcus. *Marine Drugs, 6*(1), 1–11.

Mata, T. M., Martins, A. A., & Caetano, N. S. (2010). Microalgae for biodiesel production and other applications: A review. *Renewable and Sustainable Energy Reviews, 14,* 217–232.

Maurya, R., Paliwal, Ch., Ghosh, T., Pancha, I., Chokshi, K., Mitra, M., et al. (2016). Applications of de-oiled microalgal biomass towards development of sustainable biorefinery. *Bioresource Technology, 214,* 787–796.

Menetrez, M. Y. (2012). An overview of algae biofuel production and potential environmental impact. *Environmental Science Technology, 46*(13), 7073–7085.

Michalak, I., & Chojnacka, K. (2014). Algal extracts: Technology and advances. *Engineering in Life Sciences, 14*(6), 1618–2863.

Milledge, J. J. (2011). Commercial application of microalgae other than as biofuels: A brief review. *Reviews in Environmental Science & Biotechnology, 10*(1), 31–41.

Mimouni, V., Ulmann, L., Pasquet, V., Mathieu, M., Picot, L., Bougaran, G., et al. (2012). The potential of microalgae for the production of bioactive molecules of pharmaceutical interest. *Current Pharmaceutical Biotechnology, 13*(5), 2733–2750.

Mohan, S. V., Modestra, J. A., Amulya, K., Butti, S. K., & Velvizhi, G. (2016a). A circular bioeconomy with biobased products from CO_2 sequestration. *Trends in Biotechnology, 34*(6), 506–519.

Mohan, S. V., Nikhil, G. N., Chiranjeevi, P., Nagendranatha-Reddy, Rohit, M. V., Naresh Kumar, A., et al. (2016a). Waste biorefinery models towards sustainable circular bioeconomy: Critical review and future perspectives. Bioresource Technology, 215, 2–12.

Moncada, J., Tamayo, J. A., & Cardona, C. A. (2014). Integrating first, second, and third generation biorefineries: Incorporating microalgae into the sugarcane biorefinery. *Chemical Engineering Science, 118,* 126–140.

Mondal, M., Goswami, S., Ghosh, A., Oinam, G., Tiwari, O. N., Das, P., et al. (2017). Production of biodiesel from microalgae through biological carbon capture: A review. *Biotechnology, 7*(99), 1–21.

Mulchandani, K., Kar, J. R., & Singhal, R. S. (2015). Extraction of lipids from *Chlorella saccharophila* using high-pressure homogenization followed by three phase partitioning. *Applied Biochemistry and Biotechnology, 176*(6), 1613–1626.

Mussgnug, J. H., Klassen, V., Schlüter, A., & Kruse, O. (2010). Microalgae as substrates for fermentative biogas production in a combined biorefinery concept. *Journal of Biotechnology, 150,* 51–56.

Nayak, B. K., Roy, S., & Das, D. (2014). Biohydrogen production from algal biomass (*Anabaena* sp. PCC 7120) cultivated in airlift photobioreactor. *International Journal of Hydrogen Energy, 39,* 7553–7560.

Ngamprasertsith, S., & Sawangkeaw, R. (2011). Transesterification in supercritical conditions. *RFID Technology, Security Vulnerabilities, and Countermeasures*, 75–100.

Nguyen, T. A. D., Kim, K. R., Nguyen, M. T., Kim, M. S., Kim, D., & Sim, S. J. (2010). Enhancement of fermentative hydrogen production from green algal biomass of *Thermotoga neapolitana* by various pretreatment methods. *International Journal of Hydrogen Energy, 35,* 13035–13040.

Nobre, B. P., Villalobos, F., Barragán, B. E., Oliveira, A. C., Batista, A. P., Marques, P. A. S. S., et al. (2013). A biorefinery from *Nannochloropsis* sp. microalga—Extraction of oils and pigments. Production of biohydrogen from the leftover biomass. *Bioresource Technology, 135,* 128–136.

Norsker, N. H., Barbosa, M. J., Vermue, M. H., & Wijffels, R. H. (2011). Microalgal production—A close look at the economics. *Biotechnology Advances, 29*(1), 24–27.

Nurra, C., Torras, C., Clavero, E., Ríos, S., Rey, M., Lorente, E., et al. (2014). Biorefinery concept in a microalgae pilot plant. Culturing, dynamic filtration and steam explosion fractionation. *Bioresource Technology, 163,* 136–142.

Olguin, E. (2012). Dual purpose microalgae–bacteria-based systems that treat wastewater and produce biodiesel and chemical products within a biorefinery. *Biotechnology Advances, 30*(3), 1031–1046. https://doi.org/10.1016/j.biotechadv.2012.05.001.

Orosz, M. S., & Forney, D. (2008). *A comparison of algae to biofuel conversion pathways for energy storage off-grid*. Retrieved from internet September 2016. http://web.mit.edu/mso/www/AlgaePathwayComparison.pdf.

Otsuki, T., & Zhang, F. (2004). Synthesis and tensile properties of a novel composite of Chlorella and polyethylene. *Journal of Applied Polymer Science, 92,* 812–816.

Paiva, A., Craveiro, R., Aroso, I., Martins, M., Reis, R. L., & Duarte, A. R. C. (2014). Natural deep eutectic solvents—Solvents for the 21st century. *ACS Sustainable Chemistry and Engineering, 2*(5), 1063–1071.

Park, J. Y., Park, M. S., Lee, Y. C., & Yang, J. W. (2015). Advances in direct transesterification of algal oils from wet biomass. *Bioresource Technology, 184,* 267–275.

Parniakov, O., Barba, F. J., Grimi, N., Marchal, L., Jubeau, S., Lebovka, N., et al. (2015). Pulsed electric field and pH assisted selective extraction of intracellular components from microalgae Nannochloropsis. *Algal Research, 8,* 128–134.

Pasquet, V., Farhat, F., Piot, J., Baptiste, J., Kaas, R., Patrice, T., et al. (2011). Study on the microalgal pigments ex traction process: Performance of microwave assisted extraction. *Process Biochemistry, 46*(1), 59–67.

Paul, J. H. (1982). Isolation and characterization of a Chlamydomonas L-asparaginase. *The Biochemical Journal, 203*(1), 109–115.

Pei, D., Xu, J., Zhuang, Q., Tse, H. F., & Esteban, M. A. (2010). Induced pluripotent stem cell technology in regenerative medicine and biology. *Advances in Biochemical Engineering/Biotechnology, 123*(July 2015), 127–141.

Phong, W. N., Le, C. F., Show, P. L., Chang, J. S., & Ling, T. C. (2017a). Extractive disruption process integration using ultrasonication and an aqueous two-phase system for protein recovery from Chlorella sorokiniana. *Engineering in Life Sciences, 17*(4), 357–369.

Phong, W. N., Show, P. L., The, W. H., The, T. X., Lim, H. M. Y., Nazri, N. S. B., et al. (2017b). Proteins recovery from wet microalgae using liquid biphasic flotation (LBF). *Bioresource Technology, 44*(2), 1329–1336.

Pires, J. C. M. (2017). COP21: The algae opportunity? *Renewable and Sustainable Energy Reviews, 79,* 867–877.

Posada, J. A., Brentner, L. B., Ramirez, A., Patel, M. K. (2016) Conceptual design of sustainable integrated microalgae biorefineries: Parametric analysis of energy use, greenhouse gas emissions and techno-economics. *Algal Research, 17,* 113–131.

Postma, P. R., Lam, G. P., Barbosa, M. J., Wijffels, R. H., Eppink, M. H. M., & Olivieri, G. (2016). Microalgal biorefinery for bulk and high-value products: Product extraction within cell disintegration. In D. Miklavcic (Ed.), *Handbook of electroporation* (pp. 1–20). Springer International Publishing.

Priyadarshani, I., & Sahu, D. (2012). Algae in aquaculture. *International Journal of Health Sciences and Research, 2*(1), 108–114. Retrieved from http://www.ijhsr.org/IJHSR_Vol.2_Issue.1_April2012/14.pdf.

Pulz, O., & Gross, W. (2004). Valuable products from biotechnology of microalgae. *Applied Microbiology and Biotechnology, 65*(6), 635–648.

Queiroz, M. I., Hornes, M. O., da Silva, Gonçalves, Manetti, A., Zepka, L. Q., & Jacob-Lopes, E. (2013). Fish processing wastewater as a platform of the microalgal biorefineries. *Biosystems Engineering, 115*(2), 195–202.

Quemeneur, M., Hamelin, J., Benomar, S., Guidici-Orticoni, M. T., Latrille, E., Steyer, J. P., et al. (2011). Changes in hydrogenase genetic diversity and proteomic patterns in mixed-culture dark fermentation of mono-, di- and tri-saccharides. *International Journal of Hydrogen Energy, 36,* 11654–11665.

Quinn, J. C., & Davis, R. (2015). The potentials and challenges of algae based biofuels: A review of the techno-economic, life cycle, and resource assessment modeling. *Bioresource Technology, 184,* 444–452.

Rahman, A., & Miller, C. D. (2017). Microalgae as a source of bioplastics. In R. Prasad Rastogi, D. Madamwar, & A. Pandey (Eds.), *Algae green chemistry. Recent progress in biotechnology* (pp. 189–200).Amsterdam: Elsevier.

Ranjith Kumar, R., Hanumantha Rao, P., & Arumugam, M. (2015). Lipid extraction methods from microalgae: A comprehensive review. *Frontiers in Energy Research, 2,* 1–9. https://doi.org/10.3389/fenrg.2014.00061

Reddy, M. M., Vivekanandhan, S., Misra, M., Bhatia, S. K., & Mohanty, A. K. (2013). Biobased plastics and bionanocomposites: Current status and future opportunities. *Progress in Polymer Science, 38,* 1653–1689.

Rizwan, M., Lee, J. H., & Gani, R. (2015). Optimal design of microalgae-based biorefinery: Economics, opportunities and challenges. *Applied Energy, 150,* 69–79.

Roux, J. M., Lamotte, H., & Achard, J. L. (2017). An overview of microalgae lipid extraction in a biorefinery framework. *Energy Procedia, 112,* 680–688.

Ryckebosch, E., Muylaert, K., & Foubert, I. (2012). Optimization of an analytical procedure for extraction of lipids from microalgae. *Journal of the American Oil Chemists' Society, 89*(2), 189–198.

Saadatnia, H., & Riahi, H. (2009). Cyanobacteria from paddy-fields in Iran as a biofertilizer in rice plants. *Plant Soil Environment, 55,* 207–212. Retrieved from http://www.agriculturejournals.cz/publicFiles/07199.pdf.

Saai-Anuggraha, T. S., Swaminathan, T., Sulochana, S. (2016). Microbiology of platform chemical biorefinery and metabolic engineering. In S. K. Brar, S. J. Sarma, K. Pakshirajan (Eds.), *Platform chemical biorefinery. Future Green Chemistry* (pp. 437–450). The Netherlands: Elsevier. ISBN 978-12-802980-0.

Safi, C., Charton, M., Ursu, A. V., Laroche, C., Zebib, B., Pontalier, P. Y., et al. (2014a). Release of hydro-soluble microalgal proteins using mechanical and chemical treatments. *Algal Research, 3*(1), 55–60.

Safi, C., Olivieri, G., Campos, R. P., Engelen-Smit, N., Mulder, W. J., van den Broek, L. A. M., et al. (2017). Biorefinery of microalgal soluble proteins by sequential processing and membrane filtration. *Bioresource Technology, 225,* 151–158.

Safi, C., Ursu, A. V., Laroche, C., Zebib, B., Merah, O., Pontalier, P. Y., et al. (2014b). Aqueous extraction of proteins from microalgae: Effect of different cell disruption methods. *Algal Research, 3*(1), 61–65.

Sambusiti, C., Bellucci, M., Zabaniotou, A., Beneduce, L., & Monlau, F. (2015). Algae as promising feedstocks for fermentative biohydrogen production according to a biorefinery approach: a comprehensive review. *Renewable and Sustainable Energy Reviews, 44,* 20–36.

Santibañez-Aguilar, J. E., González-Campos, J. B., Ponce-Ortega, J. M., Serna-González, M., & El-Halwagi, M. M. (2014). Optimal planning and site selection for distributed multiproduct biorefineries involving economic, environmental and social objectives. *Journal of Cleaner Production, 65,* 270–294.

Sathe, S., & Durand, P. M. (2015). A low cost, non-toxic biological method for harvesting algal biomass. *Algal Research, 11,* 169–172.

Schwenzfeier, A., Wierenga, P. A., Eppink, M. H. M., & Gruppen, H. (2014). Effect of charged polysaccharides on the techno-functional properties of fractions obtained from algae soluble protein isolate. *Food Hydrocolloids, 35,* 9–18.

Schwenzfeier, A., Wierenga, P. A., & Gruppen, H. (2011). Isolation and characterization of soluble protein from the green microalgae *Tetraselmis* sp. *Bioresource Technology, 102*(19), 9121–9127.

Shah, Md M R, Mahfuzur, R., Liang, Y., Cheng, J. J., & Daroch, M. (2016). Astaxanthin-producing green microalga *Haematococcus pluvialis*: From single cell to high value. *Frontiers in Plant Science, 7*(531), 1–28.

Shi, B., Wideman, G., & Wang, J. H. (2011). A new approach of bioCO_2 fixation by thermoplastic processing of microalgae. *Journal of Polymers and the Environment, 20,* 124–131.

Sialve, B., Bernet N., Bernard, O. (2009). Anaerobic digestion of microalgae as a necessary step to make microalgal biodiesel sustainable. *Biotechnology Advance, 27*(4), 409–416.

Silva, C. M., Ferreira, A. F., Dias, A. P., & Costa, M. (2016). A comparison between microalgae virtual biorefinery arrangements for bio-oil production based on lab-scale results. *Journal of Cleaner Production, 130,* 58–67.

Singh, J. S., Kumar, A., Rai, A. N., & Singh, D. P. (2016). Cyanobacteria: A precious bio-resource in agriculture, ecosystem, and environmental sustainability. *Frontiers in Microbiology, 7*(529), 1–19.

Sirakov, I., Velichkova, K., Stoyanova, S., Staykov, Y. (2015). The importance of microalgae for aquaculture industry. Review. *International Journal of Fisheries and Aquatic Studies, 2*(4), 81–84. Retrieved from http://www.fisheriesjournal.com/archives/2015/vol2issue4/PartB/2-4-31.pdf.

Slade, R., & Bauen, A. (2013). Micro-algae cultivation for biofuels: Cost, energy balance, environmental impacts and future prospects. *Biomass and Bioenergy, 53,* 29–38.

Solovchenko, A. E. (2015). Recent breakthroughs in the biology of astaxanthin accumulation by microalgal cell. *Photosynthesys Research, 125,* 437–449.

Spolaore, P., Joannis-Cassan, C., Duran, E., & Isambert, A. (2006). Commercial applications of microalgae. *Journal of Bioscience and Bioengineering, 101*(2), 87–96.

Sterner, C. (2013). Algenol integrated pilot-scale biorefinery for producing ethanol from hybrid algae. Report DOE/EE-0835. January 2013. Retrieved from https://www1.eere.energy.gov/bioenergy/pdfs/ibr_arra_algenol.pdf.

Stolz, P., & Obermayer, B. (2005). Manufacturing microalgae for skin care. *Cosmetics and Toiletries Magazine, 120*(3), 99–106.

Suarez Ruiz, C. A., van den Berg, C., Wijffels, R. H., & Eppink, M. H. M. (2017). Rubisco separation using biocompatible aqueous two-phase systems. *Separation and Purification Technology*, 1–8.

Suganya, T., Varman, M., Masjuki, H. H., & Renganathan, S. (2016). Macroalgae and microalgae as a potential source for commercial applications along with biofuels production: A biorefinery approach. *Renewable and Sustainable Energy Reviews, 55,* 909–941.

Sung, M. G., Lee, B., Kim, C. W., Nam, K., & Chang, Y. K. (2017). Enhancement of lipid productivity by adopting multi-stage continuous cultivation strategy in *Nannochloropsis gaditana. Bioresource Technology, 229,* 20–25.

Taher, H., Al-zuhair, S., Al-marzouqi, A. H., Haik, Y., & Farid, M. (2014). Effective extraction of microalgae lipids from wet biomass for biodiesel production. *Biomass and Bioenergy, 66,* 159–167.

Tan, C. H., Show, P. L., Chang, J. S., Ling, T. C., & Lan, J. C. W. (2014). Novel approaches of producing bioenergies from microalgae: A recent review. *Biotechnology Advances, 33*(6), 1219–1227.

Templeton, D. W., Quinn, M., Van Wychen, S., Hyman, D., & Laurens, L. M. L. (2012). Separation and quantification of microalgal carbohydrates. *Journal of Chromatography A, 1270,* 225–234.

Thomassen, G., Van Dael, M., Lemmens, B., & Van Passel, S. (2017). A review of the sustainability of algal-based biorefineries: Towards an integrated assessment framework. *Renewable and Sustainable Energy Reviews, 68*(2), 876–887. https://doi.org/10.1016/j.rser.2016.02.015.

Toledo-Cervantes, A., Estrada, J. M., Lebrero, R., & Muñoz, R. (2017). A comparative analysis of biogas upgrading technologies: Photosynthetic vs. physical/chemical processes. *Algal Research, 25,* 237–243.

Toledo-Cervantes, A., & Morales, M. (2014). Biorefinery using microalgal biomass for producing lipids, biofuels and other chemicals. In L. Torres & E. Bandala (Eds.), *Energy and environment nowadays* (pp. 17–56). Nova Science Publishers, Inc. ISBN 978-63117-399-8-1.

Tredici, M. R., Bassi, N., Prussi, M., Biondi, N., Rodolfi, L., Chini Zittelli, G., et al. (2015). Energy balance of algal biomass production in a 1-ha "Green Wall Panel" plant: how to produce algal biomass in a closed reactor achieving a high net energy ratio. *Applied Energy, 154,* 1103–1111.

Ueno, Y., Kurano, N., & Miyachi, S. (1998). Ethanol production by dark fermentation in the marine green alga, *Chlorococcum littorale. Journal of Fermentation and Bioengineering, 1998* (86), 38–43.

Uggetti, E., & Puigagut, J. (2016). Photosynthetic membrane-less microbial fuel cells to enhance microalgal biomass concentration. *Bioresource Technology, 218,* 1016–1020.

Ursu, A. V., Marcati, A., Sayd, T., Sante-Lhoutellier, V., Djelveh, G., & Michaud, P. (2014). Extraction, fractionation and functional properties of proteins from the microalgae *Chlorella vulgaris*. *Bioresource Technology, 157,* 134–139.

van der Voort, M. P. J., Vulsteke, E., & de Visser, C. L. M. (2015). *Macro-economics of algae products* (47pp). Public Output report of the EnAlgae project, Swansea, June 2015. Retrieved from http://edepot.wur.nl/347712. September 2017.

Van Reis, R., & Zydney, A. (2001). Membrane separations in biotechnology. *Current Opinion in Biotechnology, 12*(2), 208–211.

Vanthoor-Koopmans, M., Wijffels, R. H., Barbosa, M. J., & Eppink, M. H. M. (2013). Biorefinery of microalgae for food and fuel. *Bioresource Technology, 135,* 142–149.

Vigani, M., Parisi, C., Rodríguez-Cerezo, P., Barbosa, M. J., Sijtsma, L., Ploeg, M., et al. (2015). Food and feed products from microalgae: Market opportunities and challenges for the EU. *Trends in Food Science & Technology, 4,* 81–92.

Wang, H. M. D., Chen, Ch Ch., Huynh, P., & Chang, J Sh. (2015). Exploring the potential of using algae in cosmetic. *Bioresource Technology, 184,* 355–362.

Ward, A. J. (2015). The anaerobic digestion of microalgae feedstock, "life-cycle environmental impacts of biofuels and co-products". In N. R. Moheimani, et al. (Eds.), *Biomass and biofuels from microalgae, biofuel and biorefinery technologies* (pp. 331–345). Switzerland: Springer International Publishing.

Weaver, J., Husson, S. M., Murphy, L., & Wickramasinghe, S. R. (2013). Anion exchange membrane adsorbers for flow-through polishing steps: Part II. Virus, host cell protein, DNA clearance, and antibody recovery. *Biotechnology and Bioengineering, 110*(2), 500–510.

Wiesberg, I. L., Brigagão, G. V., de Medeiros, J. L., & de Queiroz Fernandes Araújo, O. (2017). Carbon dioxide utilization in a microalga-based biorefinery: Efficiency of carbon removal and economic performance under carbon taxation. *Journal of Environmental Management, 203,* 988–998.

Wijffels, R. H., & Barbosa, M. J. (2010). An outlook on microalgal biofuels. *Science, 329*(5993), 796–799.

Wijffels, R. H., Barbosa, M. J., & Eppink, M. H. M. (2010). Microalgae for the production of bulk chemicals and biofuels. *Biofuels, Bioproducts and Biorefining, 4,* 287–295.

World Metereological Organization. (2017). *Internet consult August 2017.* https://public.wmo.int/en/media/press-release/wmo-confirms-2016-hottest-year-record-about-11°c-above-pre-industrial-era.

Wu, X., Li, R., Zhao, Y., & Liu, Y. (2017). Separation of polysaccharides from *Spirulina platensis* by HSCCC with ethanol-ammonium sulfate ATPS and their antioxidant activities. *Carbohydrate Polymers, 173,* 465–472.

Xia, A., Cheng, J., Song, W., Su, H., Ding, L., Lin, R., et al. (2015). Fermentative hydrogen production using algal biomass as feedstock. *Renewable and Sustainable Energy Reviews, 51,* 209–230.

Xia, C., Zhang, J., Zhang, W., & Hu, B. (2011). A new cultivation method for microbial oil production: cell pelletization and lipid accumulation by *Mucor circinelloides*. *Biotechnology for Biofuels, 4*(15), 2–10.

Xu, Y., & Boeing, W. J. (2013). Mapping biofuel field: A bibliometric evaluation of research output. *Renewable and Sustainable Energy Reviews, 28,* 82–91.

Yaakob, Z., Ali, E., Zainal, A., Mohamad, M., & Takriff, M. S. (2014). An overview: Biomolecules from microalgae for animal feed and aquaculture. *Journal of Biological Research (Greece), 21*(1), 1–10.

Yamada, T., & Sakaguchi, K. (1982). Comparative studies on Chlorella cell walls: Induction of protoplast formation. *Archives of Microbiology, 132*(1), 10–13.

Yen, H. W., Yang, S. C., Chen, C. H., Jesisca, & Chang, J. S. (2015). Supercritical fluid extraction of valuable compounds from microalgal biomass. *Bioresource Technology, 184*, 291–296.

Yuan, J., Kendall, A., & Zhang, Y. (2015). Mass balance and life cycle assessment of biodiesel from microalgae incorporated with nutrient recycling options and technology uncertainties. *GCB Bioenergy, 7*(6), 1245–1259.

Zeller, M. A., Hunt, R., Jones, A., & Sharma, S. (2013). Bioplastics and their thermoplastic blends from Spirulina and Chlorella microalgae. *Journal of Applied Polymer Science, 130,* 3263–3275.

Zhang, F., Endo, T., Kitagawa, R., Kabeya, H., & Hirotsu, T. (2000a). Synthesis and characterization of a novel blend of polypropylene with Chlorella. *Journal of Materials Chemistry, 10,* 2666–2672.

Zhang, F., Kabeya, H., & Kitagawa, R. (2000b). An exploratory research of PVC-Chlorella composite material (PCCM) as effective utilization of Chlorella biologically fixing CO_2. *Journal of Materials Science, 5,* 2603–2609.

Zhao, G., Chen, X., Wang, L., Zhou, S., Feng, H., Chen, W. N., et al. (2013). Ultrasound assisted extraction of carbohydrates from microalgae as feedstock for yeast fermentation. *Bioresource Technology, 128,* 337–344.

Zhao, L., Peng, Y., Gao, J., & Cai, W. (2014). Bioprocess intensification: an aqueous two-phase process for the purification of C-phycocyanin from dry *Spirulina platensis*. *European Food Research and Technology, 238*(3), 451–457.

Zhou, W., Min, M., Hu, B., Ma, X., Liu, Y., Wang, Q., et al. (2013). Filamentous fungi assisted bio-flocculation: A novel alternative technique for harvesting heterotrophic and autotrophic microalgal cells. *Separation and Purification Technology, 107,* 158–165.

Zhu, L. (2015). Biorefinery as a promising approach to promote microalgae industry: An innovative framework. *Renewable and Sustainable Energy Reviews, 41,* 1376–1384.

Zhu, X., Rong, J., Chen, H., He, Ch., Hu, W., & Wang, Q. (2016). An informatics-based analysis of developments to date and prospects for the application of microalgae in the biological sequestration of industrial flue gas. *Applied Microbiology Biotechnology, 100,* 2073–2082.

Chapter 6
Life Cycle Assessment of Biofuels from Microalgae

Mariany Costa Deprá, Eduardo Jacob-Lopes and Leila Queiroz Zepka

Abstract Recently, the use of mathematical tools, such as the life cycle assessment (LCA) methodology for ecologically sound processes, with the purpose of establishing a process designer involving the limits of "cradle to grave" in an efficient and flexible way with less subjectivity, has become an ambitious challenge to be won. Therefore, to generate biofuels with low atmospheric emissions and minimal energy requirements has become crucial to commercial competitiveness. Thus, the objective of this chapter is to approach the current situation of the different scenarios of microalgal biofuels production by an evaluation of them via a life cycle assessment. The chapter is based on three main topics: (1) fundamentals for structuring a life cycle assessment, (2) biofuels data set reported in the literature, and (3) application of LCA in microalgae biofuels.

Keywords Microalgal · Biofuel · Biodiesel · Life cycle analysis

1 Introduction

The emergence of ever larger global issues, such as the energy dilemma, the warming of the climate, and the scarcity of water resources, has boosted the search for tools capable of ensuring the reliability of the results published by the industries, becoming the focus of environmental sustainability (Blanchard et al. 2017).

To this end, the application of the life cycle assessment (LCA) assumes the character of ensuring better internal management in order to promote cleaner production,

---

M. C. Deprá · E. Jacob-Lopes (✉) · L. Q. Zepka
Food Science and Technology Department, Federal University of Santa Maria, UFSM, Roraima Avenue 1000, Santa Maria, RS 97105-900, Brazil
e-mail: jacoblopes@pq.cnpq.br

L. Q. Zepka
e-mail: lqz@pq.cnpq.br

E. Jacob-Lopes et al. (eds.), *Energy from Microalgae*, Green Energy and Technology, https://doi.org/10.1007/978-3-319-69093-3_6

improve eco-efficiency, and assist in economic calculations within institutional organizations. In addition, in a broader conception, LCA also serves as the basis for reporting the data required by environmental regulatory agencies (Bicalho et al. 2017).

Given the above, the use of this mathematical tool emerges with the purpose of defining the performance of biofuels in order to assist in decision making, which provides an understanding of environmental impacts and possible increases in the efficiency of their biofuels processes, reducing costs, and promoting the marketing of its products in a sustainable way (Deprá et al. 2017).

Finally, with the objective of expanding the knowledge base and promoting the environmental importance of biofuels, the objective of the chapter was to elucidate the fundamental aspects of the implementation of an LCA as well as to report studies on microalgae biofuels.

2 Life Cycle Assessment for Biofuels

Life cycle assessment was devised in the late 1960s by the US Department of Energy. As a goal, the first studies were conducted to investigate the life cycle aspects of products and materials that address issues such as energy requirements and energy efficiency. In this context, life cycle analysis has become a key tool for assessing the sustainability potential of the processes. Therefore, this tool aims to establish an environmental approach with the objective of not restricting a systemic view of the productive chain, but rather an action on its environmental aspects and the opportunities for improvement that can be observed from a more comprehensive analysis (Curran 2006).

Afterward, in the early 1990s, the pressure of the environment and the need to use the LCA tool to be widely recognized led to the establishment of LCA standards mandated by the International Organization for Standardization (ISO) (ISO 14040 2006). This basis for implementation provided by the ISO series, in the current state, is applied in different ways and, therefore, often leads to divergent results (Gnansounou et al. 2009). Figure 1 shows the four basic steps that guide the implementation of the LCA in order to perform the analysis in a homogeneous and standardized way.

2.1 *Definition of the Purpose and Scope*

The first step in initiating a case study using the life cycle analysis tool is to establish the goal and scope definition. Depending on the purpose and scope of the LCA, there are four main options of the system boundary: cradle to gate, gate to gate, cradle to grave, and gate to tomb (Jacquemin et al. 2012). The cradle to grave option is the broader scope limit where the life cycle of a product undergoes at least

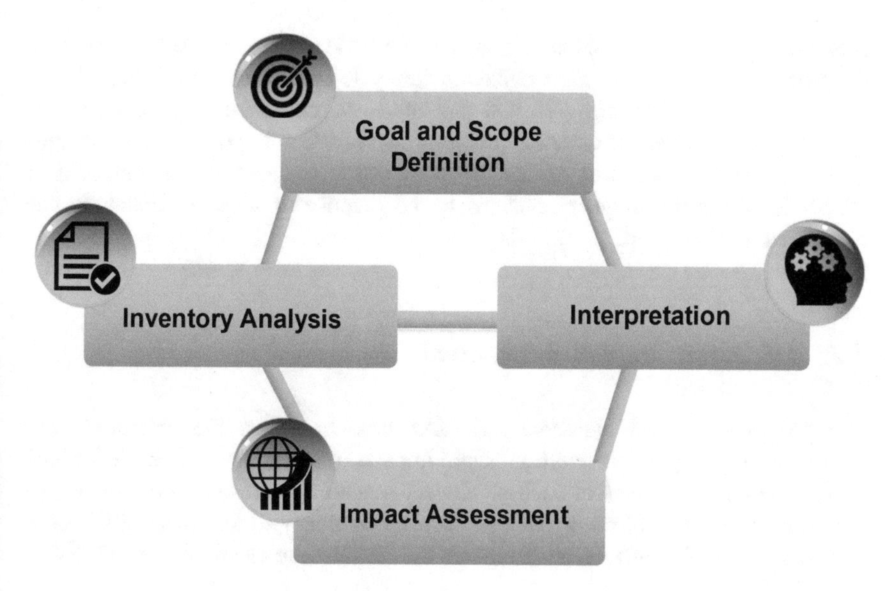

Fig. 1 Driving steps for life cycle assessment

three phases: production phase, phase of use, and phase of elimination. However, the goal usually includes the intended application, the reasons for conducting the study, the target audience, and the use of the results. Moreover, although the definition of the system is more detailed in case of project or operation improvement, the flowchart of the biofuels routes is simplified to LCA when related to the political–environmental reports (Zhang et al. 2015).

In addition, perhaps the greatest impasse of this realization is the delimitation of the process. Although this stage is inserted in all the works analyzed, they are not always structured with the same level of detail. They make the results restricted when comparisons are made.

Another important question about the starting point for reducing the constraints, the system boundary (unit operations involved), and the functional unit (calculation basis) should be clearly defined (Prox and Curran 2017). Therefore, the LCA for biofuels is configured from the process of obtaining the energy biomass, the process of extraction, transport, and use, commonly measured in units of 1 kg or 1 MJ of biofuel.

2.2 *Life Cycle Inventory*

The life cycle inventory (LCI) stage involves the compilation and quantification of inputs and outputs for each process included in the system boundary. Consequently, these inputs and outputs include the use of resources such as the release of greenhouse gases, water, and land associated with the system. In short, this stage of the process is the input to characterize the life cycle assessment (ISO 14040 2006).

In addition, as data are collected and more thoroughness is assigned to the system, new requirements or data limitations may be identified that require a change in the collection procedures so that the study objectives are still met (ISO 14041 1998). Problems that require revisions to the purpose or scope of the study can sometimes be identified.

2.3 *Life Cycle Impact Assessment*

The life cycle impact assessment (LCIA) aims to assess the magnitude and importance of the potential environmental impacts of a product/service (ISO 14042 2000). Therefore, as factors such as choice, modeling, and evaluation of impact categories can add subjectivity to the study, transparency in this stage of the LCA becomes extremely relevant, ensuring that the facilities are clearly described (Jolliet et al. 2003).

Due to this, flows are associated with the possible categories of environmental impact. The choice of an impact category is based on characterization methods according to the objectives and scope of the study. Each flow can contribute to various categories of environmental impact, as categories associated with human toxicity, acidification, and ecotoxicity (Carneiro et al. 2017).

In this sense, in order to compile and quantify the effects caused by the systemic process of producing microalgae biofuels, the quantification steps are subdivided into three categories: energy balance, water footprint, and greenhouse gas emissions.

2.3.1 Balance Energy

The concerns about energy balances are related to both the life cycle energy efficiency of biofuels and the saving of nonrenewable energy between biofuels and fossil fuels (Soccol et al. 2011).

The LCA literature defined, according to Eq. (1), the net energy ratio (NER) as the ratio of the total energy produced (energy potential of the oil or feedstock) to the energy content of the construction and materials, in addition to the energy required for all plants (Jorquera et al. 2010).

$$\mathrm{NER} = \frac{\sum \text{energy produced}}{\sum \text{energy requirements}} \tag{1}$$

Through this equation, it is possible to estimate the fossil energy needed to feed the process. The functional units used are megajoules (MJ).

2.3.2 Water Footprint

Water footprint (WF) is used to assess water use along the supply chains, sustainability of water uses within river basins, water use efficiency, water allocation equitability, and dependence of water on the supply chain. This is characterized by quantifying the freshwater consumption of a process or product per functional unit (Hoekstra 2016).

The concept of water footprint comprises three components: green WF, blue WF, and gray WF, according to Eq. (2). Green WF is defined as rainwater that is evaporated during the growing period of the culture. The blue water footprint is the volume of surface and groundwater consumed during the production of a particular product or service. Consumption includes the volume of freshwater evaporated or incorporated into a product or service. However, WF gray refers to the amount of water that cannot be reused, that is, the volume that needs treatment or that has been contaminated (Farooq et al. 2015).

$$\mathrm{WF} = \sum \mathrm{WFgreen} + \mathrm{WFblue} + \mathrm{WFgray} \tag{2}$$

Processes for microalgae biofuels, the direct withdrawal of water footprint represents the water that is consumed by each step in the process, including, for example, water for microalgae cultivation, water required to compensate evaporation of the bioreactor, water loss of the process during filtration, and the water reached during the conversion of the fuel (Mekonnen and Hoekstra 2010; Garcia and You 2015). These footprints are estimated by units of volume (m^3) per kilogram of biofuel (kg) or megajoules of biofuel (MJ) (Guieysse et al. 2013).

2.3.3 Greenhouse Gas Emissions

Absorption capacity, concentration, and residence time of the gases are used to evaluate the so-called global warming potential (GWP). In turn, the GWP is characterized as a simplified index in the Intergovernmental Panel on Climate Change (IPCC), along with the land use change coverage (LUCC) that quantifies the environmental impact generated by greenhouse gases as well as the potential of acidification, eutrophication, and depletion of the ozone layer (Forster et al. 2007).

In 2002, the United Nations Environment Program (UNEP) joined the Society of Environmental Toxicology and Chemistry (SETAC) to initiate the life cycle initiative which is an international partnership aimed at putting the cycle into practice and improving the tools of support through better data and indicators (Klöpffer 2006).

Usually, it can be quantified according to Eq. (3), sum the masses of substances that contribute to the impact (Mi), whether masses of gases (CO_2, CH_4, NO_x) contribute to these same substances impact that are published annually in reports from environmental such as the Department for Environment Food and Rural Affairs (DEFRA) (DECC 2010; Laratte et al. 2014).

$$E = \sum_{i} \mathrm{Mi} \times \mathrm{Pi} \tag{3}$$

Impact factors are reported at different times and are commonly estimated at 20, 100, or 500 years. In this way, long-lasting compounds such as carbon dioxide (CO_2) tend to remain and concentrate for a longer period in the atmosphere (around 100 years). However, compounds with smaller residence ranges, such as methane, assume heating profiles capable of emitting heating potentials 60 times larger and 23 times more potent when compared to CO_2 (Yvon-Durocher et al. 2014). In this sense, the representation of the pollutant gas and the estimated residence time should be considered according to the profile of the system to be analyzed. Moreover, among the three-time horizons cited above, the 100-year level is most commonly used as a standard time horizon for expressing GWPs (Guo and Murphy 2012).

2.4 *Interpretation*

Interpretation is the phase of the LCA where the findings of the inventory analysis and the impact assessment are combined or, in the case of life cycle inventory studies, only the results of the inventory analysis consistent with the defined objective and scope to reach conclusions and recommendations. The results of this interpretation may take the form of conclusions and recommendations to decision makers as long as they are consistent with the purpose and scope of the study (ISO 14043 2000).

In addition, this phase may involve the interactive process of reviewing the scope, as well as making assumptions made during the study. In this way, since the verification of the quality and nature of the data collection occurs through the estimation of the scenario, the parameters are modified in a systematic way always considering the initial objective defined (ILDC 2010).

3 Data Set Reported in the Literature

Microalgae are indicated as a potential alternative to traditional fuel resources because of their ability to be used for the generation biofuels. The potential fuels include biogas, biohydrogen, bioethanol, and biodiesel (Brennan and Owende

2010), each of which has advantages and disadvantages due to the processing of raw materials and limitations. However, there is a significant focus on the growth of microalgae specifically for biodiesel applications (Soh et al. 2014).

The literature in this field does not fully address the fundamental stages of life cycle application. However, it is only mentioned that the main challenges associated with the production of biofuels in the environment are related to the raw material and its significant influence on the high-energy consumption required. Moreover, literary efforts are limited. In this context, Table 1 shows surveys that report the main parameters implementing the LCA, where the main focus is to establish the net energy index and greenhouse gas emissions. Nevertheless, most studies are lacking information on the use of water resources (Živković et al. 2017).

On the other hand, the literature reports the most diverse objectives of establishing LCA. Among the most cited are lipid extraction methods and studies that reduce energy input at harvest, centrifugation, and drying, as these are the main unit operations of the microalgal biomass production unit (Chen et al. 2011; Laamanen et al. 2016).

In addition, since the main operations included in microalgae cultivation are exposed in the scope of work, the objective is to establish and provide a solid basis for the implementation of zero carbon emissions, with the purpose of consolidating integrated biorefineries to produce biofuels of microalgae (Medeiros et al. 2015; Klein et al. 2017).

As a result of the implementation of biorefineries, the distribution of environmental costs or loads in a multiproduct system provides the emergence of the allocation issue. Consequently, the biggest bottleneck in implementing the LCA is to establish the best scenario and its allocation in order to process the products without neglecting the quantification of the necessary inputs of the system (Silva et al. 2017).

It is known that microalgal biomass presents a diverse range of products to be exploited from the defatted biomass. Therefore, defining a system boundary, which makes it possible to extract all co-products from this residue, makes the system boundary unlikely and complex.

At the same time, it is necessary to estimate scenarios that determine exactly which process will be chosen. Figure 2 shows the scope of three different scenarios of a biorefinery for the production of microalgae biodiesel. Moreover, since bioenergy includes low-value but high-volume biofuels such as biodiesel. In contrast, high-value but smaller-volume co-products are designed to increase the profitability of biorefineries (Chew et al. 2017). Therefore, at the same time as bulk chemicals such as defatted biomass can be obtained, fine chemicals such as pigments can increase the economic profitability of bioprocesses (Jacob-Lopes and Franco 2010).

Table 1 Studies of life cycle assessment in microalgae biofuels

Biofuel	Cultivation system	Goal	Unit Funtional	NER	GWP ($kgCO_2eq$)	Water footprint (m^3)	References
Biodiesel	nd	Lipid extraction	1 kg	5.98–7.10	0	0.23–0.31	Collotta et al. (2017)
Biodiesel	nd	Meta-analysis	1 m^3	1.4	−1066–8222	nd	Liu et al. (2012)
Biodiesel	Hybrid	Commercial scale	1 ton	31	2137	16.3	Adesanya et al. (2014)
Biodiesel	Open ponds	Lipid extraction	1 MJ	1.13–1.82	nd	nd	Jian et al. (2015)
Biodiesel	Open ponds	Cultivation	317 GJ	30×10^4	1.810^4	12×10^4	Clarens et al. (2010)
Biodiesel	Open ponds	Pathway routes	1 MJ	0.33–92.77	0.03–1.32	nd	Dutta et al. (2016)
Ethanol	Open ponds	Biorefinery	1 MJ	1.16–2.64	89.6–233.5	nd	Quiroz-Arita et al. (2017)
Ethanol	Raceway ponds	Solvent purification	1 MJ	0.20–0.55	12.3–29.8	nd	Luo et al. (2010)

nd not defined

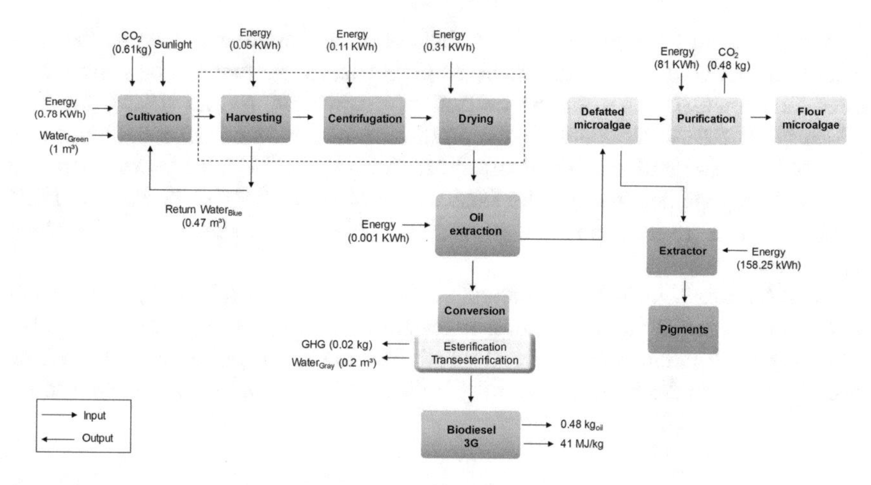

Fig. 2 Flowchart for obtaining microalgal biodiesel in an integrated scenario. Adapted from Monari et al. (2016)

Table 2 Bottlenecks of life cycle assessment implementation

Bottlenecks LCA implementation	Comment
Purpose and scope	Difficulty of delimiting cycles and establishing the inputs and outputs of process flows
Functional unit	It is very difficult to define equivalent quantities for different products. The functional unit can be a technical context, such as the amount of energy use or social functions, such as the mileage traveled by biodiesel and its different depreciation rates
Allocation	The comparative evaluation between several virtual products that have different functions. Here, it is important to know whether it is better to develop a simpler product that produces some weak pollutants and has fewer functions or if a more complex product is created with additional functions and additional pollutants
Standardization methodology	Standardize the means of computing the data. Since there are many software programs on the market, it is necessary to verify the veracity of these values stored in the databases
Environmental charges	There is a lack of understanding of impact categories and the prioritization of these categories
Interpretation	Difficulty of interpretation and comparison with the data reported in the literature

Yet, it should be noted that in the three scenarios presented, such as the generation of biodiesel, microalgae meal, or pigments, the system can be allocated in three fundamental stages in the collection, centrifugation, and drying process, which is extremely important for the production of dry biomass. In addition, it is known that these unit operations are those that require the most power from the system. Many studies that use the LCA as a tool erroneously determine the NER with less than 1, rendering the process of obtaining biodiesel by microalgae impossible.

In this context, in order for the application of the life cycle analysis does not make the microalgal biofuels process unfeasible, this methodology needs to combat some implementation bottlenecks. Table 2 lists the major obstacles that must be overcome before the LCA can be fully utilized as a standard environmental tool.

4 Application of LCA to Biofuels from Microalgae

In the case of microalgae, it is possible to use the LCA tool to apply to the process in order to define the environmental performance of biofuels. The baseline data to be quantified were adapted from Monari et al. (2016), where the energy values were expressed by hours/day and the units of volume and atmospheric emissions expressed per day.

First step: The aim of the application is to evaluate the energy demand in its environmental character in order to reveal the feasibility of this process to obtain microalgal biodiesel. The chosen alternative was to establish a simplified flow diagram of the process only for the measurement of the inputs and outputs of the procurement process, not taking into account the demands of materials required to manufacture the equipment. As a functional unit, 1 cubic meter of bioreactor volume was established, with the operating time of 24 h per day for 330 days per year. The target audience is to generate the final report, presenting perspectives on the application in a real scale of the process.

Second step: Data collection implies the quantification of system inputs. Therefore, it is necessary to fully clarify the energy demands of the equipment related to the process. Therefore, the equations related to their respective classes are applied (Box 1).

Box 1. Example of quantification of inputs and outputs of the process of obtaining microalgal biodiesel

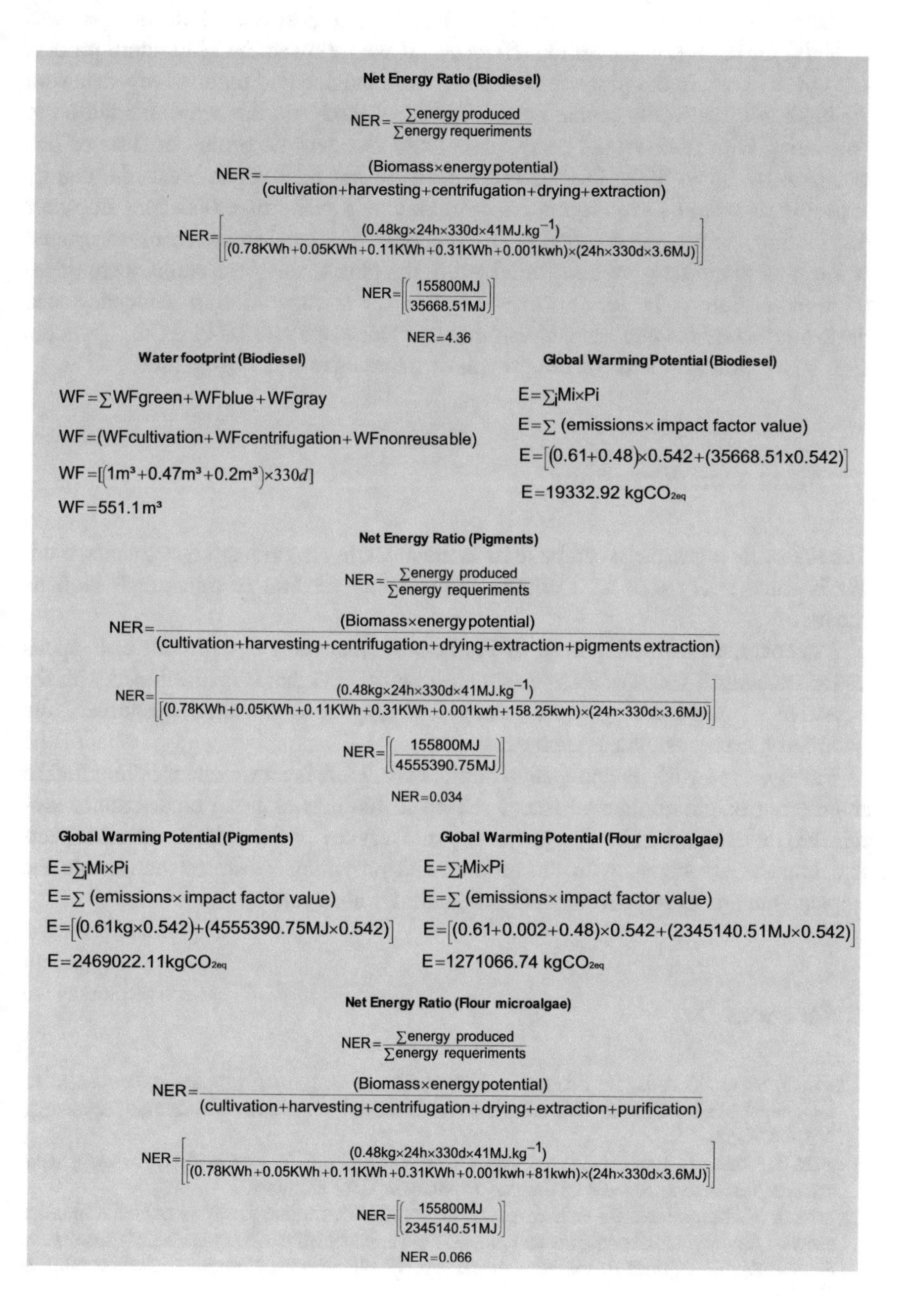

Net Energy Ratio (Biodiesel)

$$NER=\frac{\sum \text{energy produced}}{\sum \text{energy requeriments}}$$

$$NER=\frac{(\text{Biomass}\times\text{energy potential})}{(\text{cultivation}+\text{harvesting}+\text{centrifugation}+\text{drying}+\text{extraction})}$$

$$NER=\left[\frac{(0.48kg\times24h\times330d\times41MJ.kg^{-1})}{[(0.78KWh+0.05KWh+0.11KWh+0.31KWh+0.001kwh)\times(24h\times330d\times3.6MJ)]}\right]$$

$$NER=\left[\left(\frac{155800MJ}{35668.51MJ}\right)\right]$$

$$NER=4.36$$

Water footprint (Biodiesel)

$$WF=\sum WFgreen+WFblue+WFgray$$

$$WF=(WFcultivation+WFcentrifugation+WFnonreusable)$$

$$WF=[(1m^3+0.47m^3+0.2m^3)\times330d]$$

$$WF=551.1\,m^3$$

Global Warming Potential (Biodiesel)

$$E=\sum_i Mi\times Pi$$

$$E=\sum(\text{emissions}\times\text{impact factor value})$$

$$E=[(0.61+0.48)\times0.542+(35668.51\text{x}0.542)]$$

$$E=19332.92\ kgCO_{2eq}$$

Net Energy Ratio (Pigments)

$$NER=\frac{\sum \text{energy produced}}{\sum \text{energy requeriments}}$$

$$NER=\frac{(\text{Biomass}\times\text{energy potential})}{(\text{cultivation}+\text{harvesting}+\text{centrifugation}+\text{drying}+\text{extraction}+\text{pigments extraction})}$$

$$NER=\left[\frac{(0.48kg\times24h\times330d\times41MJ.kg^{-1})}{[(0.78KWh+0.05KWh+0.11KWh+0.31KWh+0.001kwh+158.25kwh)\times(24h\times330d\times3.6MJ)]}\right]$$

$$NER=\left[\left(\frac{155800MJ}{4555390.75MJ}\right)\right]$$

$$NER=0.034$$

Global Warming Potential (Pigments)

$$E=\sum_i Mi\times Pi$$

$$E=\sum(\text{emissions}\times\text{impact factor value})$$

$$E=[(0.61kg\times0.542)+(4555390.75MJ\times0.542)]$$

$$E=2469022.11kgCO_{2eq}$$

Global Warming Potential (Flour microalgae)

$$E=\sum_i Mi\times Pi$$

$$E=\sum(\text{emissions}\times\text{impact factor value})$$

$$E=[(0.61+0.002+0.48)\times0.542+(2345140.51MJ\times0.542)]$$

$$E=1271066.74\ kgCO_{2eq}$$

Net Energy Ratio (Flour microalgae)

$$NER=\frac{\sum \text{energy produced}}{\sum \text{energy requeriments}}$$

$$NER=\frac{(\text{Biomass}\times\text{energy potential})}{(\text{cultivation}+\text{harvesting}+\text{centrifugation}+\text{drying}+\text{extraction}+\text{purification})}$$

$$NER=\left[\frac{(0.48kg\times24h\times330d\times41MJ.kg^{-1})}{[(0.78KWh+0.05KWh+0.11KWh+0.31KWh+0.001kwh+81kwh)\times(24h\times330d\times3.6MJ)]}\right]$$

$$NER=\left[\left(\frac{155800MJ}{2345140.51MJ}\right)\right]$$

$$NER=0.066$$

Third step: Thus, with the interpretation of the results, it was possible to conclude the process in question; microalgae are potential raw materials for obtaining biodiesel through solvent extraction. Therefore, a positive NER of 4.36 was obtained for biodiesel generation. However, if we followed the biorefinery process and opted to obtain the pigments extracted from the defatted biomass together with the biodiesel, we would obtain negative NER of 0.034. In this way, in addition to continuing with positive net energy, we could increase the profits of sales of this chemical by up to 30%. On the other hand, to get microalgae meal, the energy expenditure would have decreased, resulting in a NER of 0.066. As for water distribution, the process resulted in an expense of 551.1 m^3 and remains unchanged before this biorefinery system. In addition, the atmospheric emissions were quantified in relation to the fossil energy required for the operation of the system and those emitted by the equipment, resulting in a value of 19,332.92 kg CO_{2eq}, yet the CO_2 absorption required for the growth of microalgae was disregarded.

5 Final Considerations

The life cycle assessment can be used in the most diverse industrial segments, but it has become a key tool as a critical parameter in relation to microalgal biofuels reports.

Moreover, to take advantage of the benefits that the LCA provides, its application involves a series of steps established by the ISO that was developed with the objective of guaranteeing great results, but they need the time, resources, and qualified human resources to be executed.

Finally, even with its application complexity, LCA is an excellent option for the monitoring of environmental issues related to biofuels and can contribute to sustainable development, thereby providing an overview of the environmental aspects and impacts associated with the product and providing subsidies that enable the implementation of improvements throughout its life cycle.

References

Adesanya, V. O., Cadena, E., Scott, S. A., & Smith, A. G. (2014). Life cycle assessment on microalgal biodiesel production using a hybrid cultivation system. *Bioresource Technology, 163,* 343–355.

Bicalho, T., Sauer, I., Rambaud, A., & Altukhova, Y. (2017). LCA data quality: A management science perspective. *Journal of Cleaner Production, 156,* 888–898.

Blanchard, R., Kumschick, S., & Richardson, D. M. (2017). Biofuel plants as potential invasive species: Environmental concerns and progress towards objective risk assessment. In *Roadmap for sustainable biofuels in southern Africa* (pp. 47–60). Nomos Verlagsgesellschaft mbH & Co. KG.

Brennan, L., & Owende, P. (2010). Biofuels from microalgae—A review of technologies for production, processing, and extractions of biofuels and co-products. *Renewable and Sustainable Energy Reviews, 14*(2), 557–577.

Carneiro, M. L. N., Pradelle, F., Braga, S. L., Gomes, M. S. P., Martins, A. R. F., Turkovics, F., et al. (2017). Potential of biofuels from algae: Comparison with fossil fuels, ethanol and biodiesel in Europe and Brazil through life cycle assessment (LCA). *Renewable and Sustainable Energy Reviews, 73,* 632–653.

Chen, C. Y., Yeh, K. L., Aisyah, R., Lee, D. J., & Chang, J. S. (2011). Cultivation, photobioreactor design and harvesting of microalgae for biodiesel production: A critical review. *Bioresource Technology, 102*(1), 71–81.

Chew, K. W., Yap, J. Y., Show, P. L., Suan, N. H., Juan, J. C., Ling, T. C., et al. (2017). Microalgae biorefinery: High value products perspectives. *Bioresource Technology, 229,* 53–62.

Clarens, A. F., Resurreccion, E. P., White, M. A., & Colosi, L. M. (2010). Environmental life cycle comparison of algae to other bioenergy feedstocks. *Environmental Science and Technology, 44*(5), 1813–1819.

Collotta, M., Champagne, P., Mabee, W., Tomasoni, G., Leite, G. B., Busi, L., et al. (2017). Comparative LCA of flocculation for the harvesting of microalgae for biofuels production. *Procedia CIRP, 61,* 756–760.

Curran, M. A. (2006). *Life cycle assessment: Principles and practice*. USA: Environmental Protection Agency (EPA). EPA/600/R-06/060. May 2006.

Department of Energy & Climate Change. (2010). *Guidelines to Defra/DECC's GHG conversion factors for company reporting*. Available at: https://www.gov.uk/government/organisations/department-of-energy-climate-change. Accessed July 2017.

Deprá, M. C., Zepka, L. Q., & Jacob-Lopes, E. (2017). Life cycle assessment as a fundamental tool to define the biofuel performance. In *Frontiers in bioenergy and biofuels*. InTech. https://doi.org/10.5772/64677.

Dutta, S., Neto, F., & Coelho, M. C. (2016). Microalgae biofuels: A comparative study on techno-economic analysis & life-cycle assessment. *Algal Research, 20,* 44–52.

Farooq, W., Suh, W. I., Park, M. S., & Yang, J. W. (2015). Water use and its recycling in microalgae cultivation for biofuel application. *Bioresource Technology, 184,* 73–81.

Forster, P., Ramaswamy, V., Artaxo, P., Berntsen, T., Betts, R., Fahey, D. W., et al. (2007). Changes in atmospheric constituents and in radiative forcing. In *Climate change 2007. The physical science basis* (Chapter 2).

Garcia, D. J., & You, F. (2015). Supply chain design and optimization: Challenges and opportunities. *Computers & Chemical Engineering, 81,* 153–170.

Gnansounou, E., Dauriat, A., Villegas, J., & Panichelli, L. (2009). Life cycle assessment of biofuels: Energy and greenhouse gas balances. *Bioresource Technology, 100*(21), 4919–4930.

Guieysse, B., Béchet, Q., & Shilton, A. (2013). Variability and uncertainty in water demand and water footprint assessments of fresh algae cultivation based on case studies from five climatic regions. *Bioresource Technology, 128,* 317–323.

Guo, M., & Murphy, R. J. (2012). LCA data quality: Sensitivity and uncertainty analysis. *Science of the Total Environment, 435,* 230–243.

Hoekstra, A. Y. (2016). A critique on the water-scarcity weighted water footprint in LCA. *Ecological Indicators, 66,* 564–573.

ILCD Handbook. (2010). *General guide for life cycle assessment—Detailed guidance* (1st ed.). Institute for Environment and Sustainability, Joint Research Centre, European Commission.

ISO—International Standard 14040. (2006). *Environmental management—Life cycle assessment. Principles and framework*. Geneva: International Organisation for Standardisation (ISO).

ISO—International Standard 14041. (1998). *Environmental management—Life cycle assessment. Goal and scope definition and Inventory analysis*. Geneva: International Organisation for Standardisation (ISO).

ISO—International Standard 14042. (2000). *Environmental management—Life cycle assessment. Life cycle impact assessment*. Geneva: International Organisation for Standardisation (ISO).

ISO—International Standard 14043. (2000). *Environmental management—Life cycle assessment. Life cycle interpretation*. Geneva: International Organisation for Standardisation (ISO).
Jacob-Lopes, E., & Franco, T. T. (2010). Microalgae-based systems for carbon dioxide sequestration and industrial biorefineries. In *Biomass*. InTech.
Jacquemin, L., Pontalier, P. Y., & Sablayrolles, C. (2012). Life cycle assessment (LCA) applied to the process industry: A review. *The International Journal of Life Cycle Assessment, 17,* 1–14.
Jian, H., Jing, Y., & Peidong, Z. (2015). Life cycle analysis on fossil energy ratio of algal biodiesel: Effects of nitrogen deficiency and oil extraction technology. *The Scientific World Journal, 2015,* 9.
Jolliet, O., Margni, M., Charles, R., Humbert, S., Payet, J., Rebitzer, G., et al. (2003). IMPACT 2002+: A new life cycle impact assessment methodology. *The International Journal of Life Cycle Assessment, 8*(6), 324–330.
Jorquera, O., Kiperstok, A., Sales, E. A., Embirucu, M., & Ghirardi, M. L. (2010). Comparative energy life-cycle analyses of microalgal biomass production in open ponds and photobioreactors. *Bioresource Technology, 101*(4), 1406–1413.
Klein, B. C., Bonomi, A., & Maciel Filho, R. (2017). Integration of microalgae production with industrial biofuel facilities: A critical review. *Renewable and Sustainable Energy Reviews, 82,* 1376–1392.
Klöpffer, W. (2006). The role of SETAC in the development of LCA. *The International Journal of Life Cycle Assessment, 11,* 116–122.
Laamanen, C. A., Ross, G. M., & Scott, J. A. (2016). Flotation harvesting of microalgae. *Renewable and Sustainable Energy Reviews, 58,* 75–86.
Laratte, B., Guillaume, B., Kim, J., & Birregah, B. (2014). Modeling cumulative effects in life cycle assessment: The case of fertilizer in wheat production contributing to the global warming potential. *Science of the Total Environment, 481,* 588–595.
Liu, X., Clarens, A. F., & Colosi, L. M. (2012). Algae biodiesel has potential despite inconclusive results to date. *Bioresource Technology, 104,* 803–806.
Luo, D., Hu, Z., Choi, D. G., Thomas, V. M., Realff, M. J., & Chance, R. R. (2010). Life cycle energy and greenhouse gas emissions for an ethanol production process based on blue-green algae. *Environmental Science and Technology, 44*(22), 8670–8677.
Medeiros, D. L., Sales, E. A., & Kiperstok, A. (2015). Energy production from microalgae biomass: Carbon footprint and energy balance. *Journal of Cleaner Production, 96,* 493–500.
Mekonnen, M. M., & Hoekstra, A. Y. (2010). *A global and high-resolution assessment of the green, blue and grey water footprint of wheat* (Vol. 42, pp. 1–94). Delft, The Netherlands: Unesco-IHE Institute for Water Education.
Monari, C., Righi, S., & Olsen, S. I. (2016). Greenhouse gas emissions and energy balance of biodiesel production from microalgae cultivated in photobioreactors in Denmark: A life-cycle modeling. *Journal of Cleaner Production, 112,* 4084–4092.
Prox, M., & Curran, M. A. (2017). Consequential life cycle assessment. In *Goal and scope definition in life cycle assessment* (pp. 145–160). Springer Netherlands.
Quiroz-Arita, C., Sheehan, J. J., & Bradley, T. H. (2017). Life cycle net energy and greenhouse gas emissions of photosynthetic cyanobacterial biorefineries: Challenges for industrial production of biofuels. *Algal Research* (in press).
Silva, J. A. M., Santos, J. J. C. S., Carvalho, M., & de Oliveira, S. (2017). On the thermoeconomic and LCA methods for waste and fuel allocation in multiproduct systems. *Energy, 127,* 775–785.
Soccol, C. R., Faraco, V., Karp, S., Vandenberghe, L. P. S., Thomaz-Soccol, V., Woiciechowski, A., et al. (2011). *Biofuels: Alternative feedstocks and conversion processes*.
Soh, L., Montazeri, M., Haznedaroglu, B. Z., Kelly, C., Peccia, J., Eckelman, M. J., et al. (2014). Evaluating microalgal integrated biorefinery schemes: Empirical controlled growth studies and life cycle assessment. *Bioresource Technology, 151,* 19–27.
Yvon-Durocher, G., Allen, A. P., Bastviken, D., Conrad, R., Gudasz, C., St-Pierre, A., et al. (2014). Methane fluxes show consistent temperature dependence across microbial to ecosystem scales. *Nature, 507*(7493), 488.

Zhang, Y., Luo, X., Buis, J. J., & Sutherland, J. W. (2015). LCA-oriented semantic representation for the product life cycle. *Journal of Cleaner Production, 86,* 146–162.

Živković, S. B., Veljković, M. V., Banković-Ilić, I. B., Krstić, I. M., Konstantinović, S. S., Ilić, S. B., et al. (2017). Technological, technical, economic, environmental, social, human health risk, toxicological and policy considerations of biodiesel production and use. *Renewable and Sustainable Energy Reviews, 79,* 222–247.

Chapter 7
The Bioeconomy of Microalgal Biofuels

Kun Peng, Jiashuo Li, Kailin Jiao, Xianhai Zeng, Lu Lin, Sharadwata Pan and Michael K. Danquah

Abstract Biofuels such as biodiesel and bioethanol, synthesized via microalgal bioprocess engineering, could be a major contributor to the purview of sustainable energy in the foreseeable future. In contrast to other biomass feedstocks like corn, sugar crops, and vegetable oil, microalgae display a number of significantly superior benefits as a raw material for biofuel manufacturing. This includes an enhanced metabolic rate of biomass production, subsistence of diverse microalgae species with sundry biochemical profiles, prospects for carbon dioxide sequestration, and either limited or near absolute monopoly from the perspective of food production modalities and logistics. However, attributing to a wide range of factors,

---

Electronic supplementary material The online version of this chapter (https://doi.org/10.1007/978-3-319-69093-3_7) contains supplementary material, which is available to authorized users.

K. Peng · J. Li
Department of New Energy Science and Engineering, School of Energy and Power Engineering, Huazhong University of Science and Technology, Wuhan 430074, China

K. Jiao · X. Zeng (✉) · L. Lin
College of Energy, Xiamen University, Xiamen 361005, China
e-mail: xianhai.zeng@xmu.edu.cn

M. K. Danquah
Department of Chemical Engineering, Faculty of Engineering and Science, Curtin University, 98009 Sarawak, Malaysia
e-mail: mkdanquah@curtin.edu.my

X. Zeng · L. Lin
Fujian Engineering and Research Center of Clean and High-Valued Technologies for Biomass, Xiamen University, Xiamen 361102, China

J. Li
State Key Laboratory of Coal Combustion, Huazhong University of Science and Technology, Wuhan 430074, China

S. Pan
School of Life Sciences Weihenstephan, Technical University of Munich, 85354 Freising, Germany

E. Jacob-Lopes et al. (eds.), *Energy from Microalgae*, Green Energy and Technology, https://doi.org/10.1007/978-3-319-69093-3_7

for instance the insipid characteristic of microalgal cultures, and the fact that microalgae cells possess trivial sizes, the process of biomass production and subsequent conversion into biofuels become prohibitively expensive. As a consequence, from an economic outlook, the large-scale production of biofuels from microalgae achieves a somewhat less appealing status, compared to the other biomass types and sources. The current chapter delivers an outline of the bioeconomy analysis for microalgae-derived biofuels. In addition, case studies on microalgal biofuel production are presented along with cost estimations and the necessary strategies to augment its commercial viability.

Keywords Techno-economic assessment · Biofuel production · Microalgae-based biofuels

1 Introduction

Biofuels are widely perceived to be significantly prospective alternatives to the traditional and non-renewable fossil fuels, attributing to their characteristics such as sustainability, and the capabilities to reduce the emission of greenhouse gases, thereby achieving the 'green' status (Demirbas 2007). Recently, global biofuel production has witnessed a rapid growth, increasing from 19.651 million tons oil equivalent (toe) in 2005 to 74.847 million toe in 2015 (BP 2016). Biofuels can be derived from a wide array of biomass materials, including agricultural crops, municipal wastes, agricultural and forestry byproducts, and aquatic products. Out of all these sources, microalgae are commonly regarded to be the most suitable feedstock, owing to its high energy intensity, high average photosynthetic efficiency (50 times that of the terrestrial plants), and high capabilities of oil production (12,000 L biodiesel per hectare) (Gao et al. 2011). In addition to these characteristics, conceivable exploitation of barren lands and water bodies makes microalgae a perfect substitute for biomass which requires high agricultural input (Hill et al. 2006; Quinn and Davis 2015).

Driven by the aforementioned advantages, both industries and academia have initiated agendas to devote time and efforts for microalgal cultivation and biofuels production, thereby leading to their considerable and continuable development. The global production of *Spirulina* biomass had increased from almost nil to nearly 3500 tons (1000 tons = 1016 tons) from 1975 to 1999 (Pulz and Gross 2004). The microalgae industry had evolved with an annual production of 7000 tons of dry matter in 2004 (Brennan and Owende 2010). The majority of the companies (~78%) contributing to the algal biofuel growth are based in the USA, followed by Europe (~13%), and auxiliary states (~9%) (Bahadar and Khan 2013). To date, the US Department of Energy (DOE) has spent about USD 85 million to develop algal biofuels through some 30 R&D initiatives or so. In addition, for the purpose of manufacturing algal oil, Aurantia, a Spanish renewable energy company, and the

Green Fuel Tech of Massachusetts (USA), have commenced a USD 92 million project alliance in 2007. It is expected that in the conceivable future, this project targets an increase of nearly 100 ha of algae greenhouses, which will yield 25,000 t of algal biomass annually (Bahadar and Khan 2013).

Although the technical feasibility of microalgae has already been proven experimentally, the microalgae-based biofuels are yet not suitable for large-scale commercial applications, even after several decades of development. The major hindrance to this end is the relatively enhanced production cost. In order to realize a 10% return rate, investigations reveal that the essential selling costs of the product per gallon of triglyceride (TAG) should be USD 18.10 for PBR and USD 8.52 for open pond manufacturing. The biodiesel production costs per gallon of diesel via hydro-treating soared to USD 9.84 and USD 20.53, while the manufacturing price per gallon for petroleum diesel was USD 2.60, clearly indicating the increases expenses associated with the former (Davis et al. 2011). US DOE reported that algal biofuels can be competitive with petroleum at approximately USD 2.38/gal (DOE 2010). It is thus obvious, that in order to seek solutions for downregulating the increased production costs, the R&D sector is dedicated to carry out frequent and elaborate analyses of economic practicalities of microalgae-based biofuel.

Techno-economic assessment (TEA) is one of the most basic and common methods applied to evaluate the feasibility of microalgae-based biofuel. TEA methods are often associated with process modeling. In 2011, Ryan Davis modeled a microalgal setup producing raw oil in the annual capacity of 10 MM gal via the Aspen Plus software, to study the cost of each process unit of fuel production. The firm inferred that the microalgal biofuel production finances would be far from being reasonable with conventional fossil fuels, in case it corresponded to construct a large-scale manufacturing setup (Davis et al. 2011). Amer et al. (2011) have reported, by comparing five microalgae to biofuels processes using the SAFEER model, that the open pond scenarios which produced either TAG or free fatty acid methyl esters, appeared to be closest to the USD 1/kg price reference, and consequently, are the most viable choices (Amer et al. 2011). In another work, Zamalloa et al. (2011) considered the anaerobic digestion of microalgae and utilized a process model and diverse indicators to conduct the analysis of uncomplicated biomethanation potential. The results highlighted the efficacy of treating electrical and heat energies equally through a feed-in price of €0.133/kWh, making the project lucrative. This stands in poor contrast to the carbon credit of €30/ton CO_2(eq), with a meagre 4% revenue returns (Zamalloa et al. 2011). Batan et al. (2016) employed a dynamic accounting model of a bounded photobioreactor microalgal facility with a manufacturing capacity of 37.85 million liters (10 million gallons) of biofuel per annum. The authors showed that the total manufacturing costs of algal raw oil and diesel per liter matched to USD 3.46 and USD 3.69, correspondingly. The financial feasibility of biofuels manufactured from microalgae relies on the entree to coproduct arcades with more incremental benefits (Batan et al. 2016). The aforementioned studies ignored the impacts of either policies or byproducts. It may be noted that the absence of these two factors could influence the accuracy of the assessments to a certain extent. Additionally, the cost of land should also be taken into consideration.

Life cycle assessment (LCA) integrated with TEA modeling is a useful tool to assess the impact of the microalgal biofuel manufacturing process over the life-cycle. In 2016, a cohesive prototype for algal biofuel synthesis was reported by Dutta et al. (2016), which assists in running a life cycle valuation and a financial practicability scrutiny, aimed at the large-scale solicitation for economic implementation of the translation routes of microalgae-derived biofuel production. The authors investigated the sustainability of microalgae-derived biofuel production of transformation routes at University of Aveiro, Portugal, and at the National Renewable Energy Laboratory, Colorado, USA, and have reported that the capital value enhancement of coproducts is predominantly noteworthy because it augments revenue which may be utilized to advance the closing fuel vending cost (Dutta et al. 2016). López-González et al. (2015) adopted concurrent differential scanning calorimetry (DSC) and thermogravimetric analysis (TGA) coupled with mass spectrometry (MS), to simulate thermochemical performance and LCA to assess environmental viability and monetary sustainability of the pyrolysis and combustion of microalgae and their oils, establishing economic feasibility of the microalgal oil pyrolysis procedure at bulky manufacturing levels (López-González et al. 2015). In contrast to the previous studies, Malik et al. (2015) used a cross-regional and fiscal input–output prototype of Australia, supplemented with engineering course statistics on algal bio-crude manufacturing to assume crossbreed life cycle evaluation for determining the primary and secondary effects of bio-crude synthesis. The results demonstrate a net carbon-negative tendency of the algal bio-crude manufacturing method. Additionally, prospects of nearly 13,000 fresh jobs along with USD 4 billion value of incentives are synonymous with manufacturing 1 million tons of bio-crude, thereby providing a boost to the economy (Malik et al. 2015). The challenge of LCA methods is that the variances in scheme restrictions and the central LCA conventions will lead to dissimilar results. It is a fact that alterable suppositions related to the coproduct distribution approaches, sourcing of electrical energy, and life cycle catalogue information vividly influence outcomes. Hence, any additional alteration in administering trails and impractical authentication of sub-processing prototypes, with small-scale statistics, will provide higher erraticism in the reported outcomes (Quinn and Davis 2015).

Supplementary explorations have been directed utilizing process modeling as the core. Delrue et al. (2012) focused on establishing a model with four assessment norms: the net energy ratio (NER), manufacturing price of biodiesel, greenhouse gas (GHG) release proportion, and water footmark, to evaluate the economic, sustainable, and energetic performance of biodiesel and other biofuel productions from microalgae. They considered three processes: hydrothermal liquefaction (HTL), oil emission, and alkane discharge and showed that HTL may be contemplated either as a substitute to wet lipid isolation, and that lipid secretion is a better choice than the typical lipid extraction process. Delrue et al. (2012) have also compared a state-of-the-art trail (hybrid raceway/PBR cultivation scheme, belt filter press for dewatering, wet lipid isolation, oil water handling and oxygen deprived residual ingestion) with a reference pathway and found that the pioneering route optimized the energy and environmental measures with a relatively high production cost for economic viability (Delrue et al. 2012, 2013). This could be improved by integration with a long-term assessment. Furthermore, research using gauges such as return on investment (ROI) and net present value (NPV), will certainly foresee long-term profits of microalgae-based biofuel industry, with more systematic consequences.

The precise objectives of this chapter are to investigate the bioeconomy of large-scale microalgal biofuels production and to identify the key factors responsible for enhanced cost. To this end, the current chapter will provide valuable information for scaling and commercialization of microalgae-based biofuels.

2 Methods

2.1 *Cost and Revenue*

According to Xin et al. (2016), the cost of algae-derived biofuels can be divided into three categories: capital investment, total fixed operating cost (TFOC), and total variable operating cost (TVOC). Both capital investment and TFOC may be obtained directly through summation of their corresponding sub-items in dollars per year. The revenue may be obtained in the similar way. However, as the units of TVOC are usually MJ/d and kg/d, TVOC items are generally estimated based on the operation time. Therefore, an estimate of the total theoretical cost could be obtained from the sum total of capital investment, TFOC and TVOC. In case there are byproducts in the production process, the actual cost equals to the difference of theoretical cost and the economic value of the byproducts. Detailed information with respect to costs and revenue is presented in the supplementary information.

Two economic indicators, namely the NPV and the ROI, are usually adopted for the economic analysis. NPV is an indicator for analyzing the profitability of an investment or a project. Alternatively, it measures the profit by computing the costs and benefits for each period of an investment or a project. NPV may be estimated as:

$$\mathrm{NPV} = \sum_{t=0}^{T} \frac{C_1 - C_0}{(1+r)^t} \tag{1}$$

where T is the time of cash flow, representing the time span during which the project is under operation and expected to have income; r stands for discount rate, i.e., the required rate of return that could be earned each period on a project with similar risk, which is set as 10% in this study; C_1 is the annual income (the benefits) and C_0 stands for annual expenditure. A positive NPV value indicates that the income brought by a project or investment has exceeded the anticipated costs, suggesting that the project or investment is acceptable, and vice versa.

ROI is used as a decision tool that allows the stakeholders to evaluate the performance of an investment or a project and compare it to others in their portfolio. The current study also uses this indicator to evaluate the efficiency of the algal biofuels production. ROI can be estimated as:

$$\mathrm{ROI} = \frac{\sum_{t=0}^{T} C_1 * P_t}{\sum_{t=0}^{T} C_0 * P_t} \tag{2}$$

where T represents the operation time, C_1 is the capital investment in dollars per year, C_0 stands for annual expenditure, and P_t is the discount factor in year t.

2.2 *Sensitivity Analysis*

The usual purpose of the sensitivity analysis is to investigate the potential changes in inputs and the corresponding effects on the economic output. Hence, the current work employs sensitivity analysis to identify the influence of each type of inputs on the basis of the economics, which may provide valued evidence for cost reduction. A typical sensitivity analysis comprises of five steps. The first step is to determine the indicators, such as payback period, ROI, NPV, and earnings before interest and tax. The second and third steps correspond to the estimation of the technical target values, and the selection of uncertainties, respectively. Particularly, in the third step, the factors that have higher probabilities of change and superior impacts on the target values of economic analysis are chosen. This is followed by the fourth step, where the influences of these uncertainties on the target value are quantified. Lastly, a comprehensive analysis is conducted and some insightful suggestions are proposed in the fifth and final step, based on the outcomes from the previous four steps.

2.3 *Data Sources*

Detailed data are elaborated in the supplementary information.

3 Results and Discussion

3.1 *Algae-Derived Ethanol*

3.1.1 Cost and Revenue

The total cost to manufacture one tonne of algae-derived ethanol is USD 6410. As depicted in Fig. 1, the highest cost is attributed to TFOC, occupying about half of

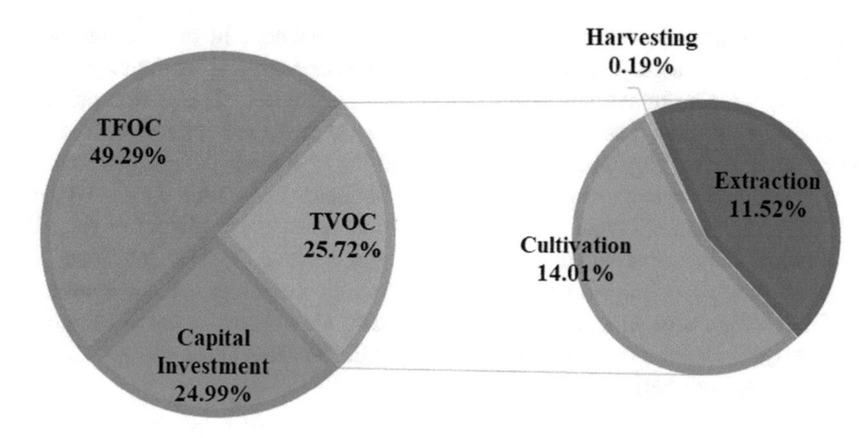

Fig. 1 Theoretical cost structure of algal ethanol

the total cost, followed by TVOC (25.72%) and TVOC (24.99%). Owing to a large number of inputs in the first year, the cost of the project amounts to be the highest during the operation period. Inevitably, this study chooses the first year to elucidate the structure of the cost (see Fig. 1). The capital investment in the first year has an economic value of USD 2.66 million, covering the costs of fixed assets involving equipment procurement and building constructions, such as photobioreactor (PBR), greenhouse, pyrolysis system, flocculation tank, centrifuge, land, storeroom. The investment for PBR is USD 1.28 million, which accounts for ~48% of the total cost. Meanwhile, the cost of the pyrolysis system and greenhouse constructions is among the principal sources of capital investments, responsible for ~17 and ~11% respectively, while the other costs possess comparatively subordinate contributions.

TFOC amounts to USD 1.56 million, with the major contributors as capital expenditures (~51%), depreciation (~17%), salaries (~14%), maintenance, insurance and taxes (~10%), among others. In a microalgal cultivation system, USD 26,763 and USD 161,457 are spent for the depreciation and maintenance of equipment annually to keep the system stable in the long run. Roughly, USD 220,000 per annum is used as the wages for the proprietors and the working personnel. The costs of cultivation, harvesting, and extraction constitute TVOC (see Fig. 1), which is USD 815,033. Cultivation, which requires a large amount of energy and nutrient input, invariably occupies the top position in TVOC. Extraction, which consumes considerable amounts of electricity and chemicals, is estimated to cost USD 365,090. Harvesting triggers the lowest cost with a fraction of 0.12%.

As a matter of fact, several factors, such as microalgal species, cultivation system, lipid content, grease content, and conversion technologies, may influence the cost estimation. Moreover, climate and season transitions (especially the changes in temperature and sunlight), having significant impacts on the mixing of nutrients and algal productivity, will also affect the total cost. In general, the cost of biofuels, such as algae-derived ethanol, is still higher than that of the fossil oils.

The revenue brought by algae-derived ethanol project is USD 8020/ton, which is higher than the cost. The revenue corresponding to the algae-derived biofuel may be credited to two segments: the products and the cost savings. The products are mainly ethanol, syngas, other liquids, and biomass, generating USD 370,615; 47,607; 1,073,089; and 1,628,041 per annum, respectively. As the wastewater-based microalgal biofuel production system recycles, nutrients like nitrogen and phosphorus may be recycled for microalgal growth to synthesize biofuels, which also gains USD 564,768 annually via the saving costs. Moreover, USD 46,883 may be earned as the carbon credit, because algae can absorb CO_2 during its growth period.

3.1.2 NPV and ROI

Assuming the discount rate to be 8%, the annual NPV is estimated to be USD 2.69 million, suggesting that the project is worth investing. The project ROI is 17.57%, which is higher than the discount rate, and implies that the algae-derived biofuel project possesses good economic benefits.

3.1.3 Sensitivity Analysis

The analysis has been performed via selection of PBR cost, salaries, operating period, chemical consumption, energy consumption, and biomass price as parameters, to evaluate their impressions on the economic cost, subject to an increase or decrease by 20% (Fig. 2).

The cost of algae-derived biofuel is most sensitive to the biomass price. The rising selling prices of biomass will greatly increase the profits of products and byproducts, ultimately lowering the actual economic cost. A 50.15% rise in the cost has been estimated, following a 20% increase in the biomass selling price. Thus, improving cultivation technology and promoting microalgae recovery efficiencies, oil extraction, and conversion rate, are a few effective approaches to enhance the economic performance.

In addition, the impacts of PBR cost, and operating period of the system on actual economic cost are significant. PBR is one of the principal project devices and incurs high expenses as well. It is anticipated that the breakthrough in the procurement of equipment, as a result of scientific and technological progress, will eventually reduce the cost of equipment input. Moreover, operation time constitutes a significant factor, having substantial connections with investments including the rental price of the land, energy, chemical and nutrient consumption, and wastewater cycling rate. A $\sim$14% increase in the tangible cost is probable, provided the operation time is cut down by 20%.

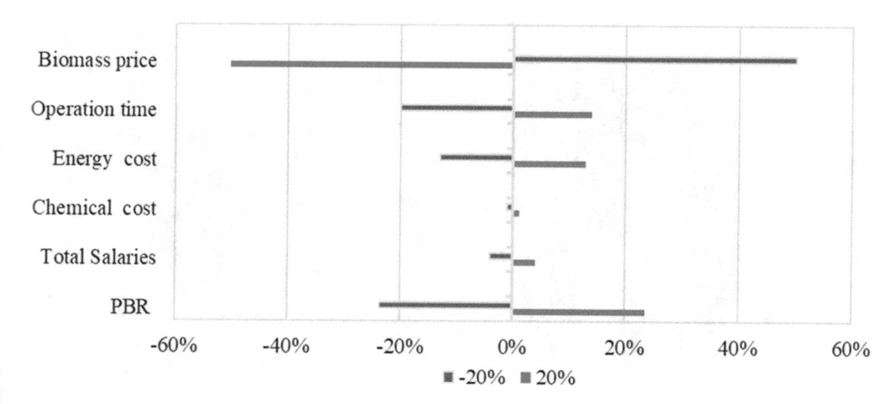

Fig. 2 Sensitivity analysis for the algal ethanol production system

Currently, the operation of the whole cultivation system is based on the electric system. As a result, the energy consumption is indispensable. Increase or decrease in energy consumption by 20% will lead to a 12.89% reduction or increase in ROI. Compared to the aforementioned factors, however, total salaries and chemical consumption do not strongly impact the cost. This is due to several reasons. Since the cultivation system is highly automated, only restricted labor is necessary for management and maintenance. It should be noted that nutrients like nitrogen and phosphorus are derived from the municipal wastewater. Consequently, a certain portion of the funds is protected as there is no need to buy extra nitrogen and phosphorus, which also hints at efficient regulation of chemical consumption.

3.2 *Algae-Derived Biodiesel*

3.2.1 Cost and Revenue

Mostly, the economic cost is a manifestation of an amalgamation of material inputs, labor, equipment repair and depreciation, and non-production inputs. It is assumed that the project will be operative for 200 days annually and will last for 15 years, according to Yang (2015). In summary, the first-year costs amount to USD 5246/ton, which is the largest among the operation period. This is because several items act as the first-year inputs. The average annual cost of the algae-derived biodiesel is USD 3523/ton. Taking the byproduct revenue into consideration, the cost will abruptly reduce to USD 960/ton.

In a similar fashion, the first-year structure is portrayed in Fig. 3. Material inputs, including energy, land, water, nutrient, catalyst, dominate the total production cost per ton of the algae-derived diesel. The cost of material inputs is up to USD 3415, corresponding to $\sim$65% of the total cost. Among the material inputs, nutrients such as nitrogenous and phosphorus fertilizer yield the largest share,

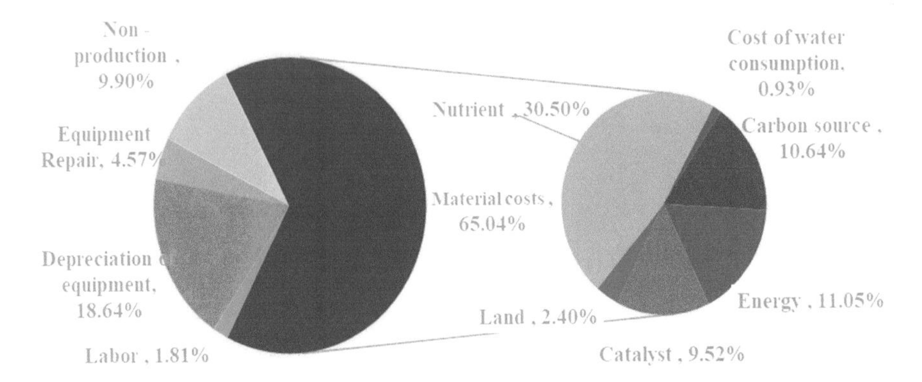

Fig. 3 Structure of algal biodiesel production costs

with >30% of the total cost. Energy consumption chiefly includes electricity, steam and transport fuel consumptions. Owing to extensive energy utilization in a variety of diverse processes, energy consumption takes the second position in material costs, with a value of USD 579.77 USD/ton ($\sim$11% of the total cost). Carbon dioxide, a necessity for algal growth, is the third largest source of material cost. To generate 1 ton of algae-derived biodiesel, USD 558.22 worth of carbon dioxide is required. In addition to the material inputs, algae-derived diesel production also demands equipment such as flocculation tank and centrifuge. Henceforth, the cost evaluation should take the equipment depreciation and repair into consideration. The results show the relative significance of these two cost sources, which contribute to nearly a quarter of the total cost.

Costs pertaining to the non-production inputs, i.e., administration, finance, and sale, are largely influenced by the management and the enterprise operation. The non-production cost is virtually unaffected and is directly correlated with the stability in the technology and management sectors. According to Yang (2015), the non-production costs account for approximately 9.9% of the total cost. Algae-derived biodiesel production mainly needs labor during the process of algae cultivation and biodiesel production. In this study, the monthly labor wage per person is assumed to be about USD 375. The total labor expenditure accounts to a small fraction ($\sim$1.18%) of the total cost.

The total annual revenue of the algae-derived biodiesel is USD 3676. The byproducts of biodiesel mainly include algal residue, pastry, glycerin, and methyl alcohol. Methyl alcohol may be used in the microalgae cultivation process to reduce the costs of chemical reagent consumption due to its relatively high price. The algal residue may be directly sold as animal feed (USD 640.03/ton) and fermentation feed (USD 447.47/ton). Meanwhile, post raw materials fermentation, the manufactured biogas (USD 212.11/ton) may be sold directly, or may be used as the feed gas for boiler gas production (USD 209.86/ton), power generation (USD 115.20/ton), and purification (USD 220.66/ton). Consequentially, to advance the economic benefits of the algae-derived biodiesel system, it is advised to vend the algal residue, yielded from microalgae biodiesel production, directly. Moreover, the algal residue may generate biogas via anaerobic fermentation. Biogas may be isolated to refined gas and CO_2. While the refined gas may be utilized as the conventional energy source for daily household use, CO_2 may be recycled for algal cultivation, which proves to be beneficial and cost effective.

3.2.2 NPV and ROI

Based on the cost and benefit data, ROI of microalgae diesel is 7.78% and the net present value (NPV) is USD 640.

3.2.3 Sensitivity Analysis

For the sensitivity analysis, an ensemble of parameters are selected: microalgae biomass per unit area, oil content, annual run time, recovery rate, oil extraction rate, CO_2 absorption rate, nutrient consumption, chemical reagent consumption, and wastewater recycling rate. The variations in microalgae diesel investment rate of return (ROI) are estimated following a ±20% alteration in any single parameter, as shown in Fig. 4.

Figure 4 indicates that the total price of the algae-derived biodiesel is maximally perceptive to nutrient utilization. There is a fluctuation in the total cost from −7.56 to 10.87%, following a ±20% alteration in the lipid content. In addition, any change in nutrient consumption strongly influences the cost. For instance, following a ±20% change in the nutrient assimilation, the total cost change varies by a margin of ±6%. Furthermore, following a ±20% change in either the recovery or the oil extraction efficiency, the total cost change will vary between −4.4 and 6%.

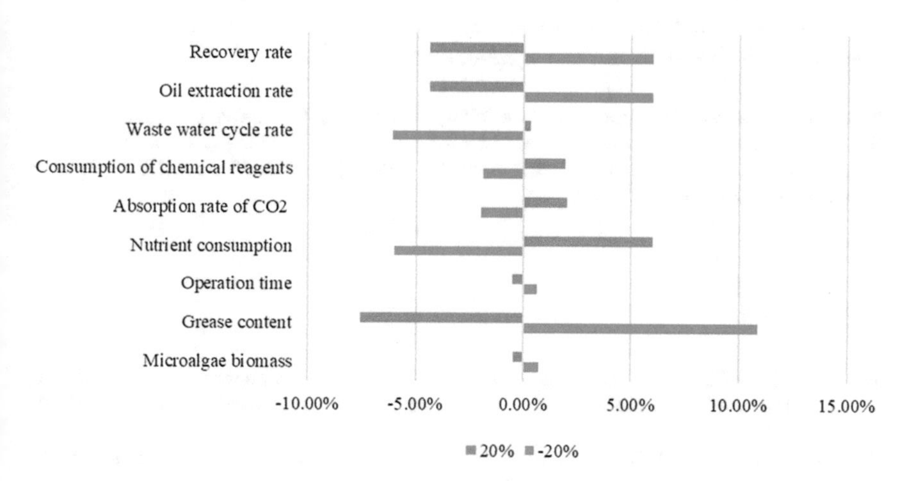

Fig. 4 Sensitivity analysis for algal biodiesel system

Thus, improving recovery and oil extraction efficiencies will definitely impact the total cost in a positive manner.

Based on the sensitivity analysis, it may be inferred that the most effective method to decrease the total price of the algae-derived biodiesel is to increase the microbial biomass and the grease content. Efforts should be made during microalgal cultivation, processing, and transformation. For instance, genetic engineering can be employed to enhance the grease content, and the algal growth rate. Other measures such as refining the carbon absorption rate and discovery of cost-effective carbon and nutrient sources are also effective ways to reduce the production cost of the microalgal biodiesel. Additionally, the choice of appropriate climate for microalgal culture, mounting the microalgal breeding time, enhancing microalgal recovery, oil extraction, and oil conversion efficiency, is all beneficial for down-regulating the cost of biodiesel production. However, efforts pertaining to the reduction of fixed investment, chemical reagent consumption, and energy consumption may lead to the cost reduction only to a certain extent, with limited effects.

4 Conclusions

The current study conducted an all-inclusive bioeconomy analysis for the algae-derived biofuel projects, such as the bioethanol and biodiesel projects. Based on the available applied data, the cost and revenue for the former is USD 6410 and USD 8020/ton, while that of the latter is USD 3523 and USD 3676/ton. Compared to the fossil fuels, the economic outputs of algae-derived biofuels are a few notches higher. NPV possesses a positive value, indicating that both the projects are worthy of probable ventures. The outcomes of sensitivity analysis imply that efficient measures such as reducing energy consumption, and increasing the microbial biomass, and grease content, are conceivable elucidations to achieve a more economically feasible status for the algae-derived biofuel.

Acknowledgements The current work is jointly financed by the research program from the Science and Technology Bureau of Xiamen City in China (3502Z20151254), and the Fundamental Research Funds for the Central Universities, HUST (2016YXMS043, 2016YXZD007), Xiamen University (20720160077), China.

References

Amer, L., Adhikari, B., & Pellegrino, J. (2011). Technoeconomic analysis of five microalgae-to-biofuels processes of varying complexity. *Bioresource Technology, 102*(20), 9350–9359.

Bahadar, A., & Khan, M. B. (2013). Progress in energy from microalgae: A review. *Renewable and Sustainable Energy Reviews, 27,* 128–148.

Batan, L. Y., Graff, G. D., & Bradley, T. H. (2016). Techno-economic and Monte Carlo probabilistic analysis of microalgae biofuel production system. *Bioresource Technology, 219,* 45–52.

BP. (2016). *Statistical Review of World Energy June 2016*. BP global.

Brennan, L., & Owende, P. (2010). Biofuels from microalgae—A review of technologies for production, processing, and extractions of biofuels and co-products. *Renewable and Sustainable Energy Reviews, 14*(2), 557–577.

Davis, R., Aden, A., & Pienkos, P. T. (2011). Techno-economic analysis of autotrophic microalgae for fuel production. *Applied Energy, 88*(10), 3524–3531.

Delrue, F., Li-Beisson, Y., Setier, P. A., Sahut, C., Roubaud, A., Froment, A. K., et al. (2013). Comparison of various microalgae liquid biofuel production pathways based on energetic, economic and environmental criteria. *Bioresource Technology, 136,* 205–212.

Delrue, F., Setier, P. A., Sahut, C., Cournac, L., Roubaud, A., Peltier, G., et al. (2012). An economic, sustainability, and energetic model of biodiesel production from microalgae. *Bioresource Technology, 111,* 191–200.

Demirbas, A. (2007). Progress and recent trends in biofuels. *Progress in Energy and Combustion Science, 33*(1), 1–18.

Dutta, S., Neto, F., & Coelho, M. C. (2016). Microalgae biofuels: A comparative study on techno-economic analysis & life-cycle assessment. *Algal Research Biomass Biofuels and Bioproducts, 20,* 44–52.

Gao, C. F., Yu, S. S., & Wu, Q. Y. (2011). Progress in microalgae based biodiesel. *Bulletin of Biology, 46*(6) (in Chinese).

Hill, J., Nelson, D., Polasky, S., & Tiffany, D. (2006). Environmental, economic, and energetic costs and benefits of biodiesel and ethanol biofuels. *Proceedings of the National Academy of Sciences of the United States of America, 103*(30), 11206–11210.

Lopez-Gonzalez, D., Puig-Gamero, M., Acién, F. G., García-Cuadra, F., Valverde, J. L., & Sanchez-Silva, L. (2015). Energetic, economic and environmental assessment of the pyrolysis and combustion of microalgae and their oils. *Renewable and Sustainable Energy Reviews, 51,* 1752–1770.

Malik, A., Lenzen, M., Ralph, P. J., & Tamburic, B. (2015). Hybrid life-cycle assessment of algal biofuel production. *Bioresource Technology, 184,* 436–443.

Pulz, O., & Gross, W. (2004). Valuable products from biotechnology of microalgae. *Applied Microbiology and Biotechnology, 65*(6), 635–648.

Quinn, J. C., & Davis, R. (2015). The potentials and challenges of algae based biofuels: A review of the techno-economic, life cycle, and resource assessment modeling. *Bioresource Technology, 184,* 444–452.

U.S.DOE. (2010). *National algal biofuels technology roadmap.*

Xin, C., Addy, M. M., Zhao, J., Cheng, Y., Cheng, S., Mu, D., et al. (2016). Comprehensive techno-economic analysis of wastewater-based algal biofuel production: A case study. *Bioresource Technology, 211,* 584–593.

Yang, Y. (2015). Economic cost analysis of biodiesel from microalgae cultivated in open pond. *Acta Energiae Solaris Sinica, 36*(2), 295–304. (in Chinese).

Zamalloa, C., Vulsteke, E., Albrecht, J., & Verstraete, W. (2011). The techno-economic potential of renewable energy through the anaerobic digestion of microalgae. *Bioresource Technology, 102*(2), 1149–1158.

Chapter 8
Biofuels from Microalgae: Biodiesel

Lucas Reijnders

Abstract It has been argued that the energy output from microalgal biofuel production should at least be 5–8 times the energy input, apart from solar irradiation driving algal photosynthesis. There is as yet no commercial production of microalgal biodiesel or large-scale demonstration project to check whether this criterion regarding the energy balance can be met in actual practice. There is, however, a set of relatively well-documented peer-reviewed scientific papers estimating energy inputs and outputs of future autotrophic microalgal biodiesel production. Energy balances for biodiesel from autotrophic microalgae grown in ponds tend to be better than for biodiesel from such microalgae grown in bioreactors. The studies regarding energy balances for biodiesel derived from microalgae grown in open ponds are considered here. None of these energy balances meets the criterion that the energy output should exceed the energy input by a factor 5–8. Estimated energy balances are variable due to divergent assumptions about microalgal varieties, applied algal and biodiesel production technologies, assumed parameters and yields and due to differences in system boundaries, allocation, and the use of credits. The studies considered here could have done better in handling uncertainties in estimated energy balances.

Keywords Biodiesel · Energy balance · Variability · Uncertainties

1 Introduction

'Microalgal biodiesel' is used here in the narrow sense: methanol- (or ethanol-) esters of microalgal fatty acids, obtained by the transesterification of triglycerides (oil) from autotrophic microalgae. To qualify as a fuel of good quality, the biodiesel

---

L. Reijnders (✉)
Institute for Biodiversity and Ecosystem Dynamics, University of Amsterdam,
Science Park 904, P.O. Box 94248, 1090 GE Amsterdam, The Netherlands
e-mail: l.reijnders@uva.nl

E. Jacob-Lopes et al. (eds.), *Energy from Microalgae*, Green Energy
and Technology, https://doi.org/10.1007/978-3-319-69093-3_8

derived from microalgal fatty acids has to be upgraded, which includes hydrogenation of unsaturated fatty acids (Reijnders 2017b).

The energy balance is defined here as energy generated by microalgal biodiesel divided by the energy input in its cradle to gate production, apart from solar irradiation driving algal photosynthesis. Energy balances can be calculated or estimated on the basis of cradle to gate life cycle assessments (see Chap. 6).

An energy balance is an important characteristic of microalgal biofuels. Chisti (2013a, b) has argued that the energy balance of algal biofuels should preferably have a value of at least 7 or 8. Reijnders (2013) has proposed an energy balance of at least 5 if the biofuel is to make a major contribution to future energy supply. These are not extreme requirements for types of energy supply derived from solar energy. For instance, recent mean cradle-to-grave energy balances for photovoltaic solar energy are >8 (Koppelaar 2017).

There is as yet no commercial production of autotrophic microalgal biodiesel or large-scale demonstration project to check whether criteria regarding the energy balance can be met in actual practice. There is, however, a set of relatively well-documented peer-reviewed scientific papers estimating energy inputs and outputs of future autotrophic microalgal biodiesel production (see Table 1). These studies are considered here. In the absence of large-scale commercial or demonstration projects and in view of limited knowledge about technologies that will actually be applied in the future, estimates of energy balances for future microalgal biofuel production may well be relatively variable and characterized by relatively large uncertainties (Reijnders and Huijbregts 2009; Colotta et al. 2016a). Variability in estimated energy balances partly originates in choices regarding microalgal varieties, production processes, process parameters, and physical output and input data (e.g., Sills et al. 2012; Naraharisetti et al. 2017). Other choices too matter significantly to energy balances estimated by life cycle assessments. These choices in part regard valuation of physical inputs. For example, flue gas, derived from fossil fuel burning and used for the production of microalgal biodiesel, may be energetically valued at zero but may also be energetically valued in positive or negative terms (Reijnders and Huijbregts 2009). Furthermore, choices as to system boundaries matter (Reijnders 2017a, b). System boundaries demarcate what is included and what is excluded in life cycle assessment. For instance, the production of water inputs and of capital goods, upgrading biodiesel, and biodiesel production wastewater treatment can in practice be included or excluded in life cycle assessments of microalgal biodiesel (Reijnders 2013, 2017b). Furthermore, there is usually more than one output from biofuel production processes. The life cycle energy inputs should be allocated to all outputs. This can be done in several ways. Allocation can take place on the basis of monetary units (prices), on the basis of physical units (energy, exergy, mass) or on the basis of substitution (embodied energy of a similar output). Different types of allocation may lead to significantly different outcomes of life cycle assessments (Reijnders 2017a). It may be noted that allocation to outputs may also be associated with additional uncertainties. For instance, allocations based on prices tend to use current prices, but these are not necessarily the same as future prices. Also, in the case of allocation by substitution, microalgal meal (one of the potential outputs of biodiesel production) has been considered a substitute for soybean meal (Maranduba et al. 2016). The equivalence of algal meal and soybean meal has however not been proven. Choices regarding the valuation of inputs, system boundaries, and allocation to outputs should be explicitly stated. It would moreover be preferable to include in assessments a sensitivity analysis indicating the impact of different choices on estimated energy balances ('scenario uncertainty'; Huijbregts et al. 2003).

There are two basic options for cultivating autotrophic algae: in open ponds and in closed bioreactors. These options can be used singly, but hybrid systems involving both bioreactors and open ponds have also been proposed (Khoo et al. 2011; Adesanya et al. 2014; Huntley et al. 2015). Published comparative life cycle studies suggest that biodiesel from microalgae cultivated in open ponds tends to have a better energy balance than biodiesel from microalgae cultivated in bioreactors (e.g., Stephenson et al. 2010; Clarens et al. 2011; Slade and Bauen 2013; Hallenbeck et al. 2016; Monari et al. 2016; Togarcheti et al. 2017). Moreover, costs of microalgal biodiesel from bioreactors are likely to be higher than the costs of microalgal biodiesel from open ponds (e.g., Dutta et al. 2016; Hallenbeck et al. 2016). For these reasons in the next sections, the focus will be on open pond-based cultivation systems for microalgae.

Table 1 Peer-reviewed life cycle assessments of energy balances associated with biodiesel derived from microalgae grown in open ponds

Authors	Estimated cradle–to-gate energy balance
Lardon et al. (2009)	<2
Batan et al. (2010)	~1
Brentner et al. (2011)	<1
Clarens et al. (2011)	<4.1
Khoo et al. (2011)	<1
Razon and Tan (2011)	<1
Shirvani et al. (2011)	<1
Xu et al. (2011)	<3
Sills et al. (2012)	<3
Frank et al. (2012)	<2
Quinn et al. (2014)	<2
Mu et al. (2014)	<2
Adhikari and Pelegrino (2015)	<1
Orfield et al. (2015)	<2
Yuan et al. (2015)	<2
Dutta et al. (2016)	<2
Crowdhury and Francetti (2017)	<3
Naraharisetti et al. (2017)	<2

2 Estimated Energy Balances

The set of relatively well-documented peer-reviewed estimates of energy balances for biodiesel derived from microalgae grown in open ponds is summarized in Table 1.

This table shows that in none of the studies considered here estimated energy balances met the criterion of a factor 5–8. As furthermore can be seen in Table 1, the variability of published energy balances for microalgal biodiesel is large (also: Quinn et al. 2014). In the following, several important causes of this variability are discussed. These regard assumptions about biomass yield and allocation and the combinations with wastewater treatment and power plants fueled by carbonaceous substances. Thereafter, the handling of uncertainties is considered.

3 Combining Microalgal Biodiesel Production with Wastewater Treatment

In studies considered here, relatively good (high) energy balances tend to be obtained when wastewater containing microalgal nutrients is used for cultivation. So far, the use of wastewater for microalgae cultivation has only been subject to small-scale investigations (Laurens et al. 2017). In the life cycle assessments considered here, the use of wastewater containing nutrients has been handled in a variety of ways. Firstly, it can be assumed that the cultivation of microalgae substitutes for conventional wastewater treatment. Based on this assumption, the input energy for conventional wastewater treatment is then subtracted from the cradle-to-gate microalgal biodiesel production (e.g., Clarens et al. 2010; Mu et al. 2014; Chowdhury and Franchetti 2017). Such subtraction does not take account of the limitations to algal purification of wastewater. For instance, wastewater tends to contain relatively large organic molecules, which cannot be metabolized by currently applied microalgae (Perez-Garcia et al. 2011). Furthermore, wastewater may well contain pathogens, predatory zooplankton, and microorganisms that might outcompete microalgae (Pittman et al. 2011; Cai et al. 2013). For this reason, wastewater may need treatment (e.g., sterilization) to minimize infection risk. Energy needed for such treatment has not been included in the life cycle assessments considered here.

In addition, or alternatively, to energy credits associated with the substitution of conventional wastewater treatment, it can be assumed that the nutrients in wastewater substitute for the inputs of synthetic nutrients and that the energy input of these nutrients present in wastewater can be valued at zero. It can be assumed too that the treatment of wastewater from microalgal biodiesel production is not within the system boundaries of life cycle assessment. Exceptions regarding the latter assumption are the studies of Lardon et al. (2009) and Clarens et al. (2011). The assumptions outlined here are contestable. One might argue from an industrial ecology perspective that nutrients derived from wastewater should have an energy value >0, as they are apparently useful for microalgal biodiesel production. Furthermore, when the substitution of conventional wastewater treatment is *within* the system boundaries, there would appear to be no good case for keeping wastewater treatment of effluents from microalgal biodiesel production, which should at least serve the recycling of nutrients (Laurens et al. 2017), *outside* the system boundaries (Stephenson et al. 2010; contrast, e.g., Mu et al. 2014). The same would appear to hold for treatment of wastewater (e.g., sterilization) to minimize infection risk for microalgal cultivation.

Furthermore, it should be noted that so far the autotrophic algal productivity of triglycerides in wastewater has been found to be low, that the supply of wastewater is unlikely to allow for producing large amounts of algal triglycerides, and that by-product usage may be problematical due to the presence of hazardous substances and pathogens (Rawat et al. 2016; Luangpipat and Chisti 2017).

4 Combination of Microalgal Biodiesel Production with Co-located Power Plants Fueled by Carbonaceous Substances and Cement Plants

Combinations of cultivating microalgae with co-located power plants (e.g., Colotta et al. 2016b) can be conducive to good energy balances if 'wastes' of power plants (flue gas, 'waste' heat) can be used as inputs in microalgal biodiesel production at an assumed zero energy value, apart from energy input needed for the transport of 'wastes' (e.g., Brentner et al. 2011; Clarens et al. 2011; Xu et al. 2011; Yuan et al. 2015). A similar assumption may be made for the use of CO_2 from co-located cement plants that generate CO_2 from calcium carbonate (Colotta et al. 2016b). The assumption that inputs of such 'wastes' may be energetically valued at zero energy input is contestable. From an industrial ecology perspective, one might argue that 'wastes' should have an energy value >0, as they are apparently useful for microalgal biodiesel production. On the other hand, one may argue from a cost perspective in favor of an energy credit (negative energy input) for using 'waste' CO_2 in microalgae cultivation as the use of flue gas may lower external costs.

It may furthermore be noted that algae can only capture a modest fraction of the flue gas emitted by power plants (Benemann 2013). The co-location with power and cement plants will severely limit the geographical scope for algal cultivation. And future power production may well shift away from burning carbonaceous fuels.

5 Assumptions About Algal Biomass Yields and Allocation

Relatively good energy balances can be achieved in studies considered here when it is assumed that biomass and lipid yields are high, with biomass yields in the order of 90 Mg ha^{-1} $year^{-1}$ or larger (Slade and Bauen 2013; Rogers et al. 2014; Naraharisetti et al. 2017). Such biomass yields are well beyond current yields in commercial cultivation of autotrophic microalgae (Reijnders 2013; Rawat et al. 2016). Whether such high yields can be achieved in future widespread commercial practice is questionable (Reijnders 2017b). A major problem for open pond cultures is the contamination with other organisms, which can lead to instability of microalgal cultivation (Rodolfi et al. 2008). This may be prevented by growing microalgae under extreme conditions such as high pH or high salt concentrations. But these conditions are not conducive to high triglyceride yields (Reijnders 2017b). And there may be problems in maintaining extreme conditions of cultivation given the vagaries of weather, such as extreme rainfall (Reijnders 2013; Chisti 2016).

On the other hand, relatively poor energy balances may be achieved when part of the outputs is considered wastes and all of the energy input is allocated to non-waste outputs (e.g., biodiesel, biogas). This is at variance with the view that mature biodiesel production systems should be designed in such a way that practically all microalgal biomass should be converted to useful outputs (e.g., Xu et al. 2011; Laurens et al. 2017).

Relatively good energy balances may be achieved when biodiesel is a product of a production facility also generating currently highly priced co-products, when the allocation is on the basis of current prices. A focus on such co-products can be noted in the current development of microalgal bioenergy (Chew et al. 2017; Laurens et al. 2017). In the case of allocation based on prices, one should evaluate what future developments in input and output prices and the impact thereof on estimated energy balances may be (cf. Kern et al. 2017). This type of evaluation has not been included in the LCAs considered here. In view of recent history, one should especially consider whether high co-product prices will be maintained when production is much increased. The price of the biodiesel co-product glycerol collapsed when the production of biodiesel based on oils from terrestrial plants was much increased (Reijnders and Huijbregts 2009).

6 Handling Uncertainties

In the introduction, it has been noted that it would be preferable to include in assessments a sensitivity analysis indicating the impact of different choices regarding the valuation of inputs, system boundaries, and allocation to outputs on estimated energy balances ('scenario uncertainty'). This type of sensitivity analysis is uncommon in the studies considered here. An exception in this respect is the study of Razon and Tan (2011). An important source of uncertainty regards future technologies, process parameters, physical inputs and outputs. The studies discussed here often considered examples of such uncertainties (e.g., Stephenson et al. 2010; Brentner et al. 2011; Clarens et al. 2011; Shirvani et al. 2011; Sills et al. 2012; Frank et al. 2012; Quinn et al. 2014; Yuan et al. 2015; Naraharisetti et al. 2017). Relatively good studies of uncertainty linked to process parameters would seem to be the studies of Clarens et al. (2011) and Sills et al. (2012) that included Monte Carlo analysis to quantify the role of uncertainty of parameters in determining the estimates of energy balances.

If compared with published methodologies for handling uncertainty in life cycle assessment (cf. Huijbregts et al. 2003; Gregory et al. 2016), the studies considered here could have done better in dealing with uncertainties.

7 Concluding Remarks

The peer-reviewed energy balances for microalgal biodiesel discussed here do not meet the criterion that the energy output should exceed the energy input by a factor 5–8. It would seem extremely likely that this would still be the case when defensible choices regarding yields, allocation, valuing the use of wastewater, and co-location with power plants would have been different as noted in this chapter. This is not favorable to a near-term widespread use of microalgal biodiesel. Moreover, microalgal biodiesel has a high near-term cost if compared with biofuels from terrestrial plants (Wijffels and Barbosa 2010; Jez et al. 2017; Zhu et al. 2017). The conclusion must be that the near-term outlook for widespread use of microalgal biodiesel is bleak (also: Chisti 2013a).

References

Adesanya, V., Cadena, E., Scott, S. A., & Smith, A. G. (2014). Life cycle assessment on microalgal biodiesel production using a hybrid cultivation system. *Bioresource Technology, 163,* 343–355.

Adhikari, B., & Pellegrino, J. (2015). Life cycle assessment of five microalgae-to-biofuels processes of varying complexity. *Journal of Renewable and Sustainable Energy, 7,* 043136 (12 pp).

Batan, L., Quinn, J., Wilson, B., & Bradley, T. (2010). Net energy and greenhouse gas emission evaluation of biodiesel derived from microalgae. *Environmental Science and Technology, 44,* 7975–7980.

Benemann, J. R. (2013). Microalgae for biofuels and animal feeds. *Biotechnology Bioengineering, 110,* 2319–2320.

Brentner, L. B., Eckelman, M. J., & Zimmerman, J. B. (2011). Combinatorial life cycle assessment to inform process design of industrial production of algal biodiesel. *Environmental Science and Technology, 45,* 7060–7067.

Cai, T., Park, S. Y., & Li, Y. (2013). Nutrient recovery from wastewater streams by microalgae: Status and prospects. *Renewable and Sustainable Energy Reviews, 10,* 360–369.
Chew, K. W., Yap, J. Y., Show, P. L., Suan, N. H., Juan, J. C., Ling, T. C., et al. (2017). Microalgae biorefinery: High value perspectives. *Bioresource Technology, 229,* 53–62.
Chisti, Y. (2013a). Constraints to commercialization of algal biofuels. *Journal of Biotechnology, 167,* 201–214.
Chisti, Y. (2013b). The problem with algal fuels. *Biotechnology Bioengineering, 110,* 2319–2328.
Chisti, Y. (2016). Large scale production of algal biomass: Raceway ponds. In F. Bux, & Y. Chisti (Eds.) *Algal Biotechnology* (pp. 21–40). Switzerland: Springer.
Chowdhury, R., & Franchetti, M. (2017). Life cycle energy demand from nutrients present in dairy waste. *Sustainable Production and Consumption, 8,* 22–27.
Clarens, A. F., Nassau, H., Resurreccion, E. P., White, M. A., & Colosi, L. M. (2011). Environmental impacts of algae-derived biodiesel and bioelectricity for transportation. *Environmental Science and Technology, 45,* 7554–7560.
Clarens, A. F., Resurreccion, E. P., White, M. A., & Colosi, L. M. (2010). Environmental life cycle comparison of algae to other bioenergy feedstocks. *Environmental Science & Technology*, 1813–1819.
Colotta, M., Busi, L., Champagne, P., Mabee, W., Tomasoni, G., & Alberti, M. (2016a). Evaluating microalgae-to-energy systems: Different approaches to life cycle assessment (LCA) studies. *Biofuels Bioproduction and Biorefining* http://doi.org/10.1002/bbb.1713.
Colotta, M., Champagne, P., Mabee, W., Tomasoni, G., Alberti, M., Busi, L., et al. (2016b). Environmental assessment of co-location alternatives for a microalgae cultivation plant: A case study of Kingston (Canada). *Energy Procedia, 95,* 29–36.
Dutta, S., Neto, F., & Coelho, M. (2016). Microalgae biofuels: A comparative study on techno-economic analysis & life cycle assessment. *Algal Research, 20,* 44–52.
Frank, E. D., Han, J., Palu-Rivera, I. A., Elgowainy, A., & Wang, M. Q. (2012). Methane and nitrous oxide emissions affect the life cycle analysis of algal biofuels. *Environmental Research Letter, 7,* 014030.
Gregory, J. F., Noshadravan, A., Olivetti, E. A., & Kirchain, R. E. (2016). A methodology for robust comparative life cycle assessments incorporating uncertainty. *Environmental Science and Technology, 50,* 6397–6405.
Hallenbeck, P. C., Grogger, M., Mraz, M., & Veverka, D. (2016). Solar biofuels production with microalgae. *Applied Energy, 179,* 136–145.
Huijbregts, M. A. J., Giliamse, W., Ragas, A. M. J., & Reijnders, L. (2003). Evaluating uncertainty in environmental life cycle assessment. A case study comparing two insulation options for a Dutch one-family dwelling. *Environmental Science and Technology, 37,* 2600–2608.
Huntley, M. E., Johnson, Z. I., Brown, S. L., Sills, D. L., Gerber, L., Archibald, I., et al. (2015). Demonstrated large-scale production of marine microalgae for fuels and feed. *Algal Research, 10,* 249–265.
Jez, S., Spinelli, D., Fierro, A., Dibenedetto, A., Aresta, M., Busi, E., et al. (2017). Comparative life cycle assessment study on environmental impact of oil production from microalgae and terrestrial oilseed crops. *Bioresource Technology, 239,* 266–275.
Kern, J. D., Hise, A. M., Characklis, G. W., Gerlach, R., Viamajala, S., & Gardner, R. (2017). Using life cycle assessment and techno-economic analysis in a real options framework to inform the design of algal biofuel production facilities. *Bioresource Technology, 225,* 418–428.
Khoo, H. H., Sharratt, P. N., Das, P., Balasubramanian, R. K., Naraharisetti, P. K., & Shaik, S. (2011). Life cycle energy and CO_2 analysis of micro-algae-to-biodiesel: Preliminary results and comparisons. *Bioresource Technology, 102,* 5800–5807.
Koppelaar, R. H. E. M. (2017). Solar-PV energy payback and net energy: Meta assessment of study quality, reproducibility, and results harmonization. *Renewable and Sustainable Energy Reviews, 72,* 1241–1255.
Lardon, L. A., Helias, A., Sialve, B., Steyer, J., & Bernard, O. (2009). Life cycle assessment of biofuel production from microalgae. *Environmental Science and Technology, 43,* 6475–6481.

Laurens, L. M. L., Chen-Glasser, M., & McMillan, J. (2017). A perspective on renewable bioenergy from photosynthetic algae as feedstock for biofuels and bioproducts. *Algal Research, 24,* 261–264.

Luangpipat, T., Chisti Y. (2017). Biomass and oil production by *Chlorella vulgaris* and four other microalgae—Effects of salinity and other factors. *Journal of Biotechnology.* http://doi.org/10.1016/j.jbiotec.2016.11.029.

Maranduba, H. L., Robra, S., Nascimento, I. A., da Cruz, R. S., Rodrigues, L. B., & de Almeida Neto, J. A. (2016). Improving the energy balance of microalgal biodiesel: Synergy with an autonomous sugarcane ethanol distillery. *Energy, 115,* 888–895.

Monari, C., Righi, S., & Olsen, S. I. (2016). Greenhouse gas emissions and energy balance of biodiesel production from microalgae cultivated in photobioreactors in Denmark: A life cycle modeling. *Journal of Cleaner Production, 112,* 4064–4092.

Mu, D., Min, M., Krohn, B., Mullins, K. A., Ruan, R., & Hill, J. (2014). Life cycle environmental impacts of wastewater-based algal biofuels. *Environmental Science and Technology, 48,* 11696–11704.

Naraharisetti, P. K., Das, P., & Sharratt, P. N. (2017). Critical factors in energy generation from microalgae. *Energy, 120,* 139–152.

Orfield, N. D., Levine, R. B., Keoleian, G. A., Miller, S. A., & Savage, P. E. (2015). Growing algae for biodiesel on direct sunlight or sugars: A comparative life cycle assessment. *ACS Sustainable Chemistry & Engineering, 3,* 386–395.

Perez-Garcia, O., Puente, Y., & Bahan, M. E. (2011). Organic carbon supplementation of sterilized municipal waste water is essential for heterotrophic growth and removing ammonium by the microalga *Chlorella vulgaris. Journal of Phycology, 47,* 190–199.

Pittman, J. K., Dean, A. P., & Osundeko, O. (2011). The potential of sustainable algal biofuel production using wastewater resources. *Bioresource Technology, 102,* 17–25.

Quinn, J. C., Smith, T. G., Downes, C. M., & Quinn, C. (2014). Microalgae to biofuels life cycle assessment—Multiple pathway evaluation. *Algal Research, 4,* 116–122.

Rawat, I., Gupta, S. K., Shriwastav, A., Singh, P., & Bux, F. (2016). Microalgae applications in wastewater treatment. In F. Bux, Y. Chisti (Eds.) *Algae Biotechnology* (pp. 249–268). Switzerland: Springer.

Razon, L. F., & Tan, R. R. (2011). Net energy analysis of the production of biodiesel and biogas from the microalgae *Haematococcus pluvialis* and *Nanochloropsis. Applied Energy, 88,* 3507–3514.

Reijnders, L. (2013). Lipid-based biofuels from autotrophic microalgae: Energetic and environmental performance. *WIRE's Energy Environmental, 2,* 73–85.

Reijnders, L. (2017a). Life cycle assessment of greenhouse gas emissions. Chapter 2–2. In W. -Y. Chen et al. (Eds.) *Handbook of climate change mitigation and adaptation* (pp. 63–91). Heidelberg: Springer (Part 1).

Reijnders, L. (2017b). Greenhouse gas balances of microalgal biofuels. In J. C. M. Pires (Ed.) *Microalgae as a source of bioenergy: Products, processes and economic.* Bentham Science Publishers.

Reijnders, L., & Huijbregts, M. A. J. (2009). *Biofuels for road transport.* Heildelberg: Springer.

Rodolfi, L., Zittelli, G. C., Bassi, N., Padovani, G., Biondi, N., Bonini, G., et al. (2008). Microalgae for oil: Strain selection, induction of lipid synthesis and outdoor mass cultivation in a low cost photobioreactor. *Biotechnology Bioengineering, 102,* 100–112.

Rogers, J. N., Rosenberg, J. N., Guzman, B. J., Oh, V. H., Mimbela, L. E., Ghassemi, A., Betenbaugh, M. J., Oyler, G. A., Donohue, M. D. (2014). A critical analysis of paddlewheel driven raceway ponds for algal biofuel production at commercial scales. *Algal Research, 4,* 76–88.

Shirvani, T., Yan, X., Inderwildi, O. R., Edwards, P. P., & King, D. A. (2011). Life cycle energy and greenhouse gas analysis for algae-derived biodiesel. *Energy & Environmental Science, 4,* 3773–3778.

Sills, D. L., Paramita, V., Franke, M. J., Johnson, M. C., Akabas, T. M., Greene, C. H., et al. (2012). Quantitative uncertainty analysis of life cycle assessment for algal biofuel production. *Environmental Science and Technology, 47,* 687–694.

Slade, R., & Bauen, A. (2013). Micro-algae cultivation for biofuels: Cost, energy balance, environmental impacts and future prospects. *Biomass and Bioenergy, 53,* 29–38.

Stephenson, A. L., Kazamis, E., Dennis, J. S., Howe, C. J., Satt, S. A., & Smith, A. G. (2010). Life cycle assessment of potential algal biodiesel production in the United Kingdom: A comparison of raceways and air lift tubular bioreactors. *Energy & Fuels, 24,* 4062–4077.

Togarcheti, S. C., Mediboyina, M. K., Chauhan, V. S., Mukherji, S. M., & Mudliar, S. N. (2017). Life cycle assessment of microalgae based biodiesel production to evaluate the impact of biomass productivity and energy source. *Resource, Conservation and Recycling, 122,* 285–294.

Wijffels, R. H., & Barbosa, M. J. (2010). An outlook on microalgal biofuels. *Science, 329,* 796–799.

Xu, L., Brilman, D. W. F., Withag, J. A. M., Brem, G., & Kersten, S. (2011). Assessment of a dry and a wet route for the production of biofuels from microalgae: Energy balance analysis. *Bioresource Technology, 102,* 5113–5122.

Yuan, J., Kendall, A., & Zhang, Y. (2015). Mass balance and life cycle assessment of biodiesel from microalgae incorporated with nutrient recycling options and technology uncertainties. *Global Change Biology Bioenergy, 7,* 1245–1259.

Zhu, L., Nugroho, Y. K., Shakeel, S. R., Li, Z., Martinkauppi, B., & Hiltunen, E. (2017). Using microalgae to produce liquid transportation biodiesel: What is next? *Renewable and Sustainable Energy Reviews, 78,* 391–400.

Chapter 9
Biofuels from Microalgae: Energy and Exergy Analysis for the Biodiesel Case

Daissy Lorena Restrepo-Serna, Mariana Ortiz-Sánchez and Carlos Ariel Cardona-Alzate

Abstract Nowadays, the microalgae have been gaining importance due to their different applications in the biofuel, food, and pharmaceutical industries. One of the applications that is commonly proposed for microalgae oil is the transformation into biodiesel through transesterification. This biodiesel is a biofuel that present energy yields similar to traditional diesel, generating an alternative to replace a fuel from petrochemical origin. The objective of this work is to analyze deeply a process for biodiesel production from microalgae oil. The process includes the cultivation, harvesting, and extraction stages for the oil. In this case, the software Aspen Plus is employed for simulation. From the results obtained (mass and energy balances), the energy, exergy, and economic and environmental analysis of the process are carried out through the development of different scenarios. Last allow to evaluate the energy, economic and environmental viability of this type of processes. As a result, this work shows the challenges to be overcome to make possible the real introduction of microalgae fuels.

Keywords Biodiesel · Biofuel · Microalgae · Exergy analysis
Energy analysis

Overview

In the development of green processes, it is considered the biofuel synthesis as a representative example. Due to the growing energetic demand, in especial for the fossil fuels, the attention for generating new alternatives has increased (Quintero et al. 2012). In last years, biodiesel has received a great interest by the scientific community due that its use leads to the reduction in harmful emissions like carbon monoxide, carbon dioxide, and particles like SO_x that are responsible of greenhouse effect (Gouveia and Oliveira 2009). Additionally, biodiesel is the only one biofuel that can be used in a conventional diesel engine without needing great modifications.

D. L. Restrepo-Serna · M. Ortiz-Sánchez · C. A. Cardona-Alzate (✉)
Chemical Engineering Department, National University of Colombia,
Manizales Campus, Km 07, Manizales, Colombia
e-mail: ccardonaal@unal.edu.co

E. Jacob-Lopes et al. (eds.), *Energy from Microalgae*, Green Energy and Technology, https://doi.org/10.1007/978-3-319-69093-3_9

Despite these characteristics, biodiesel presents certain disadvantages like a major consumption due to a less calorific power and less stability than diesel, making no possible to store it for a long time.

The biodiesel is obtained through the transesterification reaction between biological renewable sources such as vegetable oils, animal fats, and microalgae oil with an alcohol (Ma and Hanna 1999). The biodiesel production can be developed using different alkaline and organic catalysts as well as lipases obtained from animals, vegetables, or microorganism sources (Shahid and Jamal 2011). The microalgae are photosynthetic organisms that have the ability to grow very fast and live in adverse conditions. These microorganisms are present in all existent ecosystems, representing a great species variety (Shahid and Jamal 2011).

It is estimated that 50,000 microalgae species exist (Richmond 2004). Species like *Chlorella vulgaris* have aroused a major interest, due to its high protein, lipid (14–22% dry basis), and other products (Becker 1994). This specie has the ability to accumulate a great lipid quantity in absence of nitrogen, generating a fatty acid profile that can be used for biodiesel synthesis (Converti et al. 2009; Fradique et al. 2013). An advantage of microalgae is the capacity of growing in different conditions of cultivation due to its different types of growing: autotrophic, heterotrophic, and mixotrophic.

The microalgae growing includes the adaptation of its metabolism to different cultivation hostile mediums (Bumbak et al. 2011). *Desmodesmus* gene, for example has demonstrated an ordinated reproduction that exposes morphological changes in response to environmental changes like the nutrient availability, temperature an illumination (Trainor 2009).

In the present work, the use of *Chlorella protothecoides* in the biodiesel production was analyzed. For this, four steps were considered: culture and harvesting of microalgae, oil extraction and biodiesel production. To carry out the simulation of this process was used the Aspen Plus to obtain the mass and energy balances (Cerón-Salazar and Cardona-Alzate 2011), which was used in the energy, exergy, and economic and environmental analysis to understand the real viabilities of this process.

1 Microalgae Today: Applications and Uses

Microalgae have been used by indigenous populations to supply their food needing 2000 years ago. Microalgae species like *Nostoc, Arthrospira (Spirulina),* and *Aphanizomenon* were used with that aim (Spolaore et al. 2006). An example is the Azteca culture, which used microalgae like a high important food source (Venkataraman 1997). Due to the high protein content of various microalgae species, they are a non-conventional source of these proteins that can be used in foods to increase its nutritional value (Guil-Guerrero et al. 2004).

One of the markets reason for microalgae is as a source of carotenoids, being used mainly as natural food colorants and as additives for animal feed. Although the synthetic forms are less expensive than the naturals, microalgae carotenoids have the advantage of supplying natural isomers in their natural ratio (Guil-Guerrero et al. 2004; Waldenstedt et al. 2003). *D. Salina* is the microorganism most used for the production of β-carotene due to the possibilities of reaching 14% in dry weight, and its cultivation process is easy to implement (can be realized in raceway systems) (Spolaore et al. 2006). Due to the photosynthesis capacity, these microalgae are able to incorporate stable isotopes like ^{13}C, ^{15}N y ^{2}H from inorganic chemical compounds such as $^{13}CO_2$, $^{15}NO_3$, $^{2}H_2O$ that are used for protein quantification (Spolaore et al. 2006).

In recent years, due to oil crisis the studies are concentrated in the obtaining and use of microalgae lipids for energy generation, transforming it to biofuels, mainly to biodiesel (Chisti 2007). From this perspective, the microalgae cultivation has been focused on the biotechnological point of view, analyzing their implementation in biorefineries.

2 Process of Culture

2.1 Autotrophic Growth

Microalgae can implement oxygenic photosynthesis and carbon dioxide fixation through Calvin cycle. In other words, they can capture energy from light and use carbon dioxide like carbon source (Yang et al. 2000). Therefore, microalgae have the ability to mitigate carbon dioxide emissions that are produced by industry, generating high-value products (Chen et al. 2011).

2.1.1 Open Ponds Systems

The open ponds systems is the most commonly used configuration for microalgae production, due that these cultivation methods are economically feasible for high-scale biomass production (Safi et al. 2014). The cells grow under sunlight and carbon dioxide supply (Slegers et al. 2013).

These systems have some limitations due to a strict environmental control with the aim to avoid biological contaminations like bacteria or not desired species, also neutralizing pollution in the systems and water evaporation (Safi et al. 2014).

2.1.2 Photobioreactors

Although the initial investment is the highest in comparison to open ponds systems, photobioreactors permit a better contamination control, as well as better use and control of light intensity, carbon dioxide, and nutrients supply (Sforza et al. 2012).

These systems are more appropriated for cells that cannot compete and grow in difficult environments and for the obtaining of pharmaceutical and food products (Safi et al. 2014). To increase lipid productivity in microalgae, growth conditions like nitrogen on nutrients limitation are applied (Mata et al. 2010).

2.1.3 Heterotrophic Growth

The heterotrophic culture of microalgae is interesting, due to the fact that it is expected that growth rates and productivity are major in comparison with autotrophic growth. Additionally, the biomass production using these systems allows a simple and low-cost harvesting (Chen 1996). Nevertheless, one of the main limitations of this type of cultures is the high-cost attached to carbon source (glucose or acetate), in comparison with another nutrients in the medium, added to the competition of these sources with food sector (Liang et al. 2009). Given this situation, it is of great interest to find a cheap carbon source, with the aim to make these cultures competitive when used on a commercial level (Abad and Turon 2012). For this reason, many studies have been focused on the analysis of microalgae growth in agroindustrial residues such as molasses, glycerol, and pig manure (Leesing and Kookkhunthod 2011). Almost it has been demonstrated that microalgae culture can be used for wastewater treatment with high organic and salt levels (Perez-Garcia et al. 2011).

2.1.4 Mixotrophic Growth

In this type of cultures, microorganisms as microalgae can be able to get metabolic energy from both, photosynthesis and carbon organics sources (Chen et al. 2011). It has been reported a high cellular density in mixotrophic cultures for *C. vulgaris*, demonstrating that these microalgae can grow in this type of cultures (Liang et al. 2009).

3 Process of Harvesting

To consider the lipids obtained from microalgae a viable raw material, it is necessary to have in mind the harvesting method to use. This stage can represent a 20–30% of the total cost for biomass obtaining (Rawat et al. 2011) that implies a cumulative high energetic demand for biodiesel production (Dassey and Theegala 2012). These difficulties are mainly associated to the small size of the microalgae and their suspension stability in aqueous media, as a result of negative charges. Additionally, the organic material and low concentration of the streams treated that come from diluted cultures makes difficult this stage (Liu et al. 2013). The harvesting process can be classified into physical, chemical, and biological methods.

The selection of the best method depends on several factors like the final products' characteristics (Molina Grima et al. 2003), the deformation–destruction level of the algae after harvesting process thinking in posterior procedures and the economic and energetic costs (Barros et al. 2015).

3.1 Physical Methods

3.1.1 Filtration

Filtration is a dehydration method in which a pressure drop is required along the system with the object to force the fluid through a membrane (Barros et al. 2015). Certain types of filters have been used for the microalgae harvesting, depending on the cell or colonies size. Usually, a high recovery is reached if they are used in big cells, but it fails in the recovery of organisms near to bacterial size (Mohn 1980). Systems like microfiltration can be used for microalgae recovery of common microalgae with a size order of 5–6 μm (Edzwald 1993). Process like macro-filtration can be used if a previous process of flocculation is performed (Milledge and Heaven 2013). In this type of process, microalgae have the tendency to deposit in the filtrate medium, growing its thickness and flux resistance (Show and Lee 2014). Micro- and ultra-filtration processes are possible alternatives to conventional filtration, but these methods are not usually used at large-scale (Molina Grima et al. 2003).

Methods like dead-end filtration have been evaluated giving as a result that this type of processes can be energetically competitive in comparison with other harvesting methods like flocculation or vacuum filtration (Bila et al. 2012).

Cross-flow filtration methods have demonstrated to be an alternative for microalgae concentration (being concentrated at about 150 times) using less energy that required for other methods like flocculation, centrifugation, or vacuum filtration (Drexler and Yeh 2014).

3.1.2 Centrifugation

In centrifugation, forces higher than gravity are generated with the aim of performing the cell separation, allowing the separation of almost all types of microalgae (Mohn 1980). Disk stack centrifuge are the most common centrifugation systems applied in the commercial plants for the production of high-cost derived products and in the production of pilot-scale biodiesel (Molina Grima et al. 2003). The centrifugation methods are very expensive and energetically no viable if used for low-cost products like biofuels due to the volume to be processed. An alternative for the energetic consumption diminution is the application of flocculation process before decreasing the volume to be treated for centrifugation in 65% (Barros et al. 2015, Wilson 2012).

3.1.3 Flotation

The flotation technique consists in making that the microalgae float up in the water surface through air bubbles. These bubbles catch the cells making they float up with the aim to be recollected in the surface (Sharma et al. 2013). This technology has the advantage to separate low-density microalgae cultures. In some cases, it is necessary the addition of a flocculating agent to perform an effective flotation (Edzwald 1993). It has been found that for a flotation system like dissolved air flotation, the energy demand is high due to the high pressure required for the bubbles' formation (Hanotu et al. 2012).

3.1.4 Ultrasound

This method consists in the cell concentration or flocculation due to acoustic forces, without generating shear strength in the microalgae, and then the cells are recollected by methods using gravity forces like centrifugation or sedimentation. As disadvantages, the low concentration factor in comparison with other methods, and operative costs make this method unsuitable yet (Bosma et al. 2003).

3.2 Chemical Methods

3.2.1 Chemical Flocculation

Due to the negative charge present in microalgae, it does not permit the aggregation of suspended cells. These superficial charges can be neutralized adding flocculating agents, allowing the increment in particle size due to cell aggregation, increasing the sedimentation or flotation rate (Mata et al. 2010). The ideal flocculating agent must have some characteristics like non-toxic, inexpensive, and effective in low concentration compound (Molina Grima et al. 2003). For this reason, inorganic flocculating agents like $FeCl_3$ and $Al_2(SO_4)_3$ that are very effective are not ideal for an environmental suitable process (Lee et al. 2014). Another disadvantage for chemical flocculation is the possibility of cell alterations generated by flocculating agents that can interfere in the final processing of biomass like lipid extraction (Uduman et al. 2010).

4 Process of Oil Extraction

Lipid extraction from microalgae and its efficiency are an important factor in the biodiesel production process (Adam et al. 2012). Two options for lipid recovery are available, dry algal biomass and wet algal biomass extractions. In the dry algal biomass extraction, the yield is higher, but the associated cost due to biomass drying is considerable (Taher et al. 2014). On the other hand, if wet biomass is used, the cell rupture is realized in the solution where the microalgae were cultivated (Ghasemi Naghdi et al. 2016) and better energy efficient is achieved, but lipid extraction yield is low (Taher et al. 2014).

4.1 Organic Solvent Extraction

Lipid extraction with organic solvents is based on the interaction between long hydrophobic chains of fatty acids and neutral lipids through van der Waals forces, forming globules in the cytoplasm (Medina et al. 1998), which in presence of a nonpolar solvent form a solvent–lipid complex, that leave the cell due to a concentration gradient (Halim et al. 2012). Similar mechanism is applied for the extraction of polar lipids when polar solvents like alcohols are used, due to the interruption of hydrogen bonds (Pragya et al. 2013). An ideal solvent must have certain characteristics like non-toxic, cheap, volatile, and selective compound (Rawat et al. 2011).

4.2 Soxhlet Extraction

This method has the advantage that the cells are in constant contact with fresh organic solvent, avoiding the equilibrium limitation present in batch processes with solvents (Mubarak et al. 2015). It has been demonstrated that using this method, it is possible to recover almost all the microalgae lipids, been the reference method to compare with another extraction methods (Prommuak et al. 2012).

4.3 Bligh and Dyer's Method

The Bligh and Dyer's method is one of the most common methods for lipid extraction, using 1:2 chloroform/methanol (v/v); the lipids in the chloroform phase are separated (Ranjith Kumar et al. 2015). To upgrade this method, many modifications have been proposed, one of those is the addition of 1 M NaCl to avoid denatured lipids. With this method, extraction yields upon 95% from total lipid content had been obtained, with the possibility for using it in both dry and wet algal biomass (Pragya et al. 2013).

4.4 Microwave Extraction

Due to the heat generated by the interaction between water (a polar fluid) and the electric field produced by microwaves, vapor is generated next to the cell. This action breaks the cell wall and releases the compounds that are inside the microalgae (Ghasemi Naghdi et al. 2016). The heating with microwaves is very fast allowing short times for the lipid extraction (Dai et al. 2014).

4.5 Ultrasound Extraction

This method is based on the rupture of the cell through the cavitation produced from the collapse of microbubbles generated by the ultrasound waves. These bubbles are near of the cell wall and when they collapse, a shock wave is created, breaking the cell wall, releasing the internal compounds of the microalgae (Ghasemi Naghdi et al. 2016). The study of the ultrasound method has been combined with the use of organic solvents or solvent-free with the aim of has a clean and environmental process (Adam et al. 2012).

4.6 Supercritical Fluids

The extraction using supercritical fluids is a new green technology with the potential to substitute other types of extraction like the extraction with organic solvents. The use of supercritical fluids has a series of advantages due to the fluid properties like favorable mass transfer that facilities the input to the cell inside, their property variability with operational conditions and that the extracted lipids do not require a posterior stage of purification because are solvent-free (Halim et al. 2012).

4.7 Ionic Liquids Extraction

Ionic liquids are salts that have a melting point under 100 °C that due to unique properties like their thermal stability, high selectivity, solubility, and non-volatile characteristics, which are called to be an alternative for the use of organic solvents (Kim et al. 2012). It has been found that the ionic liquids with a high hydrophobicity and high acidity have better extraction yields (Yu et al. 2016).

4.8 Electroporation

Electroporation is a process in which microalgae cell wall is exposed to high intensities generated by electric fields during certain periods, destabilizing the cell wall with the object to extract the intern compounds of microalgae (Joannes et al. 2015). Electroporation is one of the most used transfection methods due to high efficiency and convenience in comparison with other transfection methods (Kang et al. 2015). This technology is called to be a promising method for cell lipid extraction due to low energy consumption, economic and possibility to be scaled-up (Toepfl et al. 2006).

5 Process of Biodiesel Production

Biodiesel production is developed using vegetable oils, animal fats, and short-chain alcohols. Methanol is the most common used alcohol and presents the best conversions, although ethanol is also used (Demirbas 2005). For the transesterification reaction advance, the use of a catalyst is needed. In last years, the biodiesel production has been investigated from the use of chemical homogeneous catalyst like acids (sulfuric acid) or basics (sodium hydroxide, sodium methoxide, potassium hydroxide, etc.); chemical heterogeneous catalyst, acids (Amberlyst 15, SO_4^{2-}), and basics (KNO_3, Al_2O_3, MgO); lipases from animal, vegetables, or microorganism sources (*Chromobacterium Viscosum*, *Candida Rugosa,* and pork pancreas) (Shahid and Jamal 2011).

6 Exergetic Analysis in Process

From the analysis of energy and especially of the exergy, it is possible to identify the zones of the process in which the main energy changes occur. In this way, it is possible to take measures in order to obtain maximum yields in terms of conversion or separation processes and above all the energy yield involved in each stage (Emets et al. 2006; Ruiz-Mercado et al. 2012; Young and Cabezas 1999).

To determine the change of exergy in the streams of a system can be used, the mathematical model presented in (Zhang et al. 2012):

$$Ex = Ex^{ph} + Ex^{ch} + Ex^{ki} + Ex^{po} \tag{1}$$

where $Ex^{ph}, Ex^{ch}, Ex^{ki}\ and\ Ex^{po}$ denote the physical, chemical, kinetic, and potential exegetical flows of each stream, respectively. Usually, the kinetic ($Ex^{ki} = mV^2/2$) and potential ($E = mhZ$) exergy not present significant values for which Eq. 1 is:

$$Ex = Ex^{ph} + Ex^{ch} \quad (2)$$

Physical exergy is defined as:

$$Ex^{ph} = \sum_i n_i ex_i^{ph} \quad (3)$$

$$ex_i^{ph} = (h_j - h_o) - T_o(s_j - s_o) \quad (4)$$

The differences $(h_j - h_o)$ *and* $(s_j - s_o)$ can be calculated from Eqs. 5 and 6:

$$(h_j - h_o) = \int_{T_o}^{T_j} CpdT \quad (5)$$

$$(s_j - s_o) = \int_{T_o}^{T_j} \frac{Cp}{T} dT - RLn\left(\frac{P}{P_o}\right) \quad (6)$$

The chemical exergy is calculated to take into account the standard exergy of each component (ex_i^{ch})

$$Ex^{ch} = \sum_i n_i \left(ex_i^{ch} + RT_o Ln\left(\frac{n_i}{\sum n_i}\right)\right) \quad (7)$$

For the equipment which presents the need to supply power from an external source, it considered another source of exergy to analyze. This is calculated by Eq. 8.

$$Ex = \int_{T_1}^{T_2} \left(1 - \frac{T_o}{T}\right) \delta Q \quad (8)$$

7 Study Case: Biodiesel from Microalgae: *Chlorella Protothecoides* Case

7.1 Process Design

7.1.1 Culture of Microalgae

In this work, it was considered a feed flow of 100,000 kg/h as calculation basis. The microalgae culture was considered (for simulation purpose) using the conditions reported by Yoo et al. (2014). The substrate employed was glucose at a concentration of 40 g/L. Before to carry out the culture, the medium was autoclaved at 121 °C for 15 min. Then the inoculum was added.

Cultivation was performed in such a way that an average residence time of 15 days was generated. Then to carry out a concentration of the biomass obtained, the mixture from the culture step was passed by a centrifugation process (Luangpipat and Chisti 2016). At this point, a cell disruption stage is carried out, in which autoclaving was used at 125 °C with 1.5 MPa for 5 min. To carry out the removal of the lipids from the previous step, a mixture of chloroform–methanol (1:1 v/v) was added in a ratio of 1:1 (Lee et al. 2010).

For the production of biodiesel, a mixture of ethanol and sodium hydroxide was realized (Bambase et al. 2007). The resulting mixture is added to the lipids previously obtained. Heating was carried out to 60 °C and passed to the transesterification reactor. This reactor is designed based on kinetics reported by Mussatto et al. (2014). In order to recover the ethanol which not reacts, a distillation step was carried out. The bottom product is mixed with hot water in a 1:3 v/v ratio. The mixture was decanted allowing the separation of the unreacted oil and a mixture of biodiesel and glycerol. This mixture was brought to a separation train in which a purity of 95% biodiesel was achieved. The obtained glycerol was mixed with the catalyst, which was sulfuric acid, which allows the precipitation of the salt generated obtaining a glycerol with a purity of 80%.

7.2 Energy and Exergy Analysis

The energy and exergetic analyses are performed from the data obtained in the simulated Aspen Plus. For the energy analysis, a quantification of the energy required by each equipment involved in each processing step was performed. Similarly for the exergetic analysis, the exergy consumed in each stage is quantified and additionally the calculation of the exergy of the currents involved in each stage of processing was carried out in order to determine the energy efficiency of the process as mentioned above. To carry out the energy and exergetic analyses of biodiesel production from microalga oil, three stages of processing are considered: culture and harvesting of microalgae, oil extraction, and biodiesel production (see Fig. 1). From this, the following results can be obtained for the analyzed case.

7.3 Economic Analysis

Using the mass and energy balances obtained by simulating the process of obtaining biodiesel, an economic analysis was carried out. For this purpose, Aspen Process Economic Analyzer software was used. In this analysis, the method of depreciation of straight line is considered. This allows to determine the economic

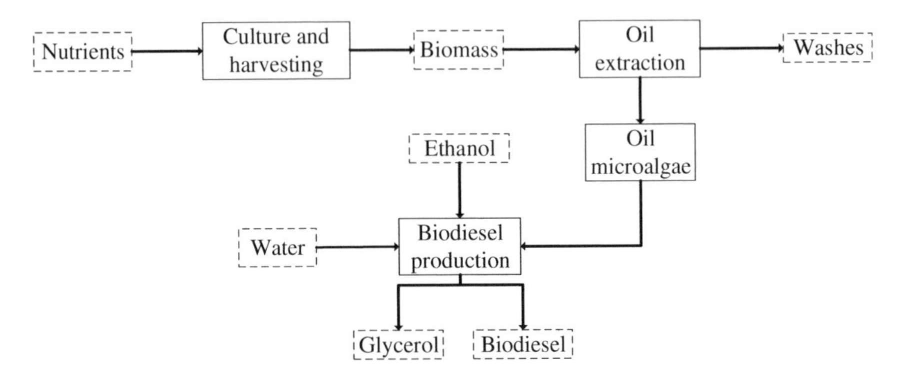

Fig. 1 Stages considered in the analysis of this work

viability of this type of processes. In Table 1 can be found economic parameters and costs of raw materials used in performing the analysis under Colombian context.

7.4 *Environmental Analysis*

By the determination of the environmental impact caused by the proposed process, the software developed by the Environmental Protection Agency (EPA): Waste Algorithm Reduction (WAR) was used. It allows to evaluate the environmental impact through eight categories: Human Toxicity Potential by Ingestion (HTPI), Human Toxicity Potential Exposure (HTPE), Terrestrial Toxicity Potential (TTP), Aquatic Toxicity Potential (ATP), Global Warming Potential (GWP), Ozone Depletion Potential (ODP), Smog Formation Potential (PCOP), and Acidification Potential (AP) (Young et al. 2000).

7.5 *Results and Discussion*

For the production of biodiesel from microalgae, there is an energy requirement of 1,003,967 MJ/h (taken into account a calculation base for feed flow of 100,000 kg/h). Here the percentage distribution of this energy is presented in Fig. 2. In this figure, it is possible to observe how the extraction of the oil of the microalgae has the highest energy consumption. One of the causes for this situation is due to the energy required for solvent evaporation once the oil extraction has been performed. On the other hand, there is the energetic requirement of the bioreactor where microalgae cultivation is carried out. Under the conditions analyzed, it is necessary that the culture and

Table 1 Investment parameters and prices used in the economic analysis

Item	Unit	Value	Ref.
Investment Parameters			
Tax rate	%	25	Dávila et al. (2014)
Interest rate	%	17	
Raw materials			
Potassium dihydrogen phosphate	USD/kg	161	Sigma-Aldrich (n.d.)
Dibasic potassium phosphate	USD/kg	199.5	
Magnesium sulfate heptahydrate	USD/kg	127.9	
Iron (II) sulfate heptahydrate	USD/kg	117.9	
Boric acid	USD/kg	92.4	
Calcium chloride dihydrate	USD/kg	149.5	
Manganese (II) chloride tetrahydrate	USD/kg	198.4	
Zinc sulfate heptahydrate	USD/kg	142.4	
Cupric sulfate pentahydrate	USD/kg	144	
Molybdenum trioxide	USD/kg	581.9	
Thiamine hydrochloride	USD/kg	426.5	
Glucose	USD/kg	165	
Glycine	USD/kg	94.3	
Sodium hydroxide	USD/kg	111.6	
Chloroform	USD/L	87.1	
Methanol	USD/L	65.4	
Ethanol	USD/L	0.55	ICIS (n.d.)
Utilities			
LP steam	USD/tonne	1.57	Moncada et al. (2015)
MP steam	USD/tonne	8.18	
HP steam	USD/tonne	9.86	
Potable water	USD/m^3	1.25	Dávila et al. (2014)
Fuel	USD/MMBTU	7.21	
Electricity	USD/kWh	0.1	
Operation			
Operator	USD/h	2.14	Dávila et al. (2014)
Supervisor	USD/h	4.29	

harvesting stage presents a consumption of 35.37% of the total energy required for the process. Finally, it is observed that the production of biodiesel presents the lowest energy consumption for this case. Thus, in energy terms, the production of microalgae oil presents a high energy consumption, due to the low yields obtained in the production of microalgae oil.

Similarly, as observed in the energy analysis, in the exergy analysis, an exergetic consumption of 19,601,822.14 MJ/h (taken into account a feed flow of

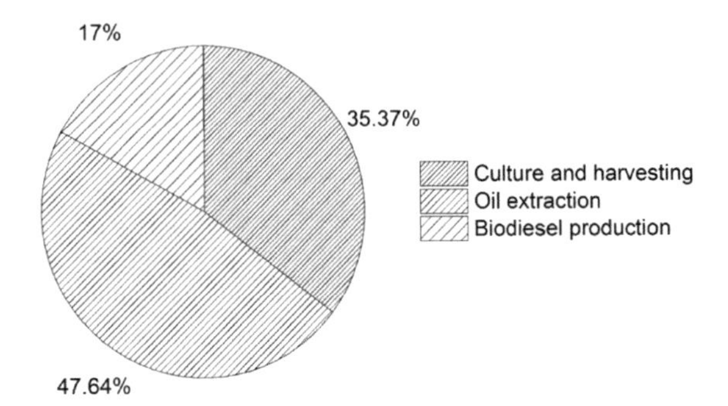

Fig. 2 Percentage distribution of energy consumption

100,000 kg/h) was evidenced, where the exergy present in the form of energy contained inside of reagents used throughout the process represents 98.88% of this percentage (see Fig. 3). On the other hand, the stage of the process that presents a greater exergetic consumption is the culture and harvesting of the microalgae. This factor is presented by the high duration of microalgae cultivation, which is why an exergetic consumption of 39.53% is present at this stage. The extraction of oil and the production of biodiesel present a percentage of 43.77 and 16.7%, respectively. Thus, the most inefficient stage of the process is the cultivation and harvesting of the microalgae. Under the process conditions analyzed, an exergetic efficiency of 37.08% was presented.

In production processes, different sources of expenses are presented, such as operating expenses, operating expenses, or operational expenses (OPEX) together with the costs of equipments (CAPEX). Similarly, another expense is the acquisition of the raw materials needed to obtain a final product such as biodiesel. Figure 4 shows the percentage distribution of expenses. In this figure, it can be

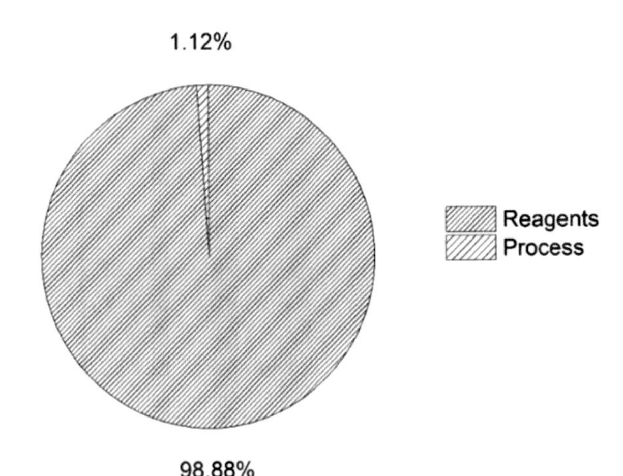

Fig. 3 Total exergy consumption by the production of biodiesel from microalgae

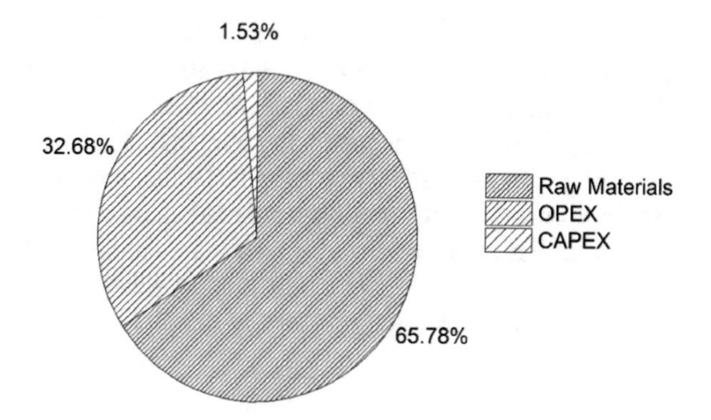

Fig. 4 Results of economic analysis by the biodiesel production from microalgae

observed that the greatest expense presented in this productive process is associated with raw materials. This is because it was necessary to use a high feed flow in the growing stage to obtain an amount of oil such that it could be used for the production of biodiesel. But even with this consideration, only the production of biodiesel does not provide a profit so that it can meet the expenses incurred throughout the process allowing a positive a profit margin. In this sense, other processes can be implemented which allow the production of other value-added products from the waste generated within the process. This alternative may lead to a decrease in OPEX, which is mainly represented by 76.14% of the cost of utilities required in the process. For this, it is necessary to carry out an energy integration of the different stages of the process, thus allowing a higher use in the energy and exergetic resources.

As for the environmental analysis, it is obtained that given the amount of waste generated in this process, a negative impact on the environment was presented. That is, the waste generated presents a higher index of contamination than the reagents used. The above can be seen in Fig. 5. Among the wastes generated are those

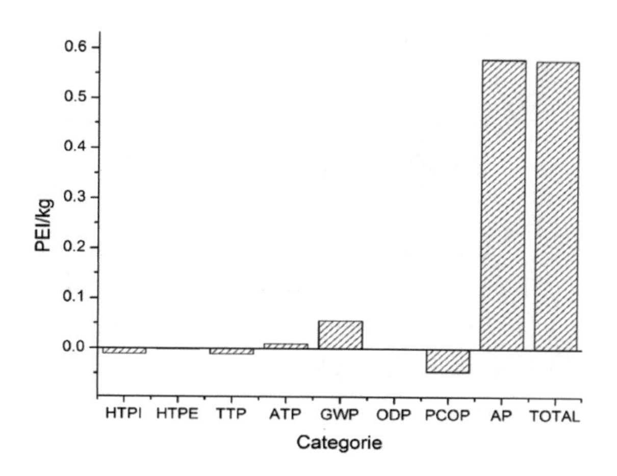

Fig. 5 Results of environmental analysis by the biodiesel production from microalgae

obtained from the extraction of the oil, which still have a potential for its use in obtaining other products through fermentation processes or chemical synthesis. On the other hand, these types of processes present a high acidification potential, which is mainly due to the high energy demand that is presented. For this, it is necessary to use a large amount of fuel to supply it. During the combustion process to obtain the required energy, there is a high release of carbon dioxide, which has a negative impact on the environment by altering the acidity of the air. As a consequence of this process can cause acid rain among other phenomena caused by the high concentration of this type of gases in the atmosphere.

8 Energy Production from Microalgae *Chlorella Protothecoides* Through the Biorefinery Concept: A New Proposal

Microalgae present a high potential to be used to obtain different value-added products, which can lead to a more sustainable production of bioenergy. For this, it is necessary to apply concepts such as biorefinery (Moncada et al. 2013; Mussatto et al. 2013). From this concept, it is possible to use the different residues that are presented in a stand-alone case of biodiesel production in other processes. Among the products with great potential to be obtained under this concept is energy from the biomass resulting from the process of lysis. In this same process, it is possible to obtain both glucose and lipids. The glucose obtained can be used in ethanol production (Quintero and Cardona 2009). Where it can be used in conjunction with lipids for the biodiesel production. Figure 6 presented this idea.

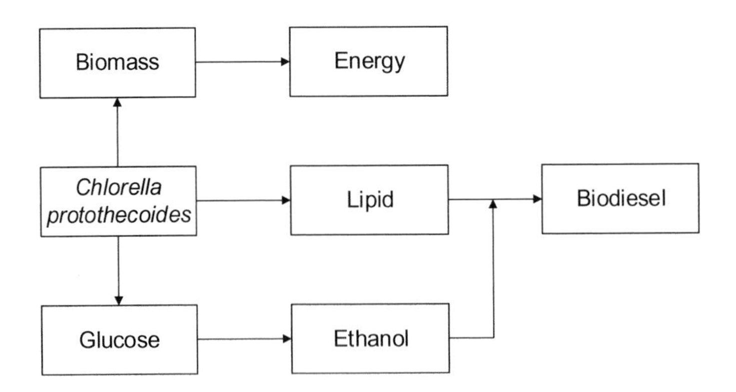

Fig. 6 Application of the *Chlorella protothecoides* under biorefinery concept

9 Conclusion

Given the high energy requirements presented by this type of processes, it is necessary to look others alternatives that allow the reduction of energy consumption in the culture and harvesting of microalgae and oil extraction stages. This can be achieved through the implementation of other technologies that allow carrying out the same process, but which require low energy consumption. In this way, a reduction in the energy requirements of the process and thus in the costs of profits can be achieved, which is reflected directly in the cost of production.

The application of energy and exergetic analysis was presented as a powerful tool for the determination of inefficiencies in this type of processes. Here the major irreversibilities are presented in the extraction of the oil due to the cellular lysis that must be made for the obtaining of the same, which makes this stage of the process present the greatest irreversibilities.

References

Abad, S., & Turon, X. (2012). Valorization of biodiesel derived glycerol as a carbon source to obtain added-value metabolites: Focus on polyunsaturated fatty acids. *Biotechnology Advances, 30*(3), 733–741.

Adam, F., Abert-Vian, M., Peltier, G., & Chemat, F. (2012). "Solvent-free" ultrasound-assisted extraction of lipids from fresh microalgae cells: A green, clean and scalable process. *Bioresource Technology, 114,* 457–465.

Bambase, M. E., Nakamura, N., Tanaka, J., & Matsumura, M. (2007). Kinetics of hydroxide-catalyzed methanolysis of crude sunflower oil for the production of fuel-grade methyl esters. *Journal of Chemical Technology and Biotechnology, 82*(3), 273–280.

Barros, A. I., Gonçalves, A. L., Simões, M., & Pires, J. C. M. (2015). Harvesting techniques applied to microalgae: A review. *Renewable and Sustainable Energy Reviews, 41,* 1489–1500.

Becker, E. (1994). Microalgae—biotechnology and microbiology. *Journal of Experimental Marine Biology and Ecology, 183,* 300–301.

Bilad, M. R., Vandamme, D., Foubert, I., Muylaert, K., & Vankelecom, I. F. J. (2012). Harvesting microalgal biomass using submerged microfiltration membranes. *Bioresource Technology, 111,* 343–352.

Bosma, R., Van Spronsen, W. A., Tramper, J., & Wijffels, R. H. (2003). Ultrasound, a new separation technique to harvest microalgae. *Journal of Applied Phycology, 15*(2–3), 143–153.

Bumbak, F., Cook, S., Zachleder, V., Hauser, S., & Kovar, K. (2011). Best practices in heterotrophic high-cell-density microalgal processes: Achievements, potential and possible limitations. *Applied Microbiology and Biotechnology.* https://doi.org/10.1007/s00253-011-3311-6.

Cerón-Salazar, I., & Cardona-Alzate, C. (2011). Integral evaluation process for obtaining pectin and essential oil from orange peel. *Inginería y Ciencia, 7*(13), 1794–9165.

Chen, C. Y., Yeh, K. L., Aisyah, R., Lee, D. J., & Chang, J. S. (2011). Cultivation, photobioreactor design and harvesting of microalgae for biodiesel production: A critical review. *Bioresource Technology, 102*(1), 71–81.

Chen, F. (1996). High cell density culture of microalgae in heterotrophic growth. *Trends in Biotechnology, 14*(11), 421–426.

Chisti, Y. (2007). Biodiesel from microalgae. *Biotechnology Advances, 25*(3), 294–306.

Converti, A., Casazza, A. A., Ortiz, E. Y., Perego, P., & Del Borghi, M. (2009). Effect of temperature and nitrogen concentration on the growth and lipid content of Nannochloropsis oculata and Chlorella vulgaris for biodiesel production. *Chemical Engineering and Processing: Process Intensification, 48*(6), 1146–1151.

Dai, Y. M., Chen, K. T., & Chen, C. C. (2014). Study of the microwave lipid extraction from microalgae for biodiesel production. *Chemical Engineering Journal, 250,* 267–273.

Dassey, A.J., Theegala, C.S. (2012). Cost Analysis of Microalgal Harvesting for Biofuel Production. *2012 Dallas, Texas, July 29 - August 1, 2012.* St. Joseph, Mich.: ASABE. https://doi.org/10.13031/2013.41701.

Dávila, J. A., Hernández, V., Castro, E., & Cardona, C. A. (2014). Economic and environmental assessment of syrup production. *Colombian case. Bioresource Technology, 161,* 84–90.

Demirbas, A. (2005). Biodiesel production from vegetable oils via catalytic and non-catalytic supercritical methanol transesterification methods. *Progress in Energy and Combustion Science.*

Drexler, I. L. C., & Yeh, D. H. (2014). Membrane applications for microalgae cultivation and harvesting: A review. *Reviews in Environmental Science & Biotechnology, 13*(4), 487–504.

Edzwald, J. K. (1993). Algae, bubbles, coagulants, and dissolved air flotation. *Water Science and Technology, 27,* 67–81.

Emets, S. V., Hoo, K. A., & Mann, U. (2006). A modified hierarchy for designing chemical processes. *Industrial and Engineering Chemistry Research, 45,* 5037–5043.

Fradique, M., Batista, A. P., Nunes, M. C., Gouveia, L., Bandarra, N. M., & Raymundo, A. (2013). Isochrysis galbana and Diacronema vlkianum biomass incorporation in pasta products as PUFA's source. *LWT - Food Science and Technology, 50*(1), 312–319.

Ghasemi Naghdi, F., González González, L. M., Chan, W., & Schenk, P. M. (2016). Progress on lipid extraction from wet algal biomass for biodiesel production. *Microbial Biotechnology, 9* (6), 718–726.

Gouveia, L., & Oliveira, A. C. (2009). Microalgae as a raw material for biofuels production. *Journal of Industrial Microbiology and Biotechnology, 36*(2), 269–274.

Guil-Guerrero, J.L., Navarro-Juárez, R., López-Martínez, J.C., Campra-Madrid, P., Rebolloso-Fuentes, M.M. (2004). Functional properties of the biomass of three microalgal species. *Journal of Food Engineering, 65*(4), 511–517.

Halim, R., Danquah, M. K., & Webley, P. A. (2012). Extraction of oil from microalgae for biodiesel production: A review. *Biotechnology Advances, 30*(3), 709–732.

Hanotu, J., Bandulasena, H. C. H., & Zimmerman, W. B. (2012). Microflotation performance for algal separation. *Biotechnology and Bioengineering, 109*(7), 1663–1673.

ICIS. (n.d.). Retrieved July 13, 2017, from http://www.icis.com/chemicals/channel-info-chemicals-a-z/.

Joannes, C., Sipaut, C.S., Dayou, J., Yasir, S.M., Mansa, R.F. (2015). Review Paper on Cell Membrane Electroporation of Microalgae using Electric Field Treatment Method for Microalgae Lipid Extraction. *IOP Conference Series: Materials Science and Engineering, 78,* 12034. https://doi.org/10.1088/1757-899X/78/1/012034.

Kang, S., Kim, K. H., & Kim, Y. C. (2015). A novel electroporation system for efficient molecular delivery into Chlamydomonas reinhardtii with a 3-dimensional microelectrode. *Scientific Reports, 5,* 15835.

Kim, Y. H., Choi, Y. K., Park, J., Lee, S., Yang, Y. H., Kim, H. J., et al. (2012). Ionic liquid-mediated extraction of lipids from algal biomass. *Bioresource Technology, 109,* 312–315.

Lee, C. S., Robinson, J., & Chong, M. F. (2014). A review on application of flocculants in wastewater treatment. *Process Safety and Environmental Protection, 92,* 489–508.

Lee, J. Y., Yoo, C., Jun, S. Y., Ahn, C. Y., & Oh, H. M. (2010). Comparison of several methods for effective lipid extraction from microalgae. *Bioresource Technology, 101*(1), S75–S77.

Leesing, R., & Kookkhunthod, S. (2011). Heterotrophic Growth of Chlorella sp. KKU-S2 for Lipid Production using Molasses as a Carbon Substrate, *9,* 87–91.

Liang, Y., Sarkany, N., & Cui, Y. (2009). Biomass and lipid productivities of Chlorella vulgaris under autotrophic, heterotrophic and mixotrophic growth conditions. *Biotechnology Letters, 31* (7), 1043–1049.

Liu, J., Zhu, Y., Tao, Y., Zhang, Y., Li, A., Li, T., et al. (2013). Freshwater microalgae harvested via flocculation induced by pH decrease. *Biotechnology for Biofuels, 6*(1), 98.

Luangpipat, T., & Chisti, Y. (2016). Biomass and oil production by Chlorella vulgaris and four other microalgae - Effects of salinity and other factors. *Journal of Biotechnology, 257,* 47–57.

Ma, F., & Hanna, M. A. (1999). Biodiesel production: a review. *Bioresource Technology, 70*(1), 1–15.

Mata, T. M., Martins, A. A., & Caetano, N. S. (2010). Microalgae for biodiesel production and other applications: A review. *Renewable and Sustainable Energy Reviews, 14*(1), 217–232.

Medina, A. R., Grima, E. M., Gimenez, A. G., & Gonzalez, M. J. I. (1998). Downstream processing of algal polyunsaturated fatty acids. *Biotechnology Advances, 16*(3), 517–580.

Milledge, J. J., & Heaven, S. (2013). A review of the harvesting of micro-algae for biofuel production. *Reviews in Environmental Science and Biotechnology, 12*(2), 165–178.

Mohn, F.H. (1980). Experiences and strategies in the recovery of biomass from mass cultures of microalgae. *Algal Biomass,* 471–547.

Molina Grima, E., Belarbi, E. H., Acién Fernández, F. G., Robles Medina, A., & Chisti, Y. (2003). Recovery of microalgal biomass and metabolites: Process options and economics. *Biotechnology Advances, 20*(7–8), 491–515.

Moncada, J., Hernández, V., Chacón, Y., Betancourt, R., & Cardona, C. A. (2015). Citrus Based Biorefineries. In D. Simmons (Ed.), *Citrus Fruits. Production, Consumption and Health Benefits* (pp. 1–26). Nova Publishers.

Moncada, J., Matallana, L. G., & Cardona, C. A. (2013). Selection of Process Pathways for Biorefinery Design Using Optimization Tools: A Colombian Case for Conversion of Sugarcane Bagasse to Ethanol, Poly-3-hydroxybutyrate (PHB), and Energy. *Industrial and Engineering Chemistry Research, 52*(11), 4132–4145.

Mubarak, M., Shaija, A., & Suchithra, T. V. (2015). A review on the extraction of lipid from microalgae for biodiesel production. *Algal Research, 7,* 117–123.

Mussatto, S. I., Moncada, J., Roberto, I. C., & Cardona, C. A. (2013). Techno-economic analysis for brewer's spent grains use on a biorefinery concept: The Brazilian case. *Bioresource Technology, 148,* 302–310.

Mussatto, S. I., Santos, J. C., Filho, W. C. R., & Silva, S. S. (2014). Purification of xylitol from fermented hemicellulosic hydrolyzate using liquid–liquid extraction and precipitation techniques. *Applied Energy, 123,* 108–120.

Perez-Garcia, O., Escalante, F.M.E., de-Bashan, L.E., & Bashan, Y. (2011). Heterotrophic cultures of microalgae: Metabolism and potential products. *Water Research,* 45, 11–36.

Pragya, N., Pandey, K. K., & Sahoo, P. K. (2013). A review on harvesting, oil extraction and biofuels production technologies from microalgae. *Renewable and Sustainable Energy Reviews, 24,* 159–171.

Prommuak, C., Pavasant, P., Quitain, A. T., Goto, M., & Shotipruk, A. (2012). Microalgal lipid extraction and evaluation of single-step biodiesel production. *Engineering Journal, 16*(5), 157–166.

Quintero, J. A., & Cardona, C. A. (2009). Ethanol dehydration by adsorption with starchy and cellulosic materials. *Industrial and Engineering Chemistry Research, 48*(14), 6783–6788.

Quintero, J. A., Felix, E. R., Rincón, L. E., Crisspín, M., Fernandez Baca, J., Khwaja, Y., et al. (2012). Social and techno-economical analysis of biodiesel production in Peru. *Energy Policy, 43,* 427–435.

Ranjith Kumar, R., Hanumantha Rao, P., & Arumugam, M. (2015). Lipid extraction methods from microalgae: A comprehensive review. *Frontiers in Energy Research, 2,* 1–9.

Rawat, I., Ranjith Kumar, R., Mutanda, T., & Bux, F. (2011). Dual role of microalgae: Phycoremediation of domestic wastewater and biomass production for sustainable biofuels production. *Applied Energy, 88*(10), 3411–3424.

Richmond, A. (2004). Handbook of microalgal culture: biotechnology and applied phycology/ edited by Amos Richmond. *Orton. Catie. Ac. Cr,* 472. https://doi.org/10.1002/9780470995280.

Ruiz-Mercado, G. J., Smith, R. L., & Gonzalez, M. A. (2012). Sustainability indicator for chemical processes: I. taxonomy. *Industrial & Engineering Chemistry Research, 51,* 2309–2328.

Safi, C., Zebib, B., Merah, O., Pontalier, P. Y., & Vaca-Garcia, C. (2014). Morphology, composition, production, processing and applications of Chlorella vulgaris: A review. *Renewable and Sustainable Energy Reviews, 35,* 265–278.

Sforza, E., Bertucco, A., Morosinotto, T., & Giacometti, G. M. (2012). Photobioreactors for microalgal growth and oil production with Nannochloropsis salina: From lab-scale experiments to large-scale design. *Chemical Engineering Research and Design, 90*(9), 1151–1158.

Shahid, E. M., & Jamal, Y. (2011). Production of biodiesel: A technical review. *Renewable and Sustainable Energy Reviews, 15*(9), 4732–4745.

Sharma, K. K., Garg, S., Li, Y., Malekizadeh, A., & Schenk, P. M. (2013). Critical analysis of current microalgae dewatering techniques. *Biofuels, 4,* 397–407.

Show, K.Y., Lee, D.J. (2014). Chapter 5 - Algal Biomass Harvesting BT - Biofuels from Algae (pp. 85–110). Amsterdam: Elsevier. https://doi.org/10.1016/B978-0-444-59558-4.00005-X.

Sigma-Aldrich. (n.d.). Retrieved July 13, 2017, from https://www.sigmaaldrich.com/us-export.html.

Slegers, P. M., Lösing, M. B., Wijffels, R. H., van Straten, G., & van Boxtel, A. J. B. (2013). Scenario evaluation of open pond microalgae production. *Algal Research, 2*(4), 358–368.

Spolaore, P., Joannis-Cassan, C., Duran, E., & Isambert, A. (2006). Commercial applications of microalgae. *Journal of Bioscience and Bioengineering, 101*(2), 87–96.

Taher, H., Al-Zuhair, S., Al-Marzouqi, A. H., Haik, Y., & Farid, M. (2014). Effective extraction of microalgae lipids from wet biomass for biodiesel production. *Biomass and Bioenergy, 66,* 159–167.

Toepfl, S., Mathys, A., Heinz, V., & Knorr, D. (2006). Review: Potential of High Hydrostatic Pressure and Pulsed Electric Fields for Energy Efficient and Environmentally Friendly Food Processing. *Food Reviews International, 22*(4), 405–423.

Trainor, F. R. (2009). Perspective: Breaking the habit. Integrating plasticity into taxonomy. *Systematics and Biodiversity, 7*(2), 95–100.

Uduman, N., Qi, Y., Danquah, M. K., Forde, G. M., & Hoadley, A. (2010). Dewatering of microalgal cultures: A major bottleneck to algae-based fuels. *Journal of Renewable and Sustainable Energy, 2,* 1–15.

Venkataraman, L. V. (1997). Spirulina platensis (Arthrospira): Physiology, Cell Biology and Biotechnologym, edited by Avigad Vonshak. *Journal of Applied Phycology, 9*(3), 295–296.

Waldenstedt, L., Inborr, J., Hansson, I., & Elwinger, K. (2003). Effects of astaxanthin-rich algal meal (Haematococcus pluvalis) on growth performance, caecal campylobacter and clostridial counts and tissue astaxanthin concentration of broiler chickens. *Animal Feed Science and Technology, 108*(1), 119–132.

Wilson, M. (2012). *Cross flow filtration for mixed-culture algae harvesting for municipal wastewater lagoons*. Master of Science Thesis, Utah State University.

Yang, C., Hua, Q., & Shimizu, K. (2000). Energetics and carbon metabolism during growth of microalgal cells under photoautotrophic, mixotrophic and cyclic light-autotrophic/dark-heterotrophic conditions. *Biochemical Engineering Journal, 6*(2), 87–102.

Yoo, S. J., Kim, J. H., & Lee, J. M. (2014). Dynamic modelling of mixotrophic microalgal photobioreactor systems with time-varying yield coefficient for the lipid consumption. *Bioresource Technology, 162,* 228–235.

Young, D. M., & Cabezas, H. (1999). Designing sustaibanle processes with simulation: The waste reduction (WAR) algorithm. *Engineering, Computers and Chemical, 23,* 1477–1491.

Young, D., Scharp, R., & Cabezas, H. (2000). The waste reduction (WAR) algorithm: Environmental impacts, energy consumption, and engineering economics. *Waste Management, 20*(8), 605–615.

Yu, Z., Chen, X., & Xia, S. (2016). The mechanism of lipids extraction from wet microalgae Scenedesmus sp. by ionic liquid assisted subcritical water. *Journal of Ocean University of China, 15*(3), 549–552.

Zhang, Y., Li, B., Li, H., & Zhang, B. (2012). Exergy analysis of biomass utilization via steam gasification and partial oxidation. *Thermochimica Acta, 538,* 21–28.

Chapter 10
Biofuels from Microalgae: Biohydrogen

Harshita Singh and Debabrata Das

Abstract Rapid industrialization and urbanization are mainly responsible for the energy crisis, environmental pollution and climate change. In addition, depletion of the fossil fuels is a major concern now. To confront these problems, it is essential to produce energy from sustainable and renewable energy sources. Hydrogen is widely considered as a clean and efficient energy carrier for the future because it does not produce carbon-based emission and has the highest energy density among any other known fuels. Due to the environmental and socioeconomic limitation associated with conventional processes for the hydrogen production, new approaches of producing hydrogen from biological sources have been greatly encouraged. From the perspective of sustainability, microalgae offer a promising source and have several advantages for the biohydrogen production. Microalgae are characterized as high rate of cell growth with superior photosynthetic efficiency and can be grown in brackish or wastewater on non-arable land. In recent years, biohydrogen production from microalgae via photolysis or being used as substrate in dark fermentation is gaining considerable interest. The present chapter describes the different methods involved in hydrogen production from microalgae. Suitability of the microalgae as a feedstock for the dark fermentation is discussed. This review also includes the challenges faced in hydrogen production from microalgae as well as the genetic and metabolic engineering approaches for the enhancement of biohydrogen production.

Keywords Microalgae · Biohydrogen · Photolysis · Dark fermentation
Genetic engineering

H. Singh · D. Das (✉)
Advance Technology Development Center, Indian Institute
of Technology, Kharagpur 721302, India
e-mail: ddas.iitkgp@gmail.com

D. Das
Department of Biotechnology, Indian Institute of Technology,
Kharagpur 721302, India

E. Jacob-Lopes et al. (eds.), *Energy from Microalgae*, Green Energy
and Technology, https://doi.org/10.1007/978-3-319-69093-3_10

1 Introduction

The exponential increase in the world population and rapid industrialization has resulted in continuous rise of the global energy demand. This has led to the depletion of fossil energy reserves, climate change and environmental pollution. To address these problems, we are compelled to find sustainable, renewable and carbon-neutral energy sources. In this regard, hydrogen is considered as a clean and efficient energy carrier. It has the highest energy density (142 kJ/g) among any other known fuels (Kumar et al. 2013), and on combustion it only produces water vapour. In addition, hydrogen can easily get converted into electricity in a fuel cell without any pollution (Batista et al. 2014) and may be used directly as a transportation fuel in an internal combustion engine (Das et al. 2014).

Different conversion technologies can be used for hydrogen production, but till today, it is produced through conventional technologies which include reforming processes, gasification and water splitting. Among the conventional processes, steam reforming of methane is widely used thermo-chemical technology and contributes 48% of the global hydrogen demand. About 30% of world hydrogen is produced by the reforming of oil/naphtha and 18% from the coal gasification (Das et al. 2014). Water electrolysis is another efficient method which produces hydrogen of very high purity and accounts for 3.9% of hydrogen production (Das et al. 2014), but this technology is challenged by the high cost of electricity. Biomass (crop residues, animal wastes, waste paper, municipal solid wastes, etc) gasification is also employed for hydrogen production, but it has the drawback of low thermal efficiency (Holladay et al. 2009). To overcome the various socioeconomic and environmental limitations associated with the currently existing industrial processes of hydrogen production, research focusing the biological hydrogen production technology has received substantial importance. This technology is not only environmentally benign but also requires less energy input as it can be carried out under ambient operating conditions (Das and Veziroglu 2001). Production of hydrogen through biological pathways is primarily controlled by the domain of bacteria and algae. In recent days, microalgae are considered valuable and tremendously potential source for the sustainable generation of biohydrogen. Interest in microalgae for hydrogen production has been ensued due to the fact that they can carry out the production of hydrogen through the process of photosynthesis utilizing most abundant natural resources, sunlight and water. Evidence of microalgal hydrogen production through biophotolysis of water was firstly put on record by Gaffron and Rubin in 1942. They studied the hydrogen metabolism in a unicellular green microalga, *Scenedesmus obliquus,* and reported the hydrogen production by this microorganism in the presence of light energy under anaerobic condition after an adaptive dark phase (Gaffron and Rubin 1942). Although photobiological hydrogen production by algae has been studied for several years, in recent decades considerable advances in this field have been made (Torzillo et al. 2015; Marquez-Reyes et al. 2015). Apart from this, over the past few years, dark fermentation utilizing microalgal substrate for biohydrogen production has also gained attention (Roy et al. 2014; Batista et al. 2014; Ortigueira et al. 2015). In comparison with other biomass, microalgae present several advantages to be used as feedstock for biohydrogen production: (1) they have higher growth rate with superior photosynthetic efficiency; (2) they can grow on non-arable land; (3) they can grow in wide variety of water sources (fresh, salt, brackish and wastewater); (4) they do not contain lignin, so no rigorous pretreatment is required (Sambusiti et al. 2015).

This chapter is aimed to describe about biohydrogen production from microalgae both by biophotolysis as well as dark fermentation. The barriers in biohydrogen production from microalgae and the molecular approaches to enhance the hydrogen production are taken into consideration.

2 Microalgal Hydrogen Production Processes

Microalgae can participate for hydrogen production mainly by two processes: (1) photolysis of water, which requires light and is closely related to the process of photosynthesis; (2) light-independent process in which microalgal biomass, rich in carbohydrate and protein, is used up as a feedstock for dark fermentation.

2.1 Biophotolysis

Biophotolysis is the action of light energy on the biological systems that results in the dissociation of substrate, usually water molecule, into hydrogen and oxygen. Unicellular green algae and cyanobacteria are organisms known to perform both oxygenic photosynthesis and biohydrogen production (Happe et al. 2000). In microalgae, the process of photolysis is closely related to the process of photosynthesis. Unlike photosynthesis, where the reductants released by the dissociation of water are consumed in the Calvin cycle or pentose phosphate pathway to reduce CO_2 for cell growth, in biophotolysis the reductants are directed for hydrogen evolution. The photosynthetic machinery of eukaryotic green algae and prokaryotic blue-green algae is similar to higher terrestrial plants. In eukaryotic microalgae, the photosynthetic machinery is embedded in the thylakoid membranes present inside an intracellular organelle, the chloroplast. In contrast, the photosynthetic apparatus of cyanobacteria lacks compartmentalization and the thylakoid membranes are present in the cytoplasm, adjacent to the plasma membrane. The thylakoid membranes contain several light-absorbing pigments such as chlorophyll a, antenna chlorophylls, carotenoid and phycobiliproteins which are arranged in two different kinds of functional arrays called photosystems (PSI and PSII). Photosystem I and photosystem II consist of distinct photochemical reaction centre, P700 and P680 respectively. Absorption of photons by the chlorophyll molecules of P700 and P680 causes their excitations and drives the electrons through thylakoid membrane to reduce ferredoxin (Fd). Under aerobic and light condition ferredoxin: $NADP^+$ oxidoreductase transfers the electrons from reduced ferredoxin to $NADP^+$ which generates NADPH. This reducing power (NADPH) is used to fix carbon for cell growth and for carbohydrates and/or lipids production. However, under some conditions, the reduced ferredoxin generated by the water splitting can be directed to reduce hydrogenase or nitrogenase enzymes for the hydrogen production. There are two types of biophotolysis for H_2 production from microalgae: direct and indirect biophotolysis.

2.1.1 Direct Biophotolysis

In this biological process, the reductive equivalents required for the hydrogen production are generated directly by the photolysis of water. Photosynthetic machinery of green algae utilizes the solar energy to split the H_2O molecule into O_2 and H_2. The electrons generated by the oxidation of H_2O molecule flows to the ferredoxin which under the optimal conditions donates the electrons directly to hydrogenase enzyme for H_2 production. In direct biophotolysis, production of H_2 at the reducing side of the PSI is associated with the simultaneous evolution of O_2 at the oxidizing side of the PSII (Melis et al. 2000). Presence of O_2 limits the hydrogen production as the hydrogenase gets deactivated at O_2 partial pressure of $<2\%$ (Ghirardi et al. 1997). Thus, in this process H_2 evolution occurs for transient period upon illumination, before the hydrogenase gets inactivated by the accumulated O_2 (Eroglu and Melis 2011). Several green microalgae such as *Chlamydomonas reinhardtii*, *Chlorella fusca*, *S. obliquus*, *Chlorococcum littorale* and *Platymonas subcordiformis* possess genomically encoded [FeFe]-hydrogenase enzyme for hydrogen generation (Eroglu and Melis 2011). Among the green algae, *C. reinhardtii* has been mostly investigated by many researchers for biohydrogen production (Melis et al. 2000; Torzillo and Seibert 2013; Tsygankov et al. 2006). In order to prevent the inactivation of H_2-evolving enzyme by O_2 and for the sustained production of H_2, different methods have been investigated such as purging the reaction mixture with inert gases (Greenbaum 1982), addition of oxygen scavenger (Randt and Senger 1985) and depletion of sulphur in the cultivation media (Melis et al. 2000).

The method of sulphur deprivation is a two-stage approach for the sustained hydrogen production by green microalgae. First stage is the growth phase in which generation of biomass takes place under suitable conditions. Second stage is the non-growth phase in which the carbohydrate-rich algal biomass is transferred to sulphur-deprived cultivation medium for H_2 production (Melis et al. 2000). Sulphur is essential for the biosynthesis of PSII protein, which is made up of sulphur-containing amino acids (cysteine or methionine). Due to the partial suppression of PSII activity, evolution of O_2 reduces and the mitochondrial respiration further helps in the depletion of oxygen, developing the essential anaerobiosis inside the cells. Anaerobic condition induces the [FeFe]-hydrogenase activity (Forestier et al. 2003) which produces H_2 by utilizing 60–90% electrons contributed

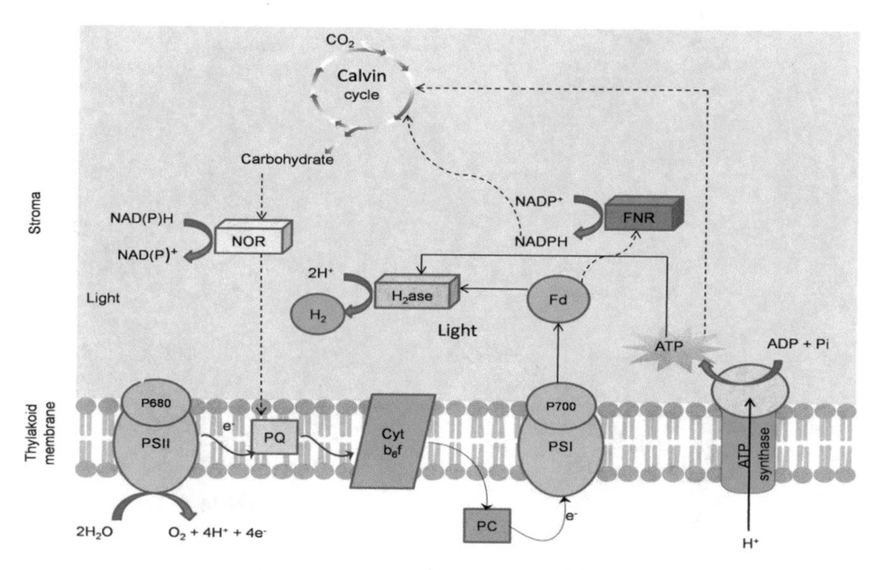

Fig. 1 Biohydrogen production via direct and indirect biophotolysis carried out by green microalgae

by the water splitting and remaining 20–30% comes from the catabolism of carbohydrate through fermentation pathway. In this way, sulphur deprivation mechanism employs both direct as well as indirect biophotolysis for H_2 generation (Fig. 1). Re-addition of sulphur in limiting amounts during the H_2 production phase helps in regenerating the depleted algal cells for another round of H_2 generation without re-establishment of aerobic condition (Kosourov et al. 2005). However, the cycling of algal suspension cultures between the sulphur deplete and sulphur replete conditions is challenging and might become simpler by using the immobilized, sulphur-deprived algal cells for sustained H_2 evolution (Laurinavichene et al. 2006).

Direct photolysis is an interesting process due to the fact that this process utilizes the most abundant natural resources, solar energy and water for the production of efficient fuel "hydrogen". However, this process suffers from the limitation of low yield and hydrogen production rate. The energy productivity via this process ranges from 0.02 to 0.12 kJ/L/h (Yu and Takahashi 2007).

2.1.2 Indirect Biophotolysis

In this process, the reductive equivalents or electrons are directly derived by the endogenously stored carbohydrates such as starch in green algae and glycogen in cyanobacteria (Fig. 2). In this method firstly, during the photosynthesis, CO_2 fixation and accumulation of carbohydrate take place. Secondly, fermentation of the carbon reserves occurs which leads to hydrogen production by the following reaction:

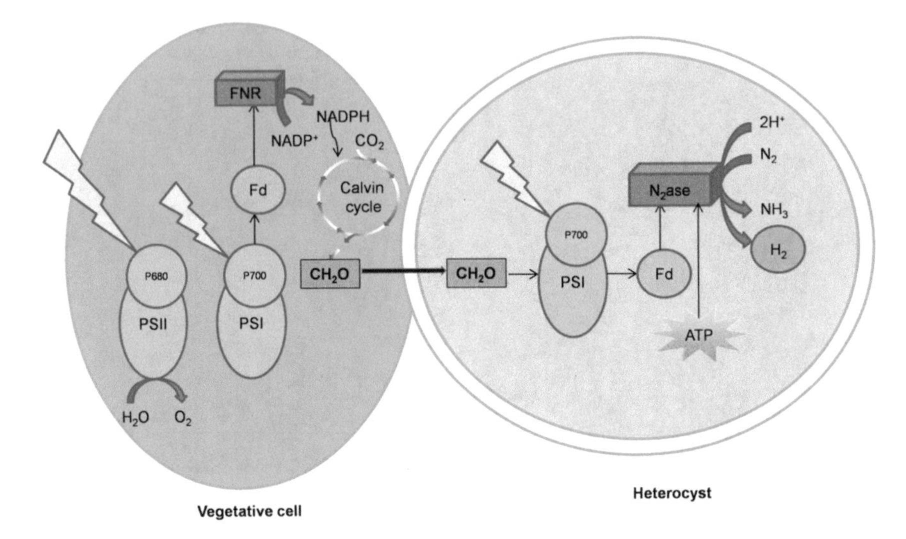

Fig. 2 Biohydrogen production via indirect biophotolysis carried out by cyanobacteria

$$12H_2O + 6CO_2 \rightarrow C_6H_{12}O_6 + 6O_2 \tag{1}$$

$$C_6H_{12}O_6 + 12H_2O \rightarrow 12H_2 + 6CO_2 \tag{2}$$

In comparison with green microalgae, H_2 production via indirect photolysis from cyanobacteria is more attractive (Yu and Takahashi 2007). In this process, the problem of H_2-producing enzyme sensitivity to O_2 is solved by the temporal or spatial separation of H_2 and O_2 evolving reactions. In spatial separation, the apparatus for photosynthesis and H_2 production is present at different locations. Temporal separation involves the reactions of O_2 and H_2 evolution to occur at different time by using light/dark cycles. In this process, during the daytime carbohydrate accumulation takes place via photosynthesis and during the night-time H_2 production occurs via fermentation of stored sugar (Miura et al. 1997).

Cyanobacteria are capable of carrying out the both, CO_2 as well as nitrogen fixation. In these organisms, nitrogen fixation occurs under anoxic conditions inside the specialized cells known as heterocysts whereas oxygenic photosynthesis and CO_2 fixation take place in the vegetative cells. Inside the heterocysts, anaerobic environment is maintained due to the absence of the O_2-evolving PSII. In addition, the O_2 impermeable cell walls of heterocysts do not allow the oxygen diffusion from the nearby vegetative cells thus helping further in creating anaerobiosis, required for the nitrogen fixation and H_2 generation by the O_2-sensitive nitrogenases (Das et al. 2014).

Nitrogen-fixing cyanobacteria known for H_2 production mostly include the genus *Anabaena, Nostoc, Calothrix* and *Oscillatoria.* In these organisms, nitrogen fixation and hydrogen evolution catalysed by the nitrogenase enzyme are described according to Eq. 3 (Eroglu and Melis 2011):

$$N_2 + 8e^- + 8H^+ \rightarrow 2NH_3 + H_2 + 16ADP + 16Pi \quad (3)$$

In the absence of N_2, nitrogenase acts as an ATP-powered hydrogenase and produces H_2 exclusively without any feedback inhibition by the following reaction (Eq. 4) (McKinlay and Harwood 2010):

$$8e^- + 8H^+ \rightarrow + 16ATP \rightarrow 2H_2 + 16ADP + 16Pi \quad (4)$$

The process of heterocyst-based H_2 production by cyanobacteria suffers the problem of low photon conversion efficiency (Benemann 2000). This is due to the high metabolic energy requirement of the nitrogenase catalysis (2 ATP per e^- transfer). Moreover, half of the energy metabolism of the cyanobacteria also accounts for the biosynthesis and maintenance of heterocysts (Benemann 2000). This energy requirement is met by the solar energy via PSI-mediated cyclic phosphorylation occurring inside the thylakoid membrane of the heterocysts. During the N_2 fixation, nitrogenase is usually accompanied by uptake hydrogenase to reutilize and retrieve the H_2/e^- for minimizing the loss of energy (Tamagnini et al. 2002). Disruption of this uptake hydrogenase activity in cyanobacteria helps in H_2 accumulation (Masukawa et al. 2002). Non-nitrogen-fixing cyanobacteria such as *Gloeobacter, Synechococcus and Synechocystis* can also generate H_2 via indirect biophotolysis. These organisms possess bidirectional [NiFe] hydrogenase which has the capability of catalysing both the synthesis and oxidation of H_2 (Tamagnini et al. 2002). The physiological function of this enzyme is still unclear but the suggested role includes: removal of excess reducing power during anaerobic fermentation and allocation of electrons to the respiratory chain by H_2 oxidation (Baebprasert et al. 2010). H_2 production by this enzyme is energetically efficient as it does not require ATP like nitrogenases. However, the rate of H_2 production by non-nitrogen-fixing cyanobacteria is comparatively lower (0.02–0.40 μmol H_2/mg chl a/h) than heterocystous cyanobacteria (0.17–0.42 μmol H_2/mg chl a/h) (Levin et al. 2004). Biohydrogen production from microalgae via photolysis is summarized in Table 1.

2.1.3 Enzymes Involved in PhotoBiological Hydrogen Production from Microalgae

The key enzymes that can catalyse the reaction of hydrogen production in microalgae are hydrogenase and nitrogenase.

Hydrogenases

Hydrogenases are the metalloenzymes and on the basis of the metal composition at their catalytic sites, they can be classified as: [FeFe]-hydrogenase, [NiFe]-hydrogenase and [Fe]-hydrogenase (Kim and Kim 2011). Among these, [Fe]-hydrogenase is found in archaea so this will not be discussed here.

Table 1 Biohydrogen production from microalgae via photolysis

Microalgae	Process condition	Hydrogen production	Reference
Anabaena variabilis PK84	Indoor helical tubular photobioreactor, Allen and Arnon medium, air +2% CO_2, 12 h light and 12 h dark cycles, 332 μE/m^2/s	19.2 mL/h/PhBR	Borodin et al. (2000)
Anabaena variabilis PK84	Outdoor tubular photobioreactor, Allen and Arnon medium, air +2% CO_2, sunlight	45.8 mL/h/PhBR	Tsygankov et al. (2002)
Chlamydomonas reinhardtii cc-124	Flat glass photobioreactor, re-addition of sulphur (25 μM) in TAP-S medium, argon sparged, 300 μE/m^2/s	5.94 μmol mg/chl/h	Kosourov et al. (2002)
Gloeocapsa alpicola	Glass bottles, cells suspended in Tris-HCl buffer, argon sparged, 24 h dark	25 μL/h/mg d. w	Troshina et al. (2002)
Anabaena PCC 7120	Indoor tubular photobioreactor, BG 11_0 medium, argon sparged, 456 μE/m^2/s	1.4 mL/h/PhBR	Lindblad et al. (2002)
Anabaena AMC 414	Indoor tubular photobioreactor, BG 11_0 medium, argon sparged, 456 μE/m^2/s	13.8 mL/h/PhBR	Lindblad et al. (2002)
Immobilized *Chlamydomonas reinhardtii* CC-1036 pf18 mt+	Flat plate photobioreactor, TAP-S medium, argon sparged, 120 μE/m^2/s	6.4 μmol/mg chl/h	Laurinavichene et al. (2006)
Synechocystis sp. PCC 6803	BG 11 medium with optimized nutrients, nitrogen atmosphere, dark condition	0.81 μmol/mgchl/h	Burrows et al. (2008)
Chlamydomonas reinhardtii strain L159I-N230Y	Flat plate photobioreactor, TAP-S medium, 70 μmol/m^2/s	5.77 mL/L/h	Torzillo et al. (2009)
Synechocystis sp. PCC 6803	Glass vial, BG11_0-S medium, 750 mM β-mercaptoethanol, argon sparged, 24 h dark	14.32 μmol/mg chla/min	Baebprasert et al. (2010)
Aphanothece halophytica	Erlenmeyer flask, nitrogen-deprived BG 11 medium, 30 μmol/m^2/s for 18 h	13.8 μmol/mgchl/h	Taikhao et al. (2013)
Nostoc PCC 7120 Δ*hupW*	Flat panel photobioreactor, BG 11 medium, alternate argon/N_2 (20/80) and 100% argon sparged, 44 μmol/m^2/s	0.71 mmol/mgchla/h	Nyberg et al. (2015)
Lyngbya sp.	Glass reactors, medium containing benzoate (600 mg/L), argon sparged, 4000 lx	17.05 μmol/gchl 1/h	Shi and Yu (2016)

[FeFe]-hydrogenases are often involved in the reduction of protons to produce H_2. These are the only type of hydrogenases found in the eukaryotic microorganisms (Vignais and Colbeau 2004). In green microalgae they are located exclusively in the stroma of the chloroplast (Eroglu and Melis 2011). These hydrogenases are monomeric or dimeric with an average molecular weight of 50 kDa. The active site cluster of the enzyme also known as H-cluster consists of six Fe atoms arranged as [4Fe-4S] sub-cluster to which [2Fe-2S] extension is covalently bridged via cysteine residue. The Fe atoms of the active site are bound to non-protein ligands, CN^- and CO groups (Peters et al. 1998, 2015). The H-cluster of the [FeFe]-hydrogenases makes them different from the other H_2-producing enzymes and results in 100-fold higher enzyme activity (Happe et al. 2002). However, in spite of high specific activity these enzymes get easily inactivated by O_2 or CO_2. The green microalgae, *C. reinhardtii,* encodes two [FeFe]-hydrogenases (HydA1 and HydA2) which are 74% similar and are expressed under anaerobic condition (Forestier et al. 2003).

[NiFe]-hydrogenases are the most numerous hydrogenases found only in prokaryotes: cyanobacteria, bacteria and archaea. The core enzyme consists of the α–β heterodimer, where the larger α-subunit possesses the NiFe bimetallic centre and the smaller β-subunit consists of the Fe–S clusters which transfer the electrons from the active site to the e^- acceptor molecule (Kim and Kim 2011). In the active site, presence of non-protein ligands (CN^- and CO groups) bound to the Fe atom is the common structural characteristic of the [FeFe]- and [NiFe]-hydrogenases (Peters et al. 2015). In cyanobacterial species, these enzymes occur in two different types: hup-encoded [NiFe]-uptake hydrogenases and hox-encoded [NiFe]-bidirectional hydrogenases. Uptake hydrogenase catalyses the oxidation of H_2 to recover the energy lost during N_2 fixation. These are found in all nitrogen-fixing cyanobacteria, but their presence in non-nitrogen-fixing cyanobacteria is still under question (Tamagnini et al. 2002). The small subunit of the enzyme does not contain the signal peptide at N-terminal; therefore, the enzyme is localized on the cytoplasmic side of either the cytoplasmic or thylakoid membrane (Tamagnini et al. 2002). In the filamentous cyanobacteria, these enzymes are found in the thylakoid membrane of the heterocysts (Tiwari and Pandey 2012). Inactivation of the gene (*hupS*) encoding the small subunit of uptake hydrogenase led to the enhanced and sustained H_2 production in *Anabaena siamensis* TISTR 8012 under high light intensity (Khetkorn et al. 2012). Bidirectional hydrogenase is the reversible enzyme that can either evolve or consume H_2 according to the existing redox state of the cell's photosynthetic membrane (Eroglu and Melis 2011). This enzyme is present in both nitrogen-fixing and non-nitrogen-fixing cyanobacteria. The enzyme is multimeric because the dimeric module of the enzyme is associated with other subunits that can bind cofactors. In cyanobacteria, during the period of adaptation to higher light intensities the reversible hydrogenases may act as an electron valve (Vignais and Colbeau 2004).

Nitrogenases

Nitrogenase is present in cyanobacteria which catalyses the nitrogen fixation by reducing the molecular nitrogen into ammonium ions that can be easily utilized by the organisms. Nitrogen fixation is ATP-requiring irreversible reaction and is essential for the maintenance of the nitrogen cycle in the atmosphere. The reduction of nitrogen to ammonia by nitrogenase is accompanied by the reduction of protons (H^+) leading to H_2 production. Nitrogenases are the metalloenzyme, and depending upon the type of metal cofactor present at the catalytic site, they can be of three types: molybdenum, iron or vanadium nitrogenases. All these three variants of nitrogenases are capable to carry out the H_2 production during the nitrogen fixation but with variable stoichiometries (Eqs. 5–7). However, in the absence of nitrogen, nitrogenases can exclusively produce the H_2 as described in Eq. 4.

$$\text{Mo-nitrogenase}: N_2 + 8e^- + 8H^+ \rightarrow 2NH_3 + H_2 \quad (5)$$

$$\text{Fe-nitrogenase}: N_2 + 21e^- + 21H^+ \rightarrow 2NH_3 + 7.5H_2 \quad (6)$$

$$\text{Fe-nitrogenase}: N_2 + 12e^- + 12H^+ \rightarrow 2NH_3 + 3H_2 \quad (7)$$

Among all the nitrogenases, the most studied one is molybdenum nitrogenase. It consists of two proteins: the larger dinitrogenase (Mo–Fe–S protein or protein I) and the smaller dinitrogenase reductase (Fe–S protein or protein II). The dinitrogenase complex has an average molecular weight of 230 kDa and is a $\alpha_2\beta_2$heterotetramer encoded by the *nifK* and *nifD* genes. The dinitrogenase reductase subunit is a homodimer of around 65 kDa and is encoded by *nifH* gene. Reductase protein receives the electron either from flavodoxin or ferredoxin (external e^- donor) and transfers it to dinitrogenase protein with concomitant hydrolysis of ATP. Hydrogen produced by the nitrogenase activity is generally consumed by the uptake hydrogenases due to which the net H_2 evolution by cyanobacteria is barely observed, at least in aerobic condition (Almon and Böger 1988).

2.2 *Dark Fermentation Using Microalgal Biomass as Feedstock*

2.2.1 Anaerobic Fermentation Process

Dark fermentation for bioH_2 production is considered as a promising technology mainly due to following reasons: process simplicity, no requirement of light energy, higher rate of H_2 evolution and potentiality to utilize wide variety of substrates (different biomass and wastewater) for H_2 production. This process involves the anaerobic breakdown of the high molecular weight organic substrates (carbohydrate, protein and lipid) into soluble metabolite products (volatile fatty acids and alcohols), H_2 and CO_2 by the facultative and obligate anaerobic bacteria. The genus of bacteria typically associated with dark fermentation includes *Clostridium*, *Klebsiella*, *Enterobacter*, *Citrobacter*, *Bacillus*, *Lactobacillus*, *Thermotoga*, *Anaerobiospirillum* and *Caldicellulosiruptor* (Xia et al. 2015). The hydrogen-producing bacteria utilizes the protons (H^+) as the electron acceptor and thus disposes the excess electrons generated by the oxidation of organic substrates in the form of H_2 (Das and Veziroglu 2001). There are two pathways for the formation of molecular H_2: NADH re-oxidation pathway and formate decomposition pathway. The H_2 production via NADH re-oxidation pathway is mediated by some specific bacteria such as *Clostridium* sp. by the following reaction (Eq. 8)

$$\mathrm{NADH} + \mathrm{H}^+ \rightarrow \mathrm{H_2} + \mathrm{NAD}^+ \tag{8}$$

This NADH is generated due to the conversion of glucose into pyruvate during the glycolysis pathway, which could be represented as follows (Eq. 9):

$$\mathrm{C_6H_{12}O_6} + 2\mathrm{NAD}^+ \rightarrow 2\mathrm{CH_3COCOOH} + 2\mathrm{NADH} + 2\mathrm{H}^+ \tag{9}$$

Pyruvate-ferredoxin oxidoreductase catalyses the breakdown of pyruvate into acetyl CoA which could be further metabolized either into acetate or butyrate (Fig. 3). In both the cases, oxidation of one mole of ferredoxin by [Fe–Fe]-hydrogenase yields one mole of H_2. Maximum H_2 yield of 4 mol/mol glucose is achieved when acetic acid is the sole metabolic end product. However, H_2 yield of only 2 mol/mol glucose is achieved when butyrate is the final product Eqs. (10 and 11):

$$\mathrm{C_6H_{12}O_6} + 2\mathrm{H_2O} \rightarrow 4\mathrm{H_2} + 2\mathrm{CO_2} + 2\mathrm{CH_3COOH} \tag{10}$$

$$\mathrm{C_6H_{12}O_6} \rightarrow 2\mathrm{H_2} + 2\mathrm{CO_2} + \mathrm{CH_3CH_2CH_2COOH} \tag{11}$$

In contrast, few facultative anaerobes such as *Escherichia coli* can carry out the H_2 evolution via formate decomposition pathway. In this case, pyruvate is converted into formate and acetyl CoA by pyruvate formate lyase. Subsequently, under the anaerobic condition the formate is cleaved into H_2 and CO_2 catalysed by formate hydrogen lyase Eqs. (12 and 13):

$$\mathrm{CH_3COCOOH} + \mathrm{HCoA} \rightarrow \mathrm{CH_3COCoA} + \mathrm{HCOO} \tag{12}$$

$$\mathrm{HCOOH} \rightarrow \mathrm{H_2} + \mathrm{CO_2} \tag{13}$$

However, when fermentation is carried out by mixed microbial consortia, glucose can undergo some other biochemical pathways which generates undesired by-products such as lactate, propionate, succinate, 2,3 butanediol, ethanol, butanol and isopropanol. Generation of these metabolites hampers the H_2 production and

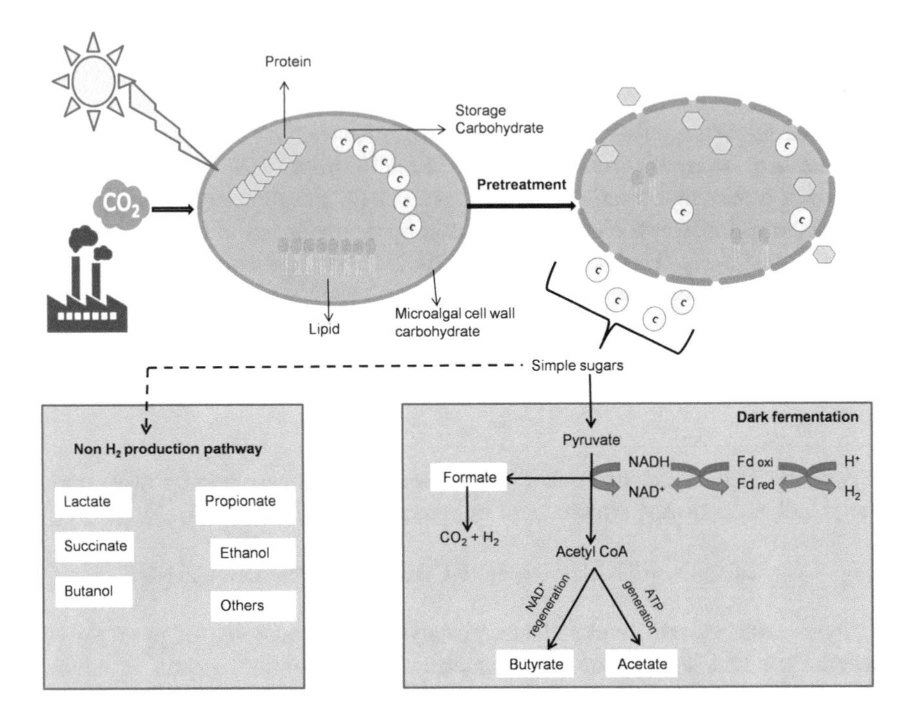

Fig. 3 Biohydrogen production via dark fermentation using algal biomass as substrate

lowers the overall yield of H_2. In such cases, the H_2 yield can be improved by inoculum pretreatment methods (enrichment of H_2-producing microorganisms) as well as by maintaining the proper operating conditions (Ghimire et al. 2015).

2.2.2 Suitability of Microalgal Biomass as a Substrate

Second-generation biofuels produced from the lignocellulosic biomass (agricultural residues and energy crops cultivated on non-arable lands) have no doubt provided the solution of the raised criticism regarding the sustainability of first-generation biofuels (biofuels produced from agricultural substrates). Nevertheless, the native recalcitrant structure of lignocellulosic biomass limits their hydrolysis by the fermentative bacteria. Indeed, to disrupt the rigid structure and to decrease the crystallinity of lignocellulosic biomass, required pretreatment methods are difficult and energy intensive. In this respect, third-generation biofuel production utilizing microalgal biomass as substrate has gained tremendous attention in recent years (Kumar et al. 2013; Nayak et al. 2014; Roy et al. 2014; Ortigueira et al. 2015; Khan et al. 2017).

Microalgal biomass offers several potential advantages to be used as an alternative to lignocellulosic feedstock for biofuel (biohydrogen) production, such as:

(1) relatively simple cell walls with no lignin therefore requires milder pretreatment, (2) high ability of CO_2 fixation, (3) higher productivity, (4) no need of arable land for mass cultivation and (5) can grow in brackish, saline and wastewaters thus reducing the freshwater footprint (Monlau et al. 2014; Sambusiti et al. 2015; Xia et al. 2015). During their growth, microalgae can synthesize and accumulate lipid, carbohydrate and protein (Monlau et al. 2014). The percentage of different components of microalgae varies according to algal species, environmental and cultivation conditions (Sambusiti et al. 2015).The first and most important task during the utilization of microalgal biomass as feedstock for dark fermentation is the selection of appropriate microalgal species having high biomass productivity and carbohydrate content (Wang et al. 2017). Microalgae store the polysaccharides either in the form of starch or glycogen.

Indeed, in microalgae the carbohydrates are also found entrapped within the cell wall mainly in the form of cellulose, hemicellulose, pectin and sulphated polysaccharide (Chen et al. 2013). Some microalgae can accumulate carbohydrate higher than 50% of their dry weight (Markou et al. 2012b). The microalgal storage and the cell wall polysaccharides upon efficient pretreatment can be released in the form of simple sugars (glucose) and contribute as a potential feedstock for dark fermentation. Appropriate cultivation condition and nutritional strategy can maximize the carbohydrate content and its productivity by altering the metabolic pathway of microalgae (Markou et al. 2012b). For instance, threefold higher carbohydrate content (39.19%) was obtained in *Scenedesmus* sp. CCNM 1077 under mixotrophic condition (glucose-supplemented medium) (Pancha et al. 2015). Ho et al. (2012) reported an increase in the carbohydrate content of *S. obliquus* CNW-N from 38.25 to 51.8%, when it was cultivated under high light intensity with nitrogen deficient condition. Moreover, Vitova et al. (2015) suggested sulphur deprivation as the most effective method of maximizing the carbohydrate content and productivity. Microalgae are also rich in macro- and microelements which are required for the growth of H_2-producing bacteria (HPB) (Sambusiti et al. 2015). The H_2 yield in dark fermentation mainly depends upon the monosaccharide content of the microalgal biomass because fermentation of lipid and protein by the HPB is thermodynamically unfavourable (Xia et al. 2015). Despite the lower potentiality of proteins for biohydrogen production, they are essential for balancing the C/N ratio of algal feedstock (Sambusiti et al. 2015). In dark fermentation, optimal C/N ratio of the substrate is an important factor for the growth and biological activity of HPB (Lay et al. 2013). The high protein content of the algal biomass results in low C/N ratio, which decreases the rate of H_2 production and limits the use of algal biomass as sole substrate. Excessive protein content leads to release and accumulation of nitrogen in the form of ammonium ion. High concentration of ammonium ion decreases the pH of the fermentation media, which may inhibit the growth of HPB or activity of enzymes participating in fermentative H_2 production. The C/N ratio of the algae can be increased by applying selected growth conditions (Montingelli et al. 2015). Moreover, an appropriate C/N ratio can be achieved by the addition of carbon-rich biomass with the microalgal biomass having high protein content. Xia et al. (2014) observed an increase in hydrogen yield, when mixed biomass of *Chlorella pyrenoidosa* and cassava starch was used as feedstock in dark fermentation.

Finally, the economic viability of the hydrogen production from microalgal feedstock is an important aspect that ought to be contemplated. Therefore, to increase the economic feasibility of the process, a biorefinery approach where microalgal biomass residues after lipid and value-added product extraction that are still rich in sugars can be utilized as feedstock under dark fermentation.

2.2.3 Pretreatment of Microalgal Biomass for Hydrogen Production

Carbohydrates in algae are entrapped within the cell wall in form of complex polymer or stored intracellularly as starch or glycogen. Therefore, when algal biomass is used as feedstock, it is necessary to carry out algal cell wall disruption followed by conversion of polymeric carbohydrates into simple fermentable sugars (glucose, arabinose, galactose, xylose and mannose), which are readily accessible for hydrogen-producing bacteria (Ho et al. 2013; Kumar et al. 2013). Efficient pretreatment of algal biomass is required to enhance the saccharification and thereafter biohydrogen yield (Xia et al. 2013). For instance, Roy et al. (2014) reported very low hydrogen production (0.03 m^3/m^3) from untreated algal biomass compared to the pretreated biomass (1.33 m^3/m^3). The effectiveness of pretreatment process depends upon the cell wall characteristics of the microalgal species. Microalgae having carbohydrate (cellulose and hemicelluloses)-based cell wall (*Chlorella kessleri* and *S. obliquus*) are difficult to be pretreated. In contrast, microalgal species having protein-based cell wall (*C. reinhardtii*, *Arthrospira platensis, Euglena gracilis*) are easily degraded (Mussgnug et al. 2010). The *Scenedesmus* sp. has one of the most resistant cell walls consisting of trilaminar structure where inner layer is composed of cellulose covered by hemicellulose. The outer layer contains sporopollenin-like biopolymer which confers high resistant to pretreatment (Miranda et al. 2012).

Nevertheless, pretreatment is an indispensable step for efficient production of hydrogen from microalgal feedstock; unoptimized pretreatment and saccharification conditions can generate sugar degradation products (furfural, hydroxymethylfurfural (HMF), formic acid, acetic acid, propionic acid and lactic acid) (Harun et al. 2014; Hernández et al. 2015; Xia et al. 2015). The accumulation of such by-products is inhibitory to the microbial growth and fermentation process thereby decreasing the overall hydrogen production (Miranda et al. 2012; Xia et al. 2015). The biomass pretreatment step is associated with high price and significantly contributes to the overall cost of biohydrogen production process (Roy et al. 2014). In order to increase the feasibility of biohydrogen production process, the selected pretreatment procedure must be simple, energy efficient, cost-effective and must enhance the polymeric carbohydrate conversion into fermentable sugars without the formation of inhibitory by-products.

To date, the pretreatment methods used for the microalgal biomass hydrolysis are mechanical, thermal, chemical, biological and combination of any two pretreatments. Mechanical pretreatment disrupts the algal cell wall by applying shear forces. Pretreatment method such as bead beating or milling is less dependent on microalgal species and can break the cell wall due to the collision of microalgal biomass with minute glass, ceramic or steel beads, under high agitation. Ultrasonication is another pretreatment method for cell wall disruption and solubilization of the organic matter. In this method, the repetitive compression and rarefaction of the sonic waves cause the formation of microbubbles which grow and then collapse, generating high pressure and heat, shear forces and free radicals thereby damaging the cell wall. Cheng et al. (2012) found that bead milling can disrupt some of the cyanobacterial cells releasing carbohydrate and protein to be utilized by hydrogenogens. However, through milling lower yield of H_2 (38.5 mLH_2/g DW) was obtained as compared to ultrasonication pretreatment (55.9 mL H_2/g DW).

Thermal pretreatment utilizes heat energy for the solubilization of microalgal biomass. Optimal range of temperature for the disintegration of organic matter varies according to substrate characteristics. For microalgal biomass, pretreatment temperature and time duration range from 65 to 180 °C and 15–60 min, respectively (Wang and Yin 2017). Thermal pretreatment by microwave heating is favoured for uniform distribution of heat and for achieving higher temperature in less time. Hydrothermal pretreatment (steam heating) at 100 and 121 °C led to increase in carbohydrates and proteins solubilization from lipid-extracted *Scenedesmus* biomass (Yang et al. 2010). However, thermal pretreatment alone is not sufficient for efficiently hydrolysing the microalgal biomass. Combination of heat and chemical pretreatment is commonly applied to improve the hydrogen yield from algal substrate. For instance, microwave and steam heating with dilute acid efficiently pretreated the biomass of *C. pyrenoidosa* with 8.6- and 9.5-fold increase in H_2 yields, respectively (Xia et al. 2013). Similarly, Roy et al. (2014) obtained high H_2 production (1.33 m^3/m^3) and reducing sugar concentration (9.6 kg/m^3) from HCl-heat pretreated biomass of *Chlorella sorokiniana*. Thermal–alkaline pretreatment of lipid-extracted *Scenedesmus* sp. biomass enhanced the H_2 production up to 168% (Yang et al. 2010).

Chemical pretreatment method involves the use of acid, alkali, solvents and oxidizing agent for the cell wall disintegration and saccharification of microalgal carbohydrates. Among the chemicals, acid and alkali reagents, generally in combination with heat, are used for the solubilization of organic matter. Liu et al. (2012) reported H_2 production of 1.42 L/L from acid (HCl)-pretreated hydrolysate of *Chlorella vulgaris* ESP6. In contrast, alkaline (NaOH)-pretreated hydrolysate was found to be less efficient for biohydrogen production. However, strong acidic conditions during the pretreatment may generate fermentative inhibitors such as furfural and HMF due to the dehydration of sugars. Moreover, the formation of inhibitory by-products can occur during the neutralization of the hydrolysate after acidic or alkaline pretreatment (Liu et al. 2012; Harun et al. 2014). Oxidizing agent such as H_2O_2 generates the nascent oxygen which helps in breaking the glycosidic bonds of complex sugars and converts it into simpler fermentable form. Roy et al. (2014) observed better H_2 production from H_2O_2-pretreated algal biomass than autoclaved and sonicated algal biomass. Diluted acid in combination with autoclaving is the most commonly used method for the pretreatment of microalgal feedstock due to its simple operation and high yield of reducing sugar (Nguyen et al. 2010; Liu et al. 2012; Kumar et al. 2013; Roy et al. 2014).

Biological pretreatment by enzymes is considered as an efficient method for the hydrolysis of microalgal biomass due to the high substrate specificity, milder operating conditions, less energy consumption and no generation of inhibitory by-products. The effectiveness of this method depends upon the substrate characteristics, enzyme dosage, temperature, pH and treatment duration. For hydrolysis, selection of enzyme is based on the microalgal cell wall composition. The commonly used enzymes for microalgal pretreatment include cellulases, α-amylases, amyloglucosidases, xylanases and proteases (Hom-Diaz et al. 2016). The pretreatment of biomass by biological method is usually carried out after physical or chemical method. Cheng et al. (2014) studied the combined effect of cellulase and glucoamylase on the reducing sugar yield from acid–heat and alkali–heat pretreated algal biomass. In combination both enzymes gave better sugar yield than cellulase alone. To increase the conversion of starch, Nguyen et al. (2010) carried out the enzymatic hydrolysis of *C. reinhardtii* biomass by utilizing Termamyl (endoglucanase) enzyme. Under the optimized enzymatic hydrolysis condition, maximum H_2 yield of 2.5 mol/mol glucose equivalent was achieved via separate hydrolysis and fermentation (SHF) process. However, pure enzymes are expensive and use of such enzymes for biomass pretreatment makes the H_2 production process economically unattractive. Therefore, beside commercial enzymes, bacterial or fungal crude enzymes can be used as cheaper alternative for microalgal biomass pretreatment. Many bacterial and fungal species possess the unique ability of producing wide variety of extracellular hydrolytic enzymes. Prajapati et al. (2015) reported that crude enzyme obtained from *Aspergillus lentulus* can efficiently solubilize the microalgal sugars. Soluble sugar concentration of 57 mg/L and 29% COD solubilization were obtained when biomass of *Chroococcus* sp. was pretreated by the fungal crude enzyme concentration of 20% v/v. Nevertheless, biological pretreatment is a green approach of obtaining high sugar yield from microalgal biomass, and the lower rate of hydrolysis makes this process time consuming and unsuitable for commercialization. Research on fermentative hydrogen production from microalgal feedstock has just started, and most of the studies have been conducted in batch systems. The main findings on biohydrogen production using microalgal biomass as substrate are presented in Table 2.

2.3 *Molecular Approaches Towards Improvement in Biohydrogen Production from Microalgae*

Production of H_2 from microalgae is an attractive process, although this renewable energy system is limited by low H_2 yield and productivity. There are several

Table 2 Biohydrogen production using microalgal biomass as substrate

Microalgae	Carbohydrate content (% w/w)	Biomass (g/L)	Pretreatment	Inoculum	Operational conditions[a]	H_2 yield	Reference
Chlamydomonas reinhardtii	–	–	Enzymatic	*Thermotoga neapolitana* DSM 4359	Serum bottles, $T = 75$ °C	311.1 mL/g glucose	Nguyen et al. (2010)
Lipid-extracted *Scenedesmus obliquus*	24.7	18	Steam heating with dilute NaOH	Anaerobic digested sludge	Serum bottles, $T = 37$ °C, pH = 6.3	45.54 mL/gVS	Yang et al. (2010)
Chlorella vulgaris	8.0	–	No pretreatment	Satellite bacteria associated with algal biomass	Glass bottles, $T = 37$ °C, pH = 8.0	10.8 mL/gVS	Lakaniemi et al. (2011)
Dunaliella tertiolecta	4.0					12.6 mL/gVS	
Arthrospira platensis	44.4	20	Microwave heating with dilute H_2SO_4 + Enzymatic	Anaerobic activated sludge	Serum bottles, $T = 35$°C, pH = 6.5	96.6 mL/gTS	Cheng et al. (2012)
Chlorella vulgaris ESP6	57.0		Steam heating with dilute HCl	*Clostridium butyricum* CGS5	Serum bottles, $T = 37$°C, pH = 7.0	81.0 mL/g TS	Liu et al. (2012)
Chlorella sorokiniana	–	10	Steam heating with dilute HCl	*Enterobacter cloacae* IIT-BT 08	Double jacketed reactor, $T = 37$ °C, pH = 6.8	201.6 mL/g COD	Kumar et al. (2013)
Nannochloropsis oceanica	21.9	25	Microwave heating with dilute H_2SO_4	Anaerobic digested sludge	Glass bottles, $T = 35$ °C, pH = 6.0	39.0 mL/gVS	Xia et al. (2013)

(continued)

Table 2 (continued)

Microalgae	Carbohydrate content (% w/w)	Biomass (g/L)	Pretreatment	Inoculum	Operational conditions[a]	H_2 yield	Reference
Lipid and pigment extracted-*Nannochloropsis* sp.	–	2.5	Milling + supercritical fluid extraction	*Enterobacter aerogenes* ATCC 13048	Glass bottles, T = 35 °C	60.6 mL/g TS	Nobre et al. (2013)
Scenedesmus obliquus	30.7	2.5	Autoclave	*Enterobacter aerogenes* ATCC 13048	Serum bottles, T = 30°C	45.1 mL/gTS	Batista et al. (2014)
		50	Autoclave	*Clostridium butyricum* DSM 10702	Serum bottles, T = 37 °C	90.3 mL/gTS	
Anabaena sp. PCC 7120	–	12	Enzymatic	Enriched thermophilic mixed culture	Double jacketed reactor, T = 60 °C, pH = 5.5	143.8 mL/g COD	Nayak et al. (2014)
Chlorella sorokiniana	14.5	14	Steam heating with dilute HCl	Enriched thermophilic mixed culture	Double jacketed reactor, T = 60 °C, pH = 6.5	333.5 mL/g hexose	Roy et al. (2014)

[a]All the hydrogen production studies were carried out in batch mode

research challenges that must be addressed to prove the technological feasibility of the process. Understanding the molecular fundamentals of H_2 production pathway and application of genetic and metabolic engineering approaches could enhance the microalgal biohydrogen production. Significant advances in the development of genetic tools have been made to overcome some of the major bottlenecks associated with microalgal H_2 production system and to improve the product yield.

2.3.1 Oxygen Sensitivity of Hydrogen-Producing Enzymes

Biophotolysis method exploits highly active microalgal hydrogenases for H_2 production. However, the extreme O_2 sensitivity of these enzymes presents a challenge for achieving sustained evolution of H_2. It has been found that the presence of O_2 irreversibly inhibits the [FeFe]-hydrogenases by attacking the [4Fe–4S] domain of the H-cluster (Stripp et al. 2009). Even O_2 not only inactivates the hydrogenases but also imposes inhibitory effect on transcription and protein maturation (Oey et al. 2016). Therefore, several studies have been conducted to increase the O_2 tolerance of the microalgal hydrogenases. Random and site-directed mutagenesis helped in obtaining mutants of *C. reinhardtii* having tenfold high O_2 tolerance (Ghirardi et al. 2000). Xu et al. (2005) developed a recombinant cyanobacterial system by transferring the O_2-tolerant hydrogenase genes from *T. roseopersicina* into *Synechococcus* sp. PCC 7942. In a different approach for O_2 sequestration, leghaemoglobin proteins (having high affinity to O_2) from legume plant (soybean) were transformed into the chloroplast of *C. reinhardtii*. This method helped in rapid consumption of O_2 and facilitated fourfold increase of H_2 production in transgenic microalgal cultures (Chen et al. 2013; Wu et al. 2010).

2.4 Photon Conversion Efficiency

For the biofuel production by utilizing the photosynthetic machinery, quantum efficiency holds paramount importance. Microalgal H_2 production system is greatly limited by the low solar conversion efficiency. Under the controlled conditions and low light intensities, algal cultures could achieve light-to-hydrogen energy conversion efficiency of up to 10% which is comparatively higher than obtained under similar conditions with solar light (<4%). In bright light, the pigments of the huge light harvesting complex (antenna system) capture more photons that can be utilized by the photosynthetic system. In such case, microalgal cells protect themselves from photodamage by dissipating (wasting) excess photons (~90%) as heat and fluorescence via a process known as 'energy-dependent non-photochemical quenching' (NPQ). This occurs at the upper layer of the algal culture; however, the cells present at lower surface may not receive sufficient light due to the 'self-shading effect' imposed by dense culture. Thus, NPQ at the top layer and the self-shading effect at lower surface result in low photon conversion efficiency.

This efficiency can be improved by modifying the antenna complex through genetic engineering. Reduction in the antenna size can minimize the energy wastage and improve the penetration of light inside the reactor. Polle et al. (2002) developed *C. reinhardtii* strain having truncated Chl antenna size of PSII, which showed better H_2 production and cellular productivity than wild strain. In other study, a truncated antenna mutant of *C. reinhardtii* under high light conditions showed 8.5-fold higher solar-to-hydrogen conversion efficiency than parent strain (Kosourov et al. 2011). To reduce the energy losses, recently simultaneous down regulation of entire LHC gene family in *C. reinhardtii* Stm3LR3 was carried out by applying RNAi technology. The mutant exhibited high photosynthetic efficiency under elevated light intensity (Mussgnug et al. 2007). Cyanobacteria (*Synechocystis* PCC 6803) lacking phycocyanin or whole phycobilisome expected to produce H_2 efficiently under photoautotrophic condition (Bernát et al. 2009).

2.5 Elimination of Uptake Hydrogenases

Another main concern to obtain adequate amount of hydrogen is the elimination of uptake hydrogenase present in the heterocyst of the nitrogen-fixing cyanobacteria. These hydrogenases catalyse the oxidation of H_2 to recover the energy lost during nitrogen fixation. In several studies, mutants developed by knockout of the uptake hydrogenase genes (*hupL* or *hupS*) resulted in higher H_2 yield. Significantly, higher amount of H_2 was obtained by the mutants of *Anabaena variabilis* developed by disruption of hup genes (Mikheeva et al. 1995; Happe et al. 2000). Khetkorn et al. (2012) demonstrated fourfold increase in hydrogen production of *A. siamensis* TISTR 8012 by the disruption of *hupS* gene. Although deletion of uptake hydrogenase helps in improving the H_2 production, *hox*-encoded [NiFe]-bidirectional hydrogenases may still reabsorb the H_2 produced by the nitrogenase due to its small Km value for H_2. In this regard, Masukawa et al. (2002) studied the effect of *hupL*, *hoxH* and *hupL*/*hoxH* deletions on photobiological H_2 production by *Anabaena* sp. PCC 7120. Compared to wild strain, the *hupL*$^-$ mutant produced H_2 at 4–7 times high rate. However, the *hoxH*$^-$ mutant did not show any improvement in H_2 production.

2.6 Substrate Utilization

In indirect biophotolysis, for H_2 production the e^- is supplied via external substrate. Strategies for improving the utilization of different substrate (sugars) by microalgae might be helpful in enhancing the biomass and biohydrogen production. In this view, modification in the transporter protein may assist the efficient transfer of external substrate inside the cell. Recently, hexose symporter (*HUP1*) gene from *C. kessleri* was heterologously expressed in *C. reinhardtii* stm6 cells, lacking the glucose transporter. The insertion of *HUP1* facilitated the import of glucose (1 mM) inside the stm6 cells. The transformed *C. reinhardtii* stm6Glc4 produced H_2 by simultaneously utilizing the water (66%) and glucose (33%) and showed fivefold increase in H_2 production than wild type (Doebbe et al. 2007).

2.7 Carbohydrate Metabolism of Microalgae

Carbohydrate-rich microalgal biomass is a suitable substrate for the fermentative H_2 production. The synthesis and accumulation of carbohydrates in microalgae occur due to CO_2 fixation, through a cyclic metabolic pathway known as Calvin cycle. CO_2 is reduced at the expense of ATP and NADPH generated during the light-dependent reaction of photosynthesis. In microalgae, the biosynthetic and catabolic pathways of energy storage molecules (starch and lipid) are closely linked. Some research findings suggest that a competition exist for the allocation of microalgal carbon between the carbohydrate and lipid synthesis (Rismani-Yazdi et al. 2011; Ho et al. 2012). Moreover, starch degradation provides main precursor (glycerol-3-phosphate, G3P) for triacylglycerol (TAG) synthesis. Thus, to enhance the biohydrogen production from microalgal feedstock, understanding and manipulating the starch metabolism become vital. The rate-limiting step in carbohydrate synthesis is catalysed by the enzyme ADP-glucose pyrophosphorylase (AGPase). An allosteric activator of AGPase is 3-phosphoglyceric acid (3-PGA) which is the intermediate product of CO_2 fixation reaction. Therefore, enhancing the photosynthetic efficiency might prove helpful to improve the carbohydrate synthesis and accumulation. In some studies, genetic modification in the RuBisCO subunits increased the photosynthetic efficiency of *Chlamydomonas* (Genkov et al. 2010; Zhu et al. 2010). An alternative strategy to enhance the microalgal starch accumulation is to decrease the starch degradation. The mechanism of microalgal carbohydrate catabolism is not completely understood, but it is well inferred in *Arabidopsis thaliana*. Phosphorolytic and/or hydrolytic enzymes play major role in starch degradation mechanism. Targeting these enzymes for gene knockout probably helps in developing microalgae with desirable phenotype (high carbohydrate content) (Radakovits et al. 2010). Except the starch stored in plastids, carbohydrates in algae are also found entrapped within the cell wall mainly in form of cellulose. The process of cellulose biosynthesis is complicated and involves several enzymatic reactions. It is synthesized by cellulase synthase utilizing UDP-glucose as precursor (Chen et al. 2013).

Due to the poor understanding of carbon partitioning between the biosynthetic pathways of energy-rich molecules, in comparison with genetic engineering, process engineering methodologies have greatly helped in the increment of microalgal carbohydrate content. However, few studies with molecular approaches have been carried out in cyanobacteria. In one such study, to enhance the cellulose yield, the genes for cellulose synthesis (*acsAB*) were transferred from *A. xylinum* into *Anabaena* sp. PCC 7120 via conjugation (Su et al. 2011). The mutant produced total extractable glucose of 0.53–0.66 mg/mL/OD750 which could be used for biohydrogen production. Recently, Patel et al. (2016) applied random mutagenesis on *Synechocystis* PCC 6803 to develop strain with high biomass and carbohydrate productivity. The mutant produced 3.6-fold more biomass and carbohydrate yield of 225 mg/L, indicating its potential to be used as fermentative feedstock.

Finally, it could be inferred that advances made in genetic and metabolic engineering have brought a major breakthrough in microalgal H_2 production process by overcoming several barriers associated with the low hydrogen yield. Indeed, there are some other problems that must be resolved to increase the overall feasibility of the process. For instance, most of the studies on photobiological H_2 production are carried out at bench-scale photobioreactors (PBRs). Due to the data scarcity, addressing several engineering issues for the scaling up of the PBR becomes challenging (Fernández-Sevilla et al. 2014). Another major problem in H_2 production is the incomplete conversion of organic substrate into H_2 and CO_2 via dark fermentation. H_2 production through this process is associated with the production of some soluble metabolites (volatile fatty acids and alcohols). This leads to low gaseous energy recovery, and the spent media rich in organic acids may pose threat to environment. To overcome this problem, an integrative system can be devised where the effluent of dark fermentation can be integrated with anaerobic digestion, photofermentation and bioelectrochemical systems (Sambusiti et al. 2015). Interestingly, volatile fatty acids rich spent media can be efficiently utilized as substrate for the mixotrophic cultivation of microalgae (Ghosh et al. 2017). Furthermore, utilization of wastewater grown and lipid/value-added product extracted microalgae as feedstock for biohydrogen production and could make the process more economically alluring.

3 Conclusion

Hydrogen production through biological routes is considered as the cleanest way of renewable energy generation. Most of the green microalgae and cyanobacteria possess novel metabolic features to carry out photobiological hydrogen evolution. Moreover, microalgal biomass has great potential to be used as substrate for fermentative biohydrogen production. Nevertheless, an efficient and economical method of biomass pretreatment is critical for carbohydrate saccharification and its utilization by hydrogen-producing bacteria. Oxygen sensitivity of hydrogenases and low photon conversion efficiency are two major bottlenecks of microalgal hydrogen production via biophotolysis, while incomplete knowledge of carbohydrate metabolism presents a challenge for developing sugar-enriched microalgal feedstock for dark fermentation. Although the application of genetic tools to enhance the biohydrogen production from microalgae is currently in its infancy, promising advances have been made to develop the genetically engineered microalgae with unprecedented precision. It is likely that further research in this direction might help in developing industrially relevant microalgal species for carbon-neutral hydrogen generation.

References

Almon, H., & Böger, P. 1988. Nitrogen and hydrogen metabolism: induction and measurement. *Methods in Enzymology*, Academic Press, 167.

Baebprasert, W., Lindblad, P., & Incharoensakdi, A. (2010). Response of H_2 production and Hox-hydrogenase activity to external factors in the unicellular cyanobacterium *Synechocystis* sp. strain PCC 6803. *International Journal of Hydrogen Energy, 35*(13), 6611–6616.

Batista, A. P., Moura, P., Marques, P. A. S. S., Ortigueira, J., Alves, L., & Gouveia, L. (2014). *Scenedesmus obliquus* as feedstock for biohydrogen production by *Enterobacter aerogenes* and *Clostridium butyricum*. *Fuel, 117,* 537–543.

Benemann, J. R. (2000). Hydrogen production by microalgae. *Journal of Applied Phycology, 12,* 291–300.

Bernát, G., Waschewski, N., & Rögner, M. (2009). Towards efficient hydrogen production: The impact of antenna size and external factors on electron transport dynamics in *Synechocystis* PCC 6803. *Photosynthesis Research, 99*(3), 205–216.

Borodin, V. B., Tsygankov, A. A., Rao, K. K., & Hall, D. O. (2000). Hydrogen production by *Anabaena variabilis* PK84 under simulated outdoor conditions. *Biotechnology and Bioengineering, 69*(5), 478–485.

Burrows, E. H., Chaplen, F. W. R., & Ely, R. L. (2008). Optimization of media nutrient composition for increased photofermentative hydrogen production by *Synechocystis* sp. PCC 6803. *International Journal of Hydrogen Energy, 33*(21), 6092–6099.

Chen, C. Y., Zhao, X. Q., Yen, H. W., Ho, S. H., Cheng, C. L., Lee, D. J., et al. (2013). Microalgae-based carbohydrates for biofuel production. *Biochemical Engineering Journal, 78,* 1–10.

Cheng, J., Liu, Y., Lin, R., Xia, A., Zhou, J., & Cen, K. (2014). Cogeneration of hydrogen and methane from the pretreated biomass of algae bloom in Taihu Lake. *International Journal of Hydrogen Energy, 39*(33), 18793–18802.

Cheng, J., Xia, A., Liu, Y., Lin, R., Zhou, J., & Cen, K. (2012). Combination of dark- and photo-fermentation to improve hydrogen production from *Arthrospira platensis* wet biomass with ammonium removal by zeolite. *International Journal of Hydrogen Energy, 37*(18), 13330–13337.

Das, D., & Veziroğlu, T. N. (2001). Hydrogen production by biological proceses: A survey of literature. *International Journal of Hydrogen Energy, 26,* 13–28.

Das, D., Khanna, N., & Dasgupta, C. N. (2014). *Biohydrogen production: Fundamentals and technology advances*. CRC Press, Taylor and Francis Group, LLC.

Doebbe, A., Rupprecht, J., Beckmann, J., Mussgnug, J. H., Hallmann, A., Hankamer, B., et al. (2007). Functional integration of the HUP1 hexose symporter gene into the genome of *C. reinhardtii*: Impacts on biological H_2 production. *Journal of Biotechnology, 131*(1), 27–33.

Eroglu, E., & Melis, A. (2011). Photobiological hydrogen production: Recent advances and state of the art. *Bioresource Technology, 102*(18), 8403–8413.

Fernández-Sevilla, J. M., Acién-Fernández, F. G., & Molina-Grima, E. (2014). Microbial bioenergy: Hydrogen production. *Advances in Photosynthesis and Respiration, 38,* 291–320.

Forestier, M., King, P., Zhang, L., Posewitz, M., Schwarzer, S., Happe, T., et al. (2003). Expression of two [Fe] -hydrogenases in *Chlamydomonas reinhardtii* under anaerobic conditions. *European Journal of Biochemistry, 270,* 2750–2758.

Gaffron, H., & Rubin, J. (1942). Fermentatinve and photochemical production of hydrogen in algae. *The Journal of General Physiology, 26*(2), 219–240.

Genkov, T., Meyer, M., Griffiths, H., & Spreitzer, R. J. (2010). Functional hybrid rubisco enzymes with plant small subunits and algal large subunits: Engineered rbcS cDNA for expression in *Chlamydomonas*. *Journal of Biological Chemistry, 285*(26), 19833–19841.

Ghimire, A., Frunzo, L., Pirozzi, F., Trably, E., Escudie, R., Lens, P. N. L., et al. (2015). A review on dark fermentative biohydrogen production from organic biomass: Process parameters and use of by-products. *Applied Energy, 144,* 73–95.

Ghirardi, M. L., Togasaki, R. K., & Seibert, M. (1997). Oxygen sensitivity of Algal H_2-production. *Applied Biochemistry and Biotechnology, 63–65,* 141–151.

Ghirardi, M. L., Zhang, L., Lee, J. W., Flynn, T., Seibert, M., Greenbaum, E., et al. (2000). Microalgae: A green source of renewable H_2. *Trends in Biotechnology, 18*(12), 506–511.

Ghosh, S., Roy, S., & Das, D. (2017). Enhancement in lipid content of *Chlorella* sp. MJ11/11 from the spent medium of thermophilic biohydrogen production process. *Bioresource Technology, 223,* 219–226.

Greenbaum, E. (1982). Photosynthetic hydrogen and oxygen production: Kinetic studies. *Science (New York), 215*(4530), 291–293.

Happe, T., Hemschemeier, A., Winkler, M., & Kaminski, A. (2002). Hydrogenases in green algae: Do they save the algae's life and solve our energy problems? *Trends in Plant Science, 7*(6), 246–250.

Happe, T., Schütz, K., & Böhme, H. (2000). Transcriptional and mutational analysis of the uptake hydrogenase of the filamentous cyanobacterium *Anabaena variabilis* ATCC 29413. *Journal of Bacteriology, 182*(6), 1624–1631.

Harun, R., Yip, J. W. S., Thiruvenkadam, S., Ghani, W. A. W. A. K., Cherrington, T., & Danquah, M. K. (2014). Algal biomass conversion to bioethanol-a step-by-step assessment. *Biotechnology Journal, 9*(1), 73–86.

Hernández, D., Riaño, B., Coca, M., & García-González, M. C. (2015). Saccharification of carbohydrates in microalgal biomass by physical, chemical and enzymatic pre-treatments as a previous step for bioethanol production. *Chemical Engineering Journal, 262,* 939–945.

Ho, S. H., Chen, C. Y., & Chang, J. S. (2012). Effect of light intensity and nitrogen starvation on CO_2 fixation and lipid/carbohydrate production of an indigenous microalga *Scenedesmus obliquus* CNW-N. *Bioresource Technology, 113,* 244–252.

Ho, S. H., Huang, S. W., Chen, C. Y., Hasunuma, T., Kondo, A., & Chang, J. S. (2013). Bioethanol production using carbohydrate rich micraolgae biomass as feedstock. *Bioresource Technology, 135,* 191–198.

Holladay, J. D., Hu, J., King, D. L., & Wang, Y. (2009). An overview of hydrogen production technologies. *Catalysis Today, 139*(4), 244–260.

Hom-Diaz, A., Passos, F., Ferrer, I., Vicent, T., & Blánquez, P. (2016). Enzymatic pretreatment of microalgae using fungal broth from *Trametes versicolor* and commercial laccase for improved biogas production. *Algal Research, 19,* 184–188.

Khan, M. I., Lee, M. G., Shin, J. H., & Kim, J. D. (2017). Pretreatment optimization of the biomass of Microcystis aeruginosa for efficient bioethanol production. *AMB Expr, 7*(19), 1–9.

Khetkorn, W., Lindblad, P., & Incharoensakdi, A. (2012). Inactivation of uptake hydrogenase leads to enhanced and sustained hydrogen production with high nitrogenase activity under high light exposure in the cyanobacterium *Anabaena siamensis* TISTR 8012. *Journal of Biological Engineering, 6*(19), 1–11.

Kim, D.-H., & Kim, M.-S. (2011). Hydrogenases for biological hydrogen production. *Bioresource Technology, 102*(18), 8423–8431.

Kosourov, S., Makarova, V., Fedorov, A. S., Tsygankov, A., Seibert, M., & Ghirardi, M. L. (2005). The effect of sulfur re-addition on H_2 photoproduction by sulfur-deprived green algae. *Photosynthesis Research, 85,* 295–305.

Kosourov, S., Tsygankov, A., Seibert, M., & Ghirardi, M. L. (2002). Sustained hydrogen photoproduction by *Chlamydomonas reinhardtii*: Effects of culture parameters. *Biotechnology and Bioengineering, 78*(7), 731–740.

Kosourov, S. N., Ghirardi, M. L., & Seibert, M. (2011). A truncated antenna mutant of *Chlamydomonas reinhardtii* can produce more hydrogen than the parental strain. *International Journal of Hydrogen Energy, 36*(3), 2044–2048.

Kumar, K., Roy, S., & Das, D. (2013). Continuous mode of carbon dioxide sequestration by *C. sorokiniana* and subsequent use of its biomass for hydrogen production by *E. cloacae* IIT-BT 08. *Bioresource Technology, 145,* 116–122.

Lakaniemi, A. M., Hulatt, C. J., Thomas, D. N., Tuovinen, O. H., & Puhakka, J. A. (2011). Biogenic hydrogen and methane production from *Chlorella vulgaris* and *Dunaliella tertiolecta* biomass. *Biotechnology for Biofuels, 4*(1), 34.

Laurinavichene, T. V., Fedorov, A. S., Ghirardi, M. L., Seibert, M., & Tsygankov, A. A. (2006). Demonstration of sustained hydrogen photoproduction by immobilized, sulfur-deprived *Chlamydomonas reinhardtii* cells. *International Journal of Hydrogen Energy, 31*(5), 659–667.

Lay, C. H., Sen, B., Chen, C. C., Wu, J. H., Lee, S. C., & Lin, C. Y. (2013). Co-fermentation of water hyacinth and beverage wastewater in powder and pellet form for hydrogen production. *Bioresource Technology, 135,* 610–615.

Levin, D. B., Pitt, L., & Love, M. (2004). Biohydrogen production: Prospects and limitations to practical application. *International Journal of Hydrogen Energy, 29*(2), 173–185.

Lindblad, P., Christensson, K., Lindberg, P., Fedorov, A., Pinto, F., & Tsygankov, A. (2002). Photoproduction of H2 by wildtype Anabaena PCC 7120 and a hydrogen uptake deficient mutant: From laboratory experiments to outdoor culture. *International Journal of Hydrogen, 27,* 1271–1281.

Liu, C. H., Chang, C. Y., Cheng, C. L., Lee, D. J., & Chang, J. S. (2012). Fermentative hydrogen production by *Clostridium butyricum* CGS5 using carbohydrate-rich microalgal biomass as feedstock. *International Journal of Hydrogen Energy, 37*(20), 15458–15464.

Márquez-Reyes, L. A., Sánchez-Saavedra, M. D. P., & Valdez-Vazquez, I. (2015). Improvement of hydrogen production by reduction of the photosynthetic oxygen in microalgae cultures of *Chlamydomonas gloeopara* and *Scenedesmus obliquus*. *International Journal of Hydrogen Energy, 40*(23), 7291–7300.

Markou, G., Angelidaki, I., & Georgakakis, D. (2012a). Microalgal carbohydrates: An overview of the factors influencing carbohydrates production, and of main bioconversion technologies for production of biofuels. *Applied Microbiology and Biotechnology, 96*(3), 631–645.

Markou, G., Chatzipavlidis, I., & Georgakakis, D. (2012b). Cultivation of *Arthrospira* (*Spirulina*) platensis in olive-oil mill wastewater treated with sodium hypochlorite. *Bioresource Technology, 112,* 234–241.

Masukawa, H., Mochimaru, M., & Sakurai, H. (2002). Disruption of the uptake hydrogenase gene, but not of the bidirectional hydrogenase gene, leads to enhanced photobiological hydrogen production by the nitrogen-fixing cyanobacterium *Anabaena* sp. PCC 7120. *Applied Microbiology and Biotechnology, 58*(5), 618–624.

McKinlay, J. B., & Harwood, C. S. (2010). Photobiological production of hydrogen gas as a biofuel. *Current Opinion in Biotechnology, 21*(3), 244–251.

Melis, A., Zhang, L., Forestier, M., Ghirardi, M. L., & Seibert, M. (2000). Sustained photobiological hydrogen gas production upon reversible inactivation of oxygen evolution in the green Alga *Chlamydomonas reinhardtii*. *Plant Physiology, 122,* 127–135.

Mikheeva, L. E., Schmitz, O., Shestakov, S. V., & Bothe, H. (1995). Mutants of the cyanobacterium *Anabaena variabilis* altered in hydrogenase activities. *Z. Natutforsch, 50,* 505–510.

Miranda, J. R., Passarinho, P. C., & Gouveia, L. (2012). Pre-treatment optimization of *Scenedesmus obliquus* microalga for bioethanol production. *Bioresource Technology, 104,* 342–348.

Miura, Y., Akano, T., Fukatsu, K., Miyasaka, H., Mizoguchi, T., Yagi, K., et al. (1997). Stably sustained hydrogen production by biophotolysis in natural day/night cycle. *Energy Conservation Management, 38,* 533–537.

Monlau, F., Sambusiti, C., Barakat, A., Quéméneur, M., Trably, E., Steyer, J. P., et al. (2014). Do furanic and phenolic compounds of lignocellulosic and algae biomass hydrolyzate inhibit anaerobic mixed cultures? A comprehensive review. *Biotechnology Advances, 32*(5), 934–951.

Montingelli, M. E., Tedesco, S., & Olabi, A. G. (2015). Biogas production from algal biomass: A review. *Renewable and Sustainable Energy Reviews, 43,* 961–972.

Mussgnug, J. H., Klassen, V., Schlüter, A., & Kruse, O. (2010). Microalgae as substrates for fermentative biogas production in a combined biorefinery concept. *Journal of Biotechnology, 150*(1), 51–56.

Mussgnug, J. H., Thomas-Hall, S., Rupprecht, J., Foo, A., Klassen, V., & McDowall, A., et al. (2007). Engineering photosynthetic light capture: Impacts on improved solar energy to biomass conversion. *Plant Biotechnology Journal, 5*(6), 802–814.

Nayak, B. K., Roy, S., & Das, D. (2014). Biohydrogen production from algal biomass (Anabaena sp. PCC 7120) cultivated in airlift photobioreactor. *International Journal of Hydrogen Energy, 39,* 7553–7560.

Nguyen, T. A. D., Kim, K. R., Nguyen, M. T., Kim, M. S., Kim, D., & Sim, S. J. (2010). Enhancement of fermentative hydrogen production from green algal biomass of Thermotoga neapolitana by various pretreatment methods. *International Journal of Hydrogen Energy, 35* (23), 13035–13040.

Nobre, B. P., Villalobos, F., Barragán, B. E., Oliveira, A. C., Batista, A. P., Marques, P. A. S. S., et al. (2013). A biorefinary from *Nannchloropsis* sp. microalga - Extraction of oils and pigments. Production of biohydrogen from the leftcover biomass. *Bioresource Tecnology, 135,* 128–136.

Nyberg, M., Heidorn, T., & Lindblad, P. (2015). Hydrogen production by the engineered cyanobacterial strain *Nostoc* PCC 7120 δhupW examined in a flat panel photobioreactor system. *Journal of Biotechnology, 215,* 35–43.

Ortigueira, J., Alves, L., Gouveia, L., & Moura, P. (2015). Third generation biohydrogen production by *Clostridium butyricum* and adapted mixed cultures from *Scenedesmus obliquus* microalga biomass. *Fuel, 153,* 128–134.

Oey, M., Sawyer, A. L., Ross, I. L., & Hankamer, B. (2016). Challenges and opportunities for hydrogen production from microalgae. *Plant Biotechnology Journal, 14*(7), 1487–1499.

Vignais, P. M., & Colbeau, A. (2004). Molecular biology of microbial hydrogenases. *Current Issues in Molecular Biology, 6,* 159–188.

Pancha, I., Chokshi, K., & Mishra, S. (2015). Enhanced biofuel production potential with nutritional stress amelioration through optimization of carbon source and light intensity in *Scenedesmus* sp. CCNM 1077. *Bioresource Technology, 179,* 565–572.

Patel, V. K., Maji, D., Pandey, S. S., Rout, P. K., Sundaram, S., & Kalra, A. (2016). Rapid budding EMS mutants of Synechocystis PCC 6803 producing carbohydrate or lipid enriched biomass. *Algal Research, 16,* 36–45.

Peters, J. W., Lanzilotta, W. N., Lemon, B. J., & Seefeldt, L. C. (1998). X-ray crystal structure of the Fe-only hydrogenase (CpI) from *Clostridium pasteurianum* to 1.8 Angstrom resolution. *Science (New York), 282*(5395), 1853–1858.

Peters, J. W., Schut, G. J., Boyd, E. S., Mulder, D. W., Shepard, E. M., Broderick, J. B., et al. (2015). [FeFe] - and [NiFe] -hydrogenase diversity, mechanism, and maturation. *Biochimica et Biophysica Acta, 1853,* 1350–1369.

Polle, J. E. W., Kanakagiri, S., Jin, E., Masuda, T., & Melis, A. (2002). Truncated chlorophyll antenna size of the photosystems—A practical method to improve microalgal productivity and hydrogen production in mass culture. *International Journal of Hydrogen Energy, 27*(11–12), 1257–1264.

Prajapati, S. K., Bhattacharya, A., Malik, A., & Vijay, V. K. (2015). Pretreatment of algal biomass using fungal crude enzymes. *Algal Research, 8,* 8–14.

Radakovits, R., Jinkerson, R. E., Darzins, A., & Posewitz, M. C. (2010). Genetic engineering of algae for enhanced biofuel production. *Eukaryotic Cell, 9*(4), 486–501.

Randt, C., & Senger, H. (1985). Participation of the two photosystems in light dependent hydrogen evolution in *Scenedesmus obliquus*. *Photochemistry and Photobiology, 42*(5), 553–557.

Rismani-Yazdi, H., Haznedaroglu, B. Z., Bibby, K., & Peccia, J. (2011). Transcriptome sequencing and annotation of the microalgae *Dunaliella tertiolecta*: Pathway description and gene discovery for production of next-generation biofuels. *BMC Genomics, 12*(1), 148.

Roy, S., Kumar, K., Ghosh, S., & Das, D. (2014). Thermophilic biohydrogen production using pre-treated algal biomass as substrate. *Biomass and Bioenergy, 61,* 157–166.

Sambusiti, C., Bellucci, M., Zabaniotou, A., Beneduce, L., & Monlau, F. (2015). Algae as promising feedstocks for fermentative biohydrogen production according to a biorefinery approach: A comprehensive review. *Renewable and Sustainable Energy Reviews, 44,* 20–36.

Shi, X. Y., & Yu, H. Q. (2016). Simultaneous metabolism of benzoate and photobiological hydrogen production by *Lyngbya* sp. *Renewable Energy, 95,* 474–477.

Stripp, S. T., Goldet, G., Brandmayr, C., Sanganas, O., Vincent, K. A., Haumann, M., et al. (2009). How oxygen attacks [FeFe] hydrogenases from photosynthetic organisms. *Proceedings of the National Academy of Sciences of the United States of America, 106*(41), 17331–17336.

Su, H. Y., Lee, T. M., Huang, Y. L., Chou, S. H., Wang, J. B., Lin, L. F., et al. (2011). Increased cellulose production by heterologous expression of cellulose synthase genes in a filamentous heterocystous cyanobacterium with a modification in photosynthesis performance and growth ability. *Botanical Studies, 52,* 265–275.

Taikhao, S., Junyapoon, S., Incharoensakdi, A., & Phunpruch, S. (2013). Factors affecting biohydrogen production by unicellular halotolerant cyanobacterium *Aphanothece halophytica. Journal of Applied Phycology, 25*(2), 575–585.

Tamagnini, P., Axelsson, R., Lindberg, P., Oxelfelt, F., Wünschiers, R., & Lindblad, P. (2002). Hydrogenases and hydrogen metabolism of cyanobacteria. *Microbiology and Molecular Biology Reviews, 66*(1), 1–20.

Tiwari, A., & Pandey, A. (2012). Cyanobacterial hydrogen production—A step towards clean environment. *International Journal of Hydrogen Energy, 37*(1), 139–150.

Torzillo, G., Scoma, A., Faraloni, C., Ena, A., & Johanningmeier, U. (2009). Increased hydrogen photoproduction by means of a sulfur-deprived *Chlamydomonas reinhardtii* D1 protein mutant. *International Journal of Hydrogen Energy, 34*(10), 4529–4536.

Torzillo, G., Scoma, A., Faraloni, C., & Giannelli, L. (2015). Advances in biotechnology of hydrogen production with the microalga *Chlamydomonas reinhardtii. Critical reviews in Biotecnology, 35*(4), 485–496.

Torzillo, G., & Seibert, M. 2013. Hydrogen production by *Chlamydomonas reinhardtii.* In *Handbook of microalgal culture: Applied phycology and biotechnology*, pp. 417–432.

Troshina, O., Serebryakova, L., Sheremetieva, M., & Lindblad, P. (2002). Production of H_2 by the unicellular cyanobacterium *Gloeocapsa alpicola* CALU 743 during fermentation. *International Journal of Hydrogen Energy, 27*(11–12), 1283–1289.

Tsygankov, A. A., Fedorov, A. S., Kosourov, S. N., & Rao, K. K. (2002). Hydrogen production by cyanobacteria in an automated outdoor photobioreactor under aerobic conditions. *Biotechnology and Bioengineering, 80*(7), 777–783.

Tsygankov, A. A., Kosourov, S. N., Tolstygina, I. V., Ghirardi, M. L., & Seibert, M. (2006). Hydrogen production by sulfur-deprived Chlamydomonas reinhardtii under photoautotrophic conditions. *International Journal of Hydrogen Energy, 31*, 1574–1584.

Vitova, M., Bisova, K., Kawano, S., & Zachleder, V. (2015). Accumulation of energy reserves in algae: From cell cycles to biotechnological applications. *Biotechnology Advances, 33*(6), 1204–1218.

Wang, J., & Yin, Y. (2017). Bihydrogen production from organic wastes. In *Green energy and technology* (pp. 123–195), Springer Nature.

Wang, Y., Ho, S.-H., Yen, H.-W., Nagarajan, D., Ren, N.-Q., & Li, S. et al. (2017). Current advances on fermantative biobutanol production using third generation feedstock. *Biotechnology Advances*. https://doi.org/10.1016/j.biotechadv.2017.06.001.

Wu, S., Huang, R., Xu, L., Yan, G., & Wang, Q. (2010). Improved hydrogen production with expression of *hemH* and *lba* genes in chloroplast of *Chlamydomonas reinhardtii. Journal of Biotechnology, 146,* 120–125.

Xia, A., Cheng, J., Ding, L., Lin, R., Song, W., Zhou, J., et al. (2014). Enhancement of energy production efficiency from mixed biomass of *Chlorella pyrenoidosa* and cassava starch through combined hydrogen fermentation and methanogenesis. *Applied Energy, 120,* 23–30.

Xia, A., Cheng, J., Lin, R., Lu, H., Zhou, J., & Cen, K. (2013). Comparison in dark hydrogen fermentation followed by photo hydrogen fermentation and methanogenesis between protein and carbohydrate compositions in *Nannochloropsis oceanica* biomass. *Bioresource Technology, 138,* 204–213.

Xia, A., Cheng, J., Song, W., Su, H., Ding, L., Lin, R., et al. (2015). Fermentative hydrogen production using algal biomass as feedstock. *Renewable and Sustainable Energy Reviews, 51,* 209–230.

Xu, Q., Yooseph, S., Smith, H.O., Venter, C.J. 2005. *Development of a novel recombinant cyanobacterial system for hydrogen production from water.* Paper presented at Genomics: GTL Program Projects, Rockville.

Yang, Z., Guo, R., Xu, X., Fan, X., & Li, X. (2010). Enhanced hydrogen production from lipid-extracted microalgal biomass residues through pretreatment. *International Journal of Hydrogen Energy, 35,* 9618–9623.

Yu, J., & Takahashi, P. (2007). Biophotolysis-based hydrogen production by cyanobacteria and green microalgae. *Trends in Applied Microbiology, 1,* 79–89.

Zhu, X.-G., Long, S. P., & Ort, D. R. (2010). Improving photosynthetic efficiency for greater yield. *Annual Review of Plant Biology, 61*(1), 235–261.

Chapter 11
Biofuels from Microalgae: Bioethanol

Reinaldo Gaspar Bastos

Abstract The industrial potential of ethanol has been tested early in 1800 to be used as an engine fuel after the invention of an internal combustion engine. Currently, there are three generations of bioethanol that have been flourished based on different feedstocks. The first-generation bioethanol is derived from fermentation of glucose contained in starch and/or sugar crops. USA and Brazil are the main producers of bioethanol worldwide utilizing corn and sugarcane, while potato, wheat, and sugar beet are the common feedstocks for bioethanol in Europe. The term "second-generation bioethanol" emerged as a boon to overcome the "food versus fuel" that occurs by the first-generation bioethanol. The second generation also referred to as "advanced biofuels" is produced by innovative processes mainly using lignocellulosic feedstock and agricultural forest residues. The emergence of the third-generation bioethanol provides more benefits as compared to the first and second generations and is focused on the use of microalgae and cyanobacteria. These organisms represent as a promising alternative feedstock due to its high lipid and carbohydrate contents, easy cultivation in a wide variety of water environment, relatively low land usage and carbon dioxide absorption. This chapter will discuss the use of microalgae for the ethanol production and the main technological routes, i.e., enzymatic hydrolysis and yeast fermentation of microalgal biomass, metabolic pathways in dark conditions, and "photofermentation."

Keywords Bioethanol · Microalgae · Carbohydrate accumulation
Hydrolysis–fermentation

R. G. Bastos (✉)
Center of Agricultural Sciences (CCA), Federal University of São Carlos (UFSCar),
Via Anhanguera, km 174, Araras, SP 13600-970, Brazil
e-mail: reinaldo.bastos@ufscar.br

E. Jacob-Lopes et al. (eds.), *Energy from Microalgae*, Green Energy
and Technology, https://doi.org/10.1007/978-3-319-69093-3_11

1 Introduction

Microalgae biomass is an interesting alternative to traditional bioethanol crops because it does not have the inherent disadvantages of bioethanol of the first or second generation. The cultivation of microalgae may occur in different culture media, without necessarily using potable water and can carry wastewater, salt water (seawater) and brackish water in its composition. Microalgae production does not compete for freshwater intended for irrigation of plantations or for human and animal consumption. In addition, microalgae cultivation can occur in small areas and in non-arable, semiarid, or desert lands, since the main factors that influence the development of microalgae are the availability of sunlight and water for cultivation (Brennan and Owende 2010). Thus, the cultivation of microalgae does not directly compete for arable land for food production nor does it increase the occurrence of burning and deforestation, the main methods for obtaining new arable land. Another advantage is that when using carbohydrates produced by certain species of microalgae, the productivity of bioethanol of the third generation (in liters per hectare per year) may be some orders of magnitude greater than the productivity of raw materials used in the production of bioethanol of the first and second generations, according to Table 1.

Historically, microalgal biomass has been largely employed in the production of several compounds for human consumption and industrial application, including sterols, amino acids, fatty acids, and carotenoids, despite being considered in the last few years for biofuel production. The interest in converting microalgae into biofuels relies on some points: productivities superior to those of conventional energy crops, after lipid and carbohydrate extraction; potentially possible to recover high-value coproducts from the debris, such as proteins and pigments; low water consumption in comparison with the irrigation of energy crops; possibility of cultivation in non-arable lands, using non-potable water, such as wastewaters and without the application of pesticides and herbicides; and improvement of air quality, due to CO_2 fixation for biomass growth.

Table 1 Bioethanol productivity from different feedstocks

Feedstocks	Bioethanol productivity (L/(ha year))
Corn straw	1050–1400
Wheat	2590
Cassava	3310
Sweet sorghum	3050–4070
Maize	3460–4020
Beet	5010–6680
Sugarcane	6190–7500
Panicum virgatum (switchgrass)	10,760
Microalgae	46,760–140,290

Adapted from Mussatto et al. (2010)

Because of their simpler structure than those of higher plants, microalgae can achieve much higher photosynthetic efficiencies than terrestrial plants. Thus, a larger share of the captured solar energy is stored through the accumulation of carbohydrates inside the cell. Similarly, microalgae biomass production occurs in relatively short times, much lower compared to terrestrial plants used in the production of the first- and secondgeneration bioethanol. The possibility of recovering the microalgal several times or continuously, depending on the type of bioreactor used for biomass production. Thus, there is an abundant and inexpensive source of biomass for the production of bioethanol. Considering the potential of microalgae use, the great diversity of species, and the different possible conditions of cultivation, the knowledge of the physiology and metabolism of these microorganisms becomes imperative for the development of new industrial processes.

The microalgae serve as raw material for different types of biofuels, among them methane, hydrogen, biodiesel, and bioethanol, which could be used together or substituting the gasoline in light vehicles (Mata et al. 2010). Since global consumption of light fossil fuels is greater than the consumption of diesel heavy vehicles, researches' efforts on microalgae bioethanol production should be increased, an economically interesting alternative.

The selection of the appropriate microalgae species for the production of biofuels is an important factor for the success of the productive process as a whole. The desirable characteristics for a microalgae to be potential organism to biofuels production are tolerate shear stresses found in the reactors (especially in closed photobioreactors), to be dominant in relation to contaminant microorganism strains, large CO_2 absorption capacity in photoautotrophic systems (high photosynthetic efficiency), tolerate large temperature variations resulting from daily and seasonal cycles, low nutrient requirement, potential of high value-added coproducts in addition to the desired product, present a short productive cycle and self-flocculation to facilitate the recovery stage of the microalgal biomass.

The use of microalgae and cyanobacteria for the production of the third-generation biofuels has many advantages over higher plants in view of producing the first- and second-generation biofuels, mainly due to their faster growth under several conditions, including in wastewater. The biochemical composition of microalgae grown under normal conditions, that is, without nutrient limitation, primarily encompasses proteins (30–50%), carbohydrates (20–40%), and lipids (8–15%). Microalgae present several compounds in their cells, such as lipids, carbohydrates, proteins, and pigments, in different concentrations. This chemical profile directly reflects the nature of the microorganism (as its species or lineage), the influence of the chosen culture conditions, and the stage of growth of the culture. In this way, the same microalgae species can present different compositions when handling the specified factors (Zepka et al. 2008). For the production of the third-generation bioethanol, one should select a microalgae species with the ability to produce high concentrations of carbohydrates instead of lipids as energy reserve compound (Mussatto et al. 2010).

Several studies have shown that limiting the amount of nitrogen in the culture medium is one of the main factors that leads to the accumulation of carbohydrates by microalgae (Dragone et al. 2011). According to Behrenset et al. (1989), microalgae in a nitrogen-deprived culture medium direct the flow of carbon to the synthesis of carbohydrates in detriment of the production of proteins. Thus, effort to increase yields of biofuels produced by microalgae is underway, including the optimization of light technologies to modify the carbon uptake pathways, aimed at a higher accumulation of biomass or specific compounds such as carbohydrates and lipids or, more recently, the use of genetic engineering for producing bioethanol, biohydrogen, and other special fermentation products (de Farias Silva and Bertucco 2016). Photosynthetic organisms are favorable for the production of biofuels, mainly because of their low cost of cultivation, but biofuel yields obtained under normal conditions are not satisfactory. In addition to the production of biodiesel, microalgae and cyanobacteria serve as attractive feedstock for the production of bioethanol, although the scientific and technological knowledge on this context is still scarce. On the contrary, studies have documented that the contents of oil and carbohydrates in microalgae cells can be increased under stress conditions, resulting, for instance, in a decrease of the protein content under nitrogen depletion (Ho et al. 2013; Wang et al. 2013). This approach could be applied to cultivate microalgae biomass richer in carbohydrates, thereby leveraging their use for the production of bioethanol, which is currently the most widely used biofuel in the world.

However, under or non-optimized growth conditions, some microalgae strains have been receiving special attention because they present the potential of industrial application for the production of bioethanol of the third generation. Hirano et al. (1997) found two with high starch: *Chlamydomonas reinhardtii* (UTEX 2247) with 45% starch (dry basis) and *C. vulgaris* (IAM C-534) with 37% starch. The microalgae yields were, respectively, 11 and 32 g dry mass/(m^2 day). Dragone et al. (2011) produced biomass of *C. vulgaris* with up to 41% starch (dry basis) under low nitrogen culture conditions. According to Doucha and Lívanský (2009), a mutant strain for the production of starch from *Chlorella* sp. can accumulate 70% starch (dry basis) under conditions of suppression of protein production.

Technologies for the first (sugar or starch feedstock) and second generations (lignocellulosic feedstock) of bioethanol basically involve two stages: the conversion of sunlight into chemical energy (such as carbohydrates and lipids) and the conversion of chemical energy into biofuel. These two stages are related to each other and result in increased production costs. As an improvement of this process, the use of a single-stage system that is capable of capturing sunlight directly and converting it into biofuel (bioethanol) would avoid one step, thereby reducing the cost of production and increasing the sustainability of the bioethanol production process. Three possible routes involving the use of microalgae and cyanobacteria biomass for bioethanol production are discussed in the literature, accordingly summarized in Fig. 1 (de Farias Silva and Bertucco 2016). The first one is the traditional process in which the biomass undergoes pretreatment steps, enzymatic hydrolysis, and yeast fermentation. The second route is the use of metabolic

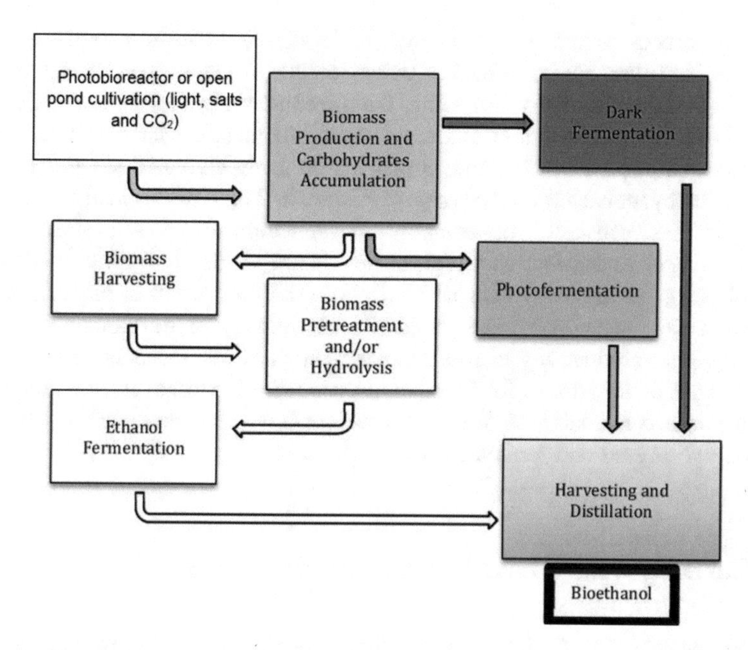

Fig. 1 Routes of bioethanol production from microalgae (adapted from de Farias Silva and Bertucco 2016)

pathways in dark conditions, redirecting photosynthesis to produce hydrogen, acids, and alcohols (such as ethanol). The third way is via "photofermentation," which is impracticable in nature. The last route requires the use of genetic engineering to redirect the preexisting biochemical pathways of microalgae for a more subjective and efficient production of bioethanol.

Photosynthesis is a vital process that drives the synthesis of all biofuels, by converting light energy into biomass, carbon storage products (carbohydrates and lipids), and a small amount of H_2. In green algae, the light-harvesting complex (LHC) (chlorophylls and carotenoids) absorbs photons from sunlight as chemical energy. This energy is used by the photosystem II (PS II) for the catalytic oxidation of water to form protons, electrons, and molecular oxygen. Low-potential electrons are transferred to the electron transport chain for the reduction of ferredoxin and then the formation of nicotinamide adenine dinucleotide phosphate (NADPH). An electrochemical gradient is formed, and the release occurs after oxidation of water in the thylakoid lumen, which is used to produce adenosine triphosphate (ATP) by ATP synthase. Photosynthetic products (NADPH and ATP) are substrates for the Calvin–Benson cycle, where CO_2 is fixed as C3 molecules that are assimilated to form sugars, lipids, and other biomolecules essential for cell growth.

Biofuels from microalgae have been the subject of intense research mainly focused on the production of biodiesel and biogas, although bioethanol and biohydrogen are also considered. The production pathways and operating conditions vary for each biofuel. Studies have already demonstrated the potential viability of industrial processes for the production of biodiesel, according to the previous chapters. However, studies aimed at consolidating a suitable process for the production of bioethanol are still ongoing. On the contrary, cyanobacteria strains have been shown to produce relevant amount of bioethanol. Markou et al. (2013) evaluated the potential of bioethanol production using carbohydrate-enriched biomass of the cyanobacterium *Arthrospira platensis*. The biomass acid hydrolysates were used as substrate for ethanolic fermentation by a salt stress-adapted *Saccharomyces cerevisiae*, with highest bioethanol yields of 16.32% ± 0.9 (gram ethanol per gram biomass) with HNO_3 0.5 N. The production of bioethanol from microalgae and cyanobacteria is a feasible technological development, as they showed higher productivity than certain crops such as sugarcane and corn (already consolidated as feedstocks for bioethanol production). Moreover, microalgae and cyanobacteria can reach 50% of their dry weight (DW) in carbohydrates, which can then be hydrolyzed and fermented with high yields.

2 Carbohydrate Accumulation by Microalgae

Microorganisms with potential for bioethanol production in this way are selected primarily in accordance with their ability to accumulate carbohydrates, which depends on environmental and nutritional conditions. The main environmental factors are light intensity, pH, salinity, and temperature, while the nutritional factors include availability and source type for nitrogen, carbon, phosphorus, sulfur, and iron (Chen et al. 2013; Markou et al. 2013).

Genera *Scenedesmus*, *Chlorella*, *Chlorococcum*, and *Tetraselmis* from *Chlorophyta* division and *Synechococcus* among other cyanobacteria have been extensively studied as feedstock for this type of bioethanol production. In general, the cultivation in a high light intensity ranged from 150 to 450 μmol/(m^2s) using a mix of CO_2 in air between 2 and 5% and mesophilic temperatures (20–30 °C) achieves around 50% of carbohydrate content under nutrient starvation, mainly nitrogen, according to Table 2 (de Farias Silva and Bertucco 2016). However, carbohydrate content could be extremely variable and the productivity depends on the cell growth too, that is, growing conditions that allow the simultaneous accumulation and growth. According to Rizza et al. (2017), generally microalgal strains that accumulated the highest levels of carbohydrates did not accumulate lipids under identical growth conditions.

The positive effect of increasing light intensity on the accumulation of starch and lipids is feasible only up to a point, usually equal to saturation of photosynthesis under given conditions in a particular species. Nutritional factors directly or indirectly influence the rate of photosynthesis and biochemical composition of microalgae. Macroelement (nitrogen, sulfur, or phosphorous) limitation is the most widely used and so far the most successful strategy for enhancing starch accumulation. For example, availability of nitrogen enhances the synthesis of proteins, pigments, and DNA, the amount of iron affects the photosynthetic electron

Table 2 Microalgae carbohydrate content in different growth conditions (adapted from Dragone et al. 2011; de Farias Silva and Bertucco 2016; Rizza et al. 2017)

Microalgae	Growth conditions	Carbohydrate (%)
Arthrospira platensis	150 μmol/(m^2s), 30 °C, bubbling air	58.0
Chlamydomonas reinhardtii UTEX 90	450 μmol/(m^2s), 23 °C, 4 days, and 130 rpm	59.7
Chlorella vulgaris KMMCC-9 UTEX26	150 μmol/(m^2/s), 20–22 °C, bubbling air	22.4
Chlorella sp. *KR-1*	80 μmol/(m^2s), 30 °C, and 10% CO_2	49.7
Chlorella sp. *TISTR 8485*	BG11 medium for 20 days	27.0
Chlorococcum sp. TISTR 8583	BG11 medium for 20 days	25.9
Scenedesmus obliquus	150 μmol/(m^2s), 25 °C, bubbling air	30.0
Scenedesmus obliquus CNW-N	210–230 μmol/(m^2s), 28 °C, 300 rpm, and 2.5% CO_2	51.8
Synechococcus elongatus PCC 7942	200 μmol/(m^2s), 28 °C, and 5% CO_2	28.0
Synechococcus sp. PCC 7002	250 μmol/(m^2s) and 1% CO_2	59.0
Tetraselmis subcordiformis FACHB-1751	150 μmol/(m^2s), 25 °C, and 3% CO_2	40.0
Ankistrodesmus sp. strain LP1	BG11 medium supplemented with 1 mM $NaNO_3$	51.3
Desmodesmus sp. strain FG	BG11 medium with 1 mM $NaNO_3$	53.5
Pseudokirchneriella sp. strain C1D	BG11 medium with 1 mM $NaNO_3$	40.5
Scenedesmus obliquus strain C1S	BG11 medium with 1 mM $NaNO_3$	29.9

transport, nitrite/nitrate and sulfate reduction, nitrogen fixation, and/or detoxification of reactive oxygen species (ROS). Sulfur involves the formation of sulfolipids, polysaccharides, and proteins, as well as in the electron transport chain. When sulfur is present at limiting concentrations, it inhibits cell division, whereas high concentrations inhibit the photosynthetic assimilation of carbon-rich compounds, such as carbohydrates. CO_2 is the most common source of carbon (autotrophic condition), and under nitrogen depletion conditions, the supplementation of CO_2 in conjunction with light intensity causes the carbon to be absorbed and converted into carbohydrates more efficiently.

According to Dragone et al. (2011), increasing microalgal starch content by nutrient limitation has been regarded as an affordable approach for the production of the third-generation bioethanol. Thus, these authors have evaluated starch accumulation in *C. vulgaris* P12 under different initial concentrations of nitrogen (0–2.2 g_{urea}/L) and iron (0–0.08 $g_{FeNa\text{-}EDTA}$/L) sources, using an experimental design. Starch accumulation occurred at nitrogen depletion conditions. Cell growth was much slower than that observed during nitrogen-supplemented cultivations. The authors proposed a two-stage cultivation process for high starch accumulation:

a first cultivation stage using nitrogen- and iron-supplemented medium, followed by a second cultivation stage in a nitrogen- and iron-free medium. The high starch content obtained (up to 41.0% of dry cell weight) suggests *C. vulgaris* P12 as a very promising feedstock for bioethanol production.

Carbohydrates are the major products derived from photosynthesis and the carbon fixation metabolism (Calvin cycle), which are either accumulated in the plastids as reserve materials (starch), or become the main component of cell walls (cellulose, pectin, and sulfated polysaccharides). However, the composition and metabolism of carbohydrates (mainly starch and cellulose) in microalgae may differ significantly from species to species. Microalgae that contain glucose-based carbohydrates are the most feasible feedstock for bioethanol production (Chen et al. 2013). The cell walls of microalgae primarily consist of an inner cell wall layer and an outer cell wall layer. The composition of the outer cell wall varies from species to species, but usually contains specific polysaccharides, such as pectin, agar, and alginate, while the inner cell wall layer is mainly composed of cellulose and other materials. Table 3 shows the compositions of the cell walls and the storage products. For some microalgae, the glucose polymers produced via cellulose/starch are the predominant component in the cell walls and stored products of microalgae. Starch and most cell wall polysaccharides can be converted into fermentable sugars for subsequent bioethanol production via microbial fermentation.

The accumulation of carbohydrates in microalgae is due to CO_2 fixation during the photosynthetic process (Fig. 2). Photosynthesis is a biological process utilizing ATP/NADPH to fix and convert CO_2 captured from the air to produce glucose and other sugars through a metabolic pathway known as the Calvin cycle. The metabolic pathways of energy-rich molecules are closely linked. Some studies demonstrated that there was a competition between lipid and starch synthesis because the major precursor for triacylglycerols synthesis is glycerol-3-phosphate (G3P), which is produced via catabolism of glucose (glycolysis). Thus, to enhance biofuels' production from microalgae-based carbohydrates, it is vital to understand and manipulate the related metabolisms to achieve higher microalgal carbohydrate

Table 3 Composition of microalgal cell wall and storage products (Chen et al. 2013)

Division	Cell wall	Storage products
Cyanophyta	Lipopolysaccharides, peptidoglycan	Cyanophycean starch
Chlorophyta	Cellulose, hemicellulose	Starch/lipid
Dinophyta	Absence or contain few cellulose	Starch
Cryptophyta	Periplast	Starch
Euglenophyta	Absence	Paramylum/lipid
Rhodophyta	Agar, carrageenan, cellulose, calcium carbonate	Floridean starch
Heterokontophyta	Naked or covered by scales or with large quantities of silica	Leucosin/lipid

accumulation via strategies like increasing glucan storage and decreasing starch degradation. The starch forms around a crystallizing nucleus and is present as an amorphous starch grain. When a chloroplast gathers enough starch, it may become an amyloplast. However, the detailed changes in enzymatic activity and metabolic flux of carbohydrate biosynthesis of microalgae are poorly understood. The manipulation of the carbohydrate metabolisms of microalgae by genetic engineering has also been proposed. With the development of genetic engineering of microalgae, and a better understanding of the biochemistry of microalgae carbohydrate metabolisms, superior strains for carbohydrate accumulation could be developed.

Except the starch in plastids, microalgal extracellular coverings (cell wall) are another carbohydrate-rich part, which could be transformed to biofuel. However, the compositions of microalgal extracellular coverings are diverse by species. Among them, cellulose is one of the main fermentable carbohydrates in most of green algae. Cellulose synthesis is a complicated process that includes many enzymatic reactions. The starting substrate for cellulose synthesis is UDP-glucose, which is formed from the reaction of UDP and fructose catalyzed by sucrose synthase. Despite the understanding of main carbohydrate metabolism in microalgae, in-depth knowledge on its regulation is still lacking. It is important to integrate updated information of genomic sequences, transcriptomes, proteomes, and metabolomes data at systems level to meet the challenges on economic biofuels production from microalgae.

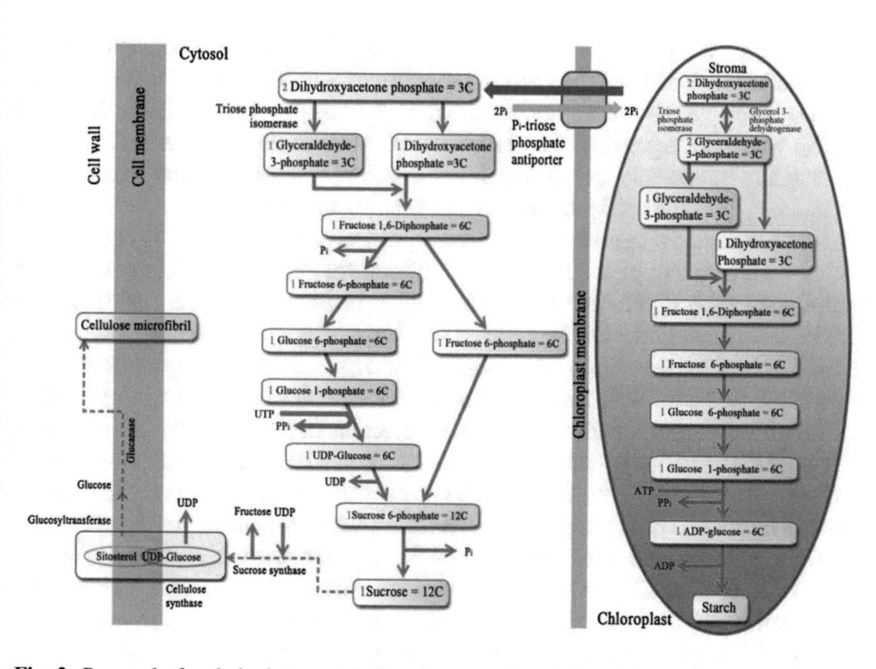

Fig. 2 Proposal of carbohydrate metabolism in green algae (adapted from Chen et al. 2013)

Although systems study of microalgae on carbohydrate metabolisms is currently in its infant stage, omics studies on microalgae have made significant progress. Such a strategy will open a door for efficient carbohydrate metabolic regulation and genetic engineering of microalgae for biofuels' production.

3 Technologies of Microalgal Carbohydrates to Bioethanol

The main technological routes for bioethanol production by microalgal biomass involve hydrolysis–yeast fermentation, the use of metabolic pathways in dark conditions, and "photofermentation."

The hydrolysis of biomass is the most used method for the use of microalgal carbohydrates. Hydrolysis–fermentation of microalgal biomass is based on the production of microalgae biomass succeeded by pretreatment steps, involving breakdown of the cell structure and hydrolysis of the biomass, and frequently by the addition of enzymes. The treated biomass is then fermented with yeasts or bacteria to obtain ethanol. The main drawbacks of this route are the multistep processes required, which demands more energy, and the use of enzymes and yeasts, which accounts for a considerable proportion of the costs. On the contrary, the hydrolysis/fermentation process converts biomass at the highest rate, because of the well-known high efficiency of enzymes and yeasts in converting biomass into products.

Markou et al. (2013) studied the potential of bioethanol production using carbohydrate-enriched biomass of the cyanobacteria *A. platensis*. For the saccharification of the carbohydrate-enriched biomass, four acids (H_2SO_4, HNO_3, HCl, and H_3PO_4) were investigated. The hydrolysates then were used as substrate for ethanol fermentation by a salt stress-adapted *Saccharomyces cerevisiae* strain. According to the authors, the highest bioethanol yield was observed at acid concentration of 0.5 N. At this concentration, fermentation of hydrolysates with HCl as catalyst had the lowest bioethanol yield (13.41% gram of ethanol per gram dry biomass), while hydrolysates with H_2SO_4 and HNO_3 as catalysts had bioethanol yield of 16.27 and 16.32%, respectively. *Chlorella* biomass was hydrolyzed in the presence of 2% HCl and 2.5% $MgCl_2$, a sugar concentration of nearly 12%, and a sugar recovery of about 83% was obtained. Fermentation experiments demonstrated that glucose in the *Chlorella* biomass hydrolysates was converted into ethanol by *S. cerevisiae* with a yield of 0.47 g/g, which is 91% of the theoretical yield (Zhou et al. 2011).

Rizza et al. (2017) researched *Desmodesmus* sp. strain for production of biomass fermentable. Hydrolyzed preparations were brought to pH 5.5–6.0 with $Mg(OH)_2$ crystals and used directly or after concentration by freeze-drying for ethanol fermentation. A detailed time-course analysis of the increase in biomass and accumulation of total carbohydrates and proteins indicated that *Desmodesmus* sp. strain FG grew robustly, its reaching ratios of carbohydrates to protein over 2.

Microalgae biomass at 100 g/L was hydrolyzed according to the optimized conditions to yield soluble carbohydrates preparations. These preparations were inoculated with *S. cerevisiae* cells and accumulated about 23 g ethanol per liter, representing approximately 81% of the maximum theoretical. These results indicated that microalgae biomass could be converted into ethanol by baker's yeast efficiently as commercial grade dextrose and that other nutrients, usually used to improve fermentation, such as the N-source, were already present in the hydrolyzed microalgal biomass. Both almost complete exhaustion of carbohydrates from the fermentation broth and high conversion efficiency of carbohydrates into ethanol indicated very high enrichment of fermentable sugars in the biomass of the strains selected in this study and in their corresponding hydrolysates. It also indicated that sugar loss and/or generation of fermentation inhibitors from microalgal biomass remained at negligible levels after the optimized saccharification treatment. These results contribute to support the potential of microalgae biomass as an alternative feedstock for bioethanol and the value of bioprospecting programs to identified candidate strains among natural biodiversity.

Yuan et al. (2016) evaluated liquid hot water pretreatment prior to enzymatic hydrolysis of *Scenedesmus* sp. The concentration and recovery of total sugars and glucose at 100 °C were 0.85 and 0.26 g/L, respectively, while 13.4 and 0.16 g/L at 200 °C. These results indicated that the increase of temperature could accelerate the motions of solvent molecules (sulfuric acid) and improve the liberation of sugars. Thus, according to these authors, liquid hot water pretreatment could greatly enhance the enzymatic efficiency and could be regarded as an ideal method for glucose recovery from microalgae.

Mixed microalgae cultures could be considered as an attractive research area compared to traditional pure culture to dominate cultivation contamination risk and enhance economic feasibility of large-scale biofuel production. In this sense, Shokrkar et al. (2017) evaluate the effect of different pretreatment strategies including acidic, alkaline, and enzymatic hydrolysis on the sugar extraction from mixed microalgae. According to these authors, total carbohydrates content of microalgal biomass increased about 20.1% in the absence of nitrogen (about 36% in terms of volatile suspended solids amount). Dilute acids decompose cellulose, and starch in the biomass to release simple sugars. Hydrolysis kinetic depends on the type of substrate, temperature, acid concentration, and reaction time. Results showed that the mixture of dilute sulfuric acid and $MgSO_4$ exhibited a higher sugar yield than dilute acid. Among all pretreatments used, the enzymatic treatment with thermostable enzymes showed the highest recovery of 0.951 g of extracted glucose per gram of total sugar. Moreover, the enzymatic pretreatment of wet microalgae was compared with dried ones at identical operational conditions and dried biomass concentration of 50 g/L, and similar sugar yields were achieved which would be advantageous to reduce the need for drying of the microalgae biomass. Fermentation of the acidic and enzymatic treated samples to ethanol using *Saccharomyces cerevisiae* showed yield of 0.38 and 0.46 g/g glucose, corresponding to 76 and 92% of the theoretical values, respectively. These authors reported that bioethanol yield after enzymatic hydrolysis of mixed microalgae culture is higher than that of acid hydrolysis. Carbohydrates in microalgae biomass are mainly cellulose and starch. Cellulose molecules are glucose polymers linked together by β-1,4 glucosidic bonds, as opposed to the α-1,4 and α-1,6 glucosidic bonds for starch. In the enzymatic pretreatment of algae, β-glucosidase/cellulase hydrolyzed β-1,4 glucosidic bonds of algal cellulose, whereas α-amylase liquefied algal starch to oligosaccharides through the hydrolysis of the α-1,4 glucosidic linkages, and then amyloglucosidase hydrolyzed α-1,4 and α-1,6 glucosidic bonds of oligosaccharides into glucose. Therefore, it is desirable to use three enzymes in the enzymatic pretreatment of microalgae, thus improving the hydrolysis yields even further.

Another process known as "dark fermentation" refers to the conversion of organic substrates into biohydrogen (de Farias Silva and Bertucco 2016). Fermentative and hydrolytic microorganisms hydrolyze complex organic polymers into monomers, which are subsequently converted into a mixture of organic acids of low molecular weight and alcohols, mainly acetic acid and ethanol. Various microalgae and cyanobacteria that are capable of expelling ethanol through the cell wall by means of intracellular process in the absence of light include *C. reinhardtii*, *Chlamydomonas moewusii*, *C. vulgaris*, *Oscillatoria limnetica*, *Oscillatoria limosa*, *Gleocapsa alpícola*, *Cyanothece* sp., *Chlorococcum littorale*, and *Spirulina* sp. and *Synechococcus* sp. However, dark fermentation is disadvantageous in terms of hydrogen productivity, because approximately 80–90% of the initial chemical oxygen demand (COD) remains in the form of acids and alcohols after the process. Even under optimal operating conditions, typical yields vary only between 1 and 2 mol H_2 per mol of glucose. The production of ethanol is favored by the accumulation of carbohydrates in the microalgae cells through photosynthesis, and then, the microalgae are forced to synthesize ethanol through fermentative metabolism directly from their carbohydrate and lipid reserves when switching the growth to dark conditions. However, it can be concluded that dark fermentation of microalgae is not an efficient process for the production of bioethanol.

"Photofermentation" is a process of growing interest principally after the announcement of the installation of industrial plants where modified cyanobacteria are used to produce bioethanol directly. The "photofermentative" route (simply, photanol) is a natural mechanism of converting sunlight into products of fermentation through a highly efficient metabolic pathway. Photanol is not only limited to ethanol production, but it is also used for a large number of naturally occurring products resulting from glycolysis-based fermentation (Rai and Singh 2016). Thus, several cyanobacteria species can be genetically modified by introducing specific fermentation cassettes through molecular engineering procedures, and then tested as a fermentative organism. *Synechococcus* sp. is a unicellular cyanobacterium living in freshwater that has been relatively well characterized. It is capable of tolerating insertion of foreign DNA to be transformed and replicated using shuttle vectors between *Escherichia coli* and cyanobacteria, or insertion of foreign DNA into the chromosome through homologous recombination at selected active sites. *Synechocystis* sp. PCC 6803 was the first photosynthetic organism that had its genome sequenced and one of the best characterized cyanobacteria.

Thermosynechococcus is also naturally transformable. The metabolic pathway of ethanol synthesis is briefly summarized: After fixation of inorganic carbon by Calvin cycle, it forms phosphoglycerate that is converted into pyruvate by two enzymes (pyruvate decarboxylase and alcohol dehydrogenase), and finally into ethanol. Therefore, the "photofermentation" process for obtaining ethanol includes two stages: photosynthesis and fermentation. Each stage has its key factors that determine the efficiency of the process and the metabolic needs of the cyanobacteria. In any case, this route requires the use of genetically modified microorganisms.

Figures 3 and 4 present the schematic diagram of the assumed fermentative pathways operating in dark-incubated wild type *Chlamydomonas reinhardtii* and mutant PFL1-deficient strain 48F5 (Philipps et al. 2011). In fermenting *C. reinhardtii* wild type cells (CC-124), pyruvate from glycolytic glucose oxidation, serves as substrate for several enzymes. Pyruvate formate lyase (PFL1) cleaves pyruvate into formate and acetyl CoA. Acetyl CoA is converted to acetate by the successive action of phosphotransacetylase (PTA) and acetate kinase (ACK), resulting in ATP production, or to ethanol by a bifunctional aldehyde/alcohol dehydrogenase (ADH1), resulting in oxidation of NAD(P)H. Pyruvate decarboxylase (PDC) decarboxylates pyruvate yielding acetaldehyde, which is further reduced to ethanol by alcohol dehydrogenase (ADH). Another pathway leads to D-lactate production by the action of D-lactate dehydrogenase (D-LDH). Pyruvate ferredoxin oxidoreductase (PFR1) oxidatively decarboxylates pyruvate, resulting in reduced ferredoxin (FDX), CO_2, and acetyl CoA. The latter can probably be metabolized by PTA and ACK or ADH1 (indicated by a dotted line). Reduced FDX could then function as an electron donor for the hydrogenase (HYD1), resulting in hydrogen evolution in the dark.

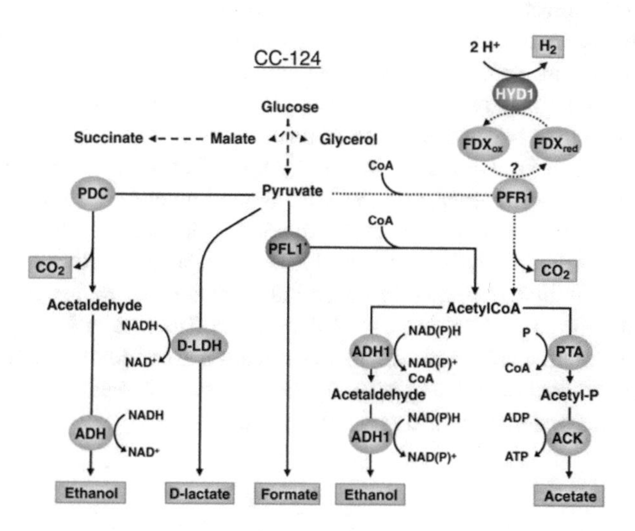

Fig. 3 Fermentative pathways of wild type *Chlamydomonas reinhardtii* and mutant PFL1-deficient strain 48F5

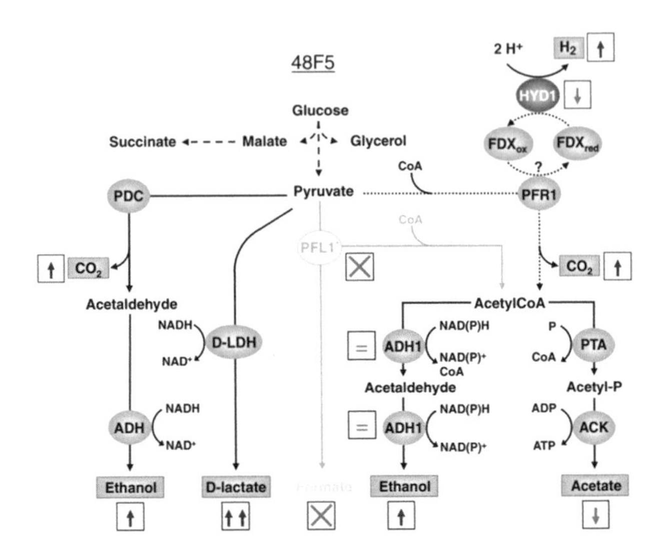

Fig. 4 Fermentative pathways of *C. reinhardtii* mutant PFL1-deficient strain 48F5

Pathways could be resulting in the other products as malate, succinate, and glycerol, which have been reported to be fermentative products of *C. reinhardtii.*

In contrast, the PFL1 pathway is not active in strain 48F5 because of disruption of the PFL1 gene (indicated by light gray lines and text, and red crosses near PFL1 and below formate in Fig. 4). Instead, the dark-incubated PFL1 mutant generated more H_2, CO_2, ethanol, and D-lactate than the wild type, while acetate secretion was reduced. Strain 48F5 also showed reduced in vitro hydrogenase activity and reduced HYD1 transcript and HYD1 protein levels. The amounts of ADH1 were almost identical in the wild type and the PFL1 mutant. Red downward arrows indicate a reduction, green upward arrows indicate an increase, and orange equal symbols indicate unchanged results. The double upward arrows for D-lactate indicate a more than twofold increase in this metabolite.

Costa et al. (2015) reported in their study the effect of inoculum concentration and carbon source to *C. reinhardtii*, as well as the influence of hybrid system and coculture (*C. reinhardtii* and *R. capsulatus*) on the photofermentative ethanol production. Maximum ethanol content (19.94 g/L) and productivity (0.17 g/(Lh)) were achieved by hybrid system in which the effluent of *C. reinhardtii* containing organic acids was used as substrate to *R. capsulatus*. The results from this work are beneficial to comprehend the potentiality of microalgae and photosynthetic bacteria to synthesize ethanol concerning several strategies such as media composition and different culture systems (hybrid and cocultivation).

4 Cases and Outlook for Commercial Production

It is broadly accepted that microalgal-based biofuels' economics would be largely improved if obtained in the frame of biomass biorefineries for the production of multiple commodities and higher value products. According to the US Energy Information Administration (EIA 2017), world biofuel production will increase from approximately 1.3 million barrels per day in 2010 to approximately 3.0 million barrels per day in 2040 (Kim et al. 2017). Fermentations run in study of (Rizza et al. 2017) yielded as coproducts 0.06 kg dry edible yeast *S. cerevisiae* per 1 kg dry *Desmodesmus* sp. biomass and the spent fermentation broth that would be used as animal feed supplements or other biotechnological applications. It is presumed that CO_2 produced as a fermentation product (at least 0.22 kg/kg of dry *Desmodesmus* biomass) could be recycled into microalgae to increase productivity and reduce the C footprint of bioethanol production, as previously reported in the literature (Stewart and Hessami 2005).

Moreover, sufficient carbohydrate content and efficient biomass harvest are required for economical bioethanol production from microalgae. Kim et al. (2017) studied the red algae *P. cruentum*, which is one of the most promising candidate organisms for producing fatty acids, lipids, carbohydrates, and pigments, from seawater and freshwater. In this, research was compared to the separate hydrolysis and fermentation, and simultaneous saccharification and fermentation methods. After optimizing each process, these authors designed an overall mass balance for bioethanol production: 100 g of seawater microalgae consists of 16.9 g glucose, 5.3 g of galactose, and 4.7 g of xylose, whereas 100 g of freshwater microalgae consists of 16.6 g glucose, 5.5 g galactose, and 6.4 g xylose. Saccharification and fermentation processing (5% substrate loading, w/v) of microalgae was conducted with pectinase (4.8 mg/g), cellulase (7.2 mg/g), and *S. cerevisiae* at 37 °C for 12 h, resulting in ethanol production of 5.58 and 5.90 g, respectively (Fig. 5). These results suggest that freshwater is a more efficient candidate for bioethanol production than seawater biomass.

Algenol is an American company owner of the first industrial plant for bioethanol production from engineered microorganisms. *Cyanobacterium* sp. with plasmids of a heterologous alcohol dehydrogenase gene (from *Synechocystis*) and pyruvate decarboxylase gene (from Zymomonas) (Piven et al. 2015). These high photosynthetic efficiency values can be ascribed not only to the species used but also to the geometrical characteristics of the photobioreactors (vertical bags) and to the cultivation under continuous conditions. The main limitations reported about this process are the fixed carbon/ethanol ratio, incidence of light, contaminants, and CO_2 supply time.

Other companies have been research of bioethanol production from microalgae, according to review (de Farias Silva and Bertucco 2016). In 2011, Joule Unlimited started a project to build an industrial plant using an engineered cyanobacterium from light, carbonic gas, water, and salts, with authorization of the Environmental Protection Agency (EPA) in 2014. This company claims to have an efficient system

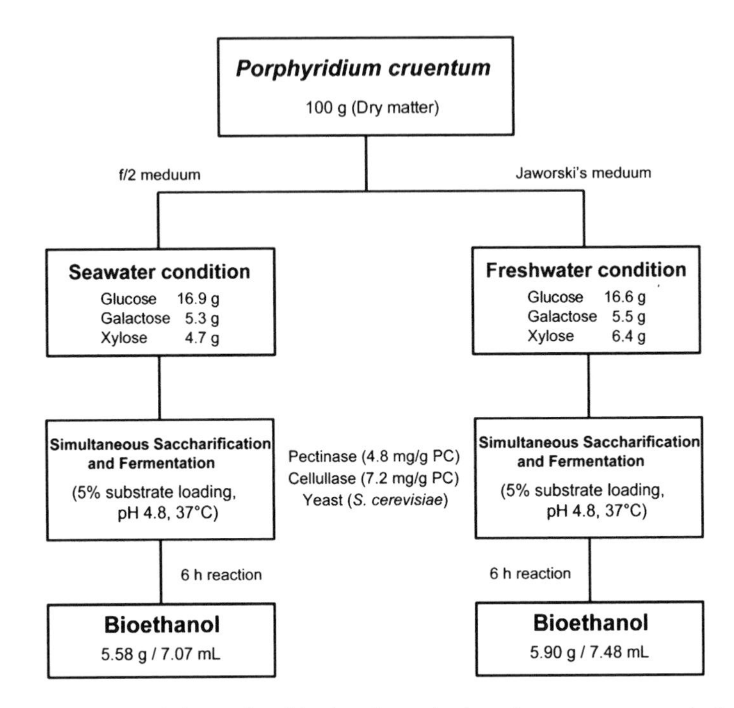

Fig. 5 Overall mass balance for bioethanol production from seawater and freshwater *Porphyridium cruentum* (Kim et al. 2017)

to directly produce biofuels such as alkanes and ethanol from CO_2. It was reported that a photosynthetic efficiency of 6–7% was achieved, in comparison with algal open-pond values of 1.5%, both in outdoor conditions. The system proposed is based on a reactor called SolarConverter® (a horizontal thin film plastic using CO_2 in a closed system and outdoor), where the mixing, culture density, and geometry (depth and surface area) have been studied to optimize the capture and conversion of CO_2 by an appropriate combination of the light and dark areas with the reactor. The company estimated ethanol productivity >230,000 L/(ha year) with a production cost of US$ 0.16/L of ethanol with subsidies (US$ 0.32/L without subsidies).

The costs of bioethanol production from sugarcane (Brazil, 0.16–0.22 US$/L) are lower than those from corn (USA, 0.25–0.40 US$/L), sugar beet (Europe, 0.43–0.73 US$/L), and lignocellulosic materials (USA, 0.43–0.93 US$/L) (Gupta and Verma 2015). It is quite difficult to estimate the economics of bioethanol from genetically engineered cyanobacteria. Algenol announced a production cost of approximately 0.79 US$/L, and potential application of this method of bioethanol production will be increased with the continuous decrease.

5 Conclusions

According to available scientific literature and company initiatives, it is clear that the bioethanol production from microalgae should focus on not only the increase of the carbohydrate content but also the higher productivity of biomass. Technical and economic evaluations are necessary to verify the gains and losses of energy involved in the production of ethanol from microbiological biomass. The relevance of genetically engineered microorganisms with traditional processes must also be discussed, as it is well known that enzymes and yeasts can efficiently produce bioethanol with high productivity. The main technological bottlenecks of hydrolysis and fermentation seem to be being solved by several researches in this area, even helped by the production of the second-generation ethanol. Finally, more studies are necessary, particularly for better understanding of carbohydrate accumulation (hydrolysis and fermentation), as well as metabolic pathway of dark and photofermentation, which appears indeed as a highly promising technological application in the future.

References

Behrens, P. W., Bingham, S. E., Hoeksema, S. D., Cohoon, D. L., & Cox, J. C. (1989). Studies on the incorporation of CO_2 into starch by *Chlorella vulgaris*. *Journal of Applied Phycology, 1,* 123–130.

Brennan, l, & Owende, P. (2010). Biofuels from microalgae—A review of technologies for production, processing, and extractions of biofuels and co-products. *Renewable and Sustainable Energy Reviews, 14,* 557–577.

Chen, C. Y., Zhao, X. Q., Yen, H. W., Ho, S. H., Cheng, C. L., Lee, D. J., et al. (2013). Microalgae-based carbohydrates for biofuel production. *Biochemical Engineering Journal, 78,* 1–10.

Costa, R. L., Oliveira, T. V., Ferreira, J. S., Cardoso, V. L., & Batista, F. R. (2015). Prospective technology on bioethanol production from photofermentation. *Bioresource Technology, 181,* 330–337.

de Farias Silva, C. E., & Bertucco, A. (2016). Bioethanol from microalgae and cyanobacteria: A review and technological outlook. *Process Biochemistry, 51*(11), 1833–1842.

Doucha, J., & Lívanský, K. (2009). Outdoor open thin-layer microalgal photobioreactor: potential productivity. *Journal of Applied Phycology, 21,* 111–117.

Dragone, G., Fernandes, B. D., Abreu, A. P., Vicente, A. A., & Teixeira, J. A. (2011). Nutrient limitation as a strategy for increasing starch accumulation in microalgae. *Applied Energy, 88* (10), 3331–3335.

Gupta, A., & Verma, J. P. (2015). Sustainable bio-ethanol production from agro-residues: A review. *Renewable and Sustainable Energy Reviews, 41,* 550–567.

Hirano, A., Ueda, R., Hirayama, S., & Ogushi, Y. (1997). CO2 fixation and ethanol production with microalgal photosynthesis and intracellular anaerobic fermentation. *Energy, 22,* 137–142.

Ho, S. H., Huang, S. W., Chen, C. Y., Hasunuma, T., Kondo, A., & Chang, J. S. (2013). Bioethanol production using carbohydrate-rich microalgae biomass as feedstock. *Bioresource Technology, 135,* 191–198.

Kim, H. M., Oh, C. H., & Bae, H. J. (2017). Comparison of red microalgae (*Porphyridium cruentum*) culture conditions for bioethanol production. *Bioresource Technology, 233,* 44–50.

Markou, G., Angelidaki, I., Nerantzis, E., & Georgakakis, D. (2013). Bioethanol production by carbohydrate-enriched biomass of *Arthrospira* (Spirulina) platensis. *Energies, 6*(8), 3937–3950.

Mata, T. M., Martins, A. A., & Caetano, N. S. (2010). Microalgae for biodiesel production and other applications: A review. *Renewable and Sustainable Energy Reviews, 14,* 217–232.

Mussatto, S. I., Dragone, G., Guimarães, P. M., Silva, J. P., Carneiro, L. M., Roberto, I. C., et al. (2010). Technological trends, global market, and challenges of bio-ethanol production. *Biotechnology Advances, 28,* 817–830.

Philipps, G., Krawietz, D., Hemschemeier, A., & Happe, T. (2011). A pyruvate formate lyase-deficient *Chlamydomonas reinhardtii* strain provides evidence for a link between fermentation and hydrogen production in green algae. *Plant Journal, 66*(2), 330–340.

Piven, I., Friedrich A., Dühring, U., Uliczka, F., Baier, K., Inaba, M., et al. (2015). *Cyanobacterium sp. host cell and vector for production of chemical compounds in cyanobacterial cultures.* U.S.Patent 8846369 B2.

Rai, P. K., & Singh, S. P. (2016). Integrated dark- and photo-fermentation: Recent advances and provisions for improvement. *International Journal of Hydrogen Energy, 41*(44), 19957–19971.

Rizza, L. S., Smachetti, M. E. S., Nascimento, M., Salerno, G. L., & Curatti, L. (2017). Bioprospecting for native microalgae as an alternative source of sugars for the production of bioethanol. *Algal Research, 22,* 140–147.

Shokrkar, H., Ebrahimi, S., & Zamani, M. (2017). Bioethanol production from acidic and enzymatic hydrolysates of mixed microalgae culture. *Fuel, 200,* 380–386.

Stewart, C., & Hessami, M. A. (2005). A study of methods of carbon dioxide capture and sequestration—The sustainability of a photosynthetic bioreactor approach. *Energy Conversion and Management, 46*(3), 403–420.

U.S. Energy Information Administration (EIA). (2017). *Annual energy outlook.* https://www.eia.gov/outlooks/aeo/.

Wang, L., Li, Y., Sommerfeld, M., & Hu, Q. (2013). A flexible culture process for production of the green microalga *Scenedesmus dimorphus* rich in protein, carbohydrate or lipid. *Bioresource Technology, 129,* 289–295.

Yuan, T., Li, X., Xiao, S., Guo, Y., Zhou, W., Xu, J., et al. (2016). Microalgae pretreatment with liquid hot water to enhance enzymatic hydrolysis efficiency. *Bioresource Technology, 220,* 530–536.

Zepka, L. Q., Jacob-Lopes, E., Goldbeck, R., & Queiroz, M. I. (2008). Production and biochemical profile of the microalgae *Aphanothece microscopica Nägeli* submitted to different drying conditions. *Chemical Engineering and Processing, 47,* 1305–1310.

Zhou, N., Zhang, Y., Wu, X., Gong, X., & Wang, Q. (2011). Hydrolysis of chlorella biomass for fermentable sugars in the presence of HCl and $MgCl_2$. *Bioresource Technology, 102*(21), 10158–10161.

Chapter 12
Biofuels from Microalgae: Biomethane

Fabiana Passos, Cesar Mota, Andrés Donoso-Bravo, Sergi Astals, David Jeison and Raúl Muñoz

Abstract The high cost of axenic microalgae cultivation in photobioreactors limits nowadays the potential uses of microalgal biomass as a feedstock for the production of biodiesel or bioethanol. In this context, microalgae-based wastewater treatment (WWT) has emerged as the leading method of cultivation for supplying microalgae at low cost and low environmental impacts, while achieving sewage treatment. Nonetheless, the year-round dynamics in microalgae population and cell composition when grown in WWTPs restrict the use of this low-quality biomass to biogas production via anaerobic digestion. Although the macromolecular composition of the microalgae produced during wastewater treatment is similar to that of sewage sludge, the recalcitrant nature of microalgae cell walls requires an optimisation of pretreatment technologies for enhancing microalgae biodegradability. In addition, the low C/N ratio, the high water content and the suspended nature of microalgae suggest that microalgal biomass will also benefit from anaerobic co-digestion with carbon-rich substrates, which constitutes a field for further research. Photosynthetic microalgae growth can also support an effective CO_2 capture and H_2S oxidation from biogas, which would generate a high-quality biomethane complying with most

F. Passos · C. Mota
Department of Sanitary and Environmental Engineering, Federal University of Minas Gerais, Campus Pampulha, Antônio Carlos Avenue, 6.627, Belo Horizonte, MG, Brazil

A. Donoso-Bravo
Inria Chile, Apoquindo Avenue, Floor 12, Las Condes, Santiago 2827, Chile

S. Astals
Advanced Water Management Centre, The University of Queensland, St. Lucia, QLD 4072, Australia

D. Jeison
Biochemical Engineering School, Pontificia Universidad Católica de Valparaíso, Avenida Brasil 2085, Valparaíso 2362803, Chile

R. Muñoz (✉)
Department of Chemical Engineering and Environmental Technology, University of Valladolid, Dr. Mergelina, S/N, Valladolid, Spain
e-mail: mutora@iq.uva.es

E. Jacob-Lopes et al. (eds.), *Energy from Microalgae*, Green Energy and Technology, https://doi.org/10.1007/978-3-319-69093-3_12

international regulations for injection into natural gas grids or use as autogas. This book chapter will critically review the most recent advances in biogas production from microalgae, with a special focus on pretreatment technologies, co-digestion opportunities, modelling strategies, biogas upgrading and process microbiology.

Keywords Anaerobic co-digestion · Biogas upgrading · Microbiology Modelling · Pretreatments

1 Introduction

During the last decade, microalgae production and bioconversion have been widely investigated for bioenergy generation purposes. Nonetheless, energy and life cycle assessments of theoretical and pilot-scale studies have consistently shown that such technology is only feasible if microalgae are grown in open ponds fed with wastewater (Sialve et al. 2009). In this context, high rate algal ponds (HRAPs) have been proved efficient in removing organic matter and nutrients from contaminated effluents (Park et al. 2011), and cost-effective alternatives when compared to activated sludge processes (no external input of aeration is required due to the natural occurrence of photosynthesis).

The microalgae-bacteria biomass produced in such systems may be valorised through anaerobic digestion (AD) with the concomitant production of biogas. This process is already well known and has long been used to produce bioenergy from organic residues such as sewage sludge, agricultural and industrial by-products. In fact, AD may convert microalgae-based wastewater treatment plants (WWTPs) into net energy producers by converting methane into heat and electricity that may be subsequently used in biomass pretreatment and wastewater biodegradation (Passos and Ferrer 2014). Additionally, the mineralisation of microalgae containing organic nitrogen and phosphorus may convert microalgae into a stabilised biosolid fertilizer (Solé-Bundó et al. 2017).

Nonetheless, this technology platform has some bottlenecks that hinder its viability at full-scale. The main issues are: (i) low microalgae production rates due to carbon or light limitation, (ii) costly biomass concentration and (iii) slow biodegradability in anaerobic digesters. Some of these challenges may be overcome by applying pretreatment or co-digestion technologies. Pretreatment can be used to enhance microalgae anaerobic biodegradability by weakening or disrupting microalgae cell wall structure; co-digestion improves the process biogas yield by improving the organic loading rate while controlling ammonia concentration. On the other hand, mathematical models and reactor design and operation strategies need to be carefully reviewed for a better understanding and optimisation of process performance. Finally, the biogas produced during the AD of microalgae should be upgraded prior to its combustion on-site, injection into natural gas grids or used as autogas.

This chapter aims at presenting and discussing the main topics involved in microalgae AD, i.e. the microbiology involved, pretreatment technologies, co-digestion with other substrates, design and operational considerations, process modelling and biogas upgrading to biomethane.

2 The Role of Microbiology in the Anaerobic Digestion of Microalgae

AD of microalgae is a spontaneous process in which organic matter from microalgal cells is converted to biogas through reactions catalysed by naturally occurring microorganisms. Like most biological processes, AD is affected by a variety of factors such as "substrate type", environmental, physical, biological and chemical conditions. Microalgal biomass is composed mainly of organic compounds (mostly lipids, carbohydrates and protein), as well as nitrogen, phosphorus and oligonutrients such as zinc, cobalt and iron. The average composition of microalgae can be expressed as $CO_{0.48}H_{1.83}N_{0.11}P_{0.01}$ (Grobbelaar 2004). The content of proteins, lipids and carbohydrates in microalgae is strongly species dependent (Table 1) and varies from 6 to 52%, from 7 to 23% and from 5 to 23%, respectively (Brown et al. 1997).

Two of the most important factors determining the methane yield in anaerobic digestion of microalgal biomass are the composition of microalgae cell wall and its contribution to the total cell mass. Cell wall composition is recognised as the limiting factor in hydrolysis of microalgae (Chen and Oswald 1998. Microalgae cell wall comprises 12–36% of total cell mass (w/w) (Table 2) and may contain biopolymers (e.g. algaenan, cellulose, sporopollenin, glucosamine, proline and carotenoids) and/or structures (such as trilaminar outer wall or trilaminar sheath—TLS) that are resistant to anaerobic degradation (Kadouri et al. 1988; Brown 1997; Derenne et al. 1992; Gelin et al. 1997; Okuda 2002; Simpson et al. 2003). Cell walls recalcitrant to microbial attack may prevent microalgal intracellular organic

Table 1 Gross composition of several microalgae species

Microalgae species	Proteins (%)	Lipids (%)	Carbohydrates (%)
Euglena gracilis	39–61	14–20	14–18
Chlamydomonas reinhardtii	48	21	17
Chlorella pyrenoidosa	57	2	26
Chlorella vulgaris	51–58	14–22	12–17
Dunaliella salina	57	6	32
Spirulina maxima	60–71	6–7	13–16
Spirulina platensis	46–63	4–9	8–14
Scenedesmus obliquus	50–56	12–14	10–17

Adapted from Sialve et al. (2009)

content from being converted to biogas, which affects the final methane yield. However, a variety of pretreatments (below discussed) have been shown to be effective at breaking microalgae cell walls and increasing methane yield (Angelidaki and Ahring 2000; Alzate et al. 2012).

Cultivation of microalgae under nitrogen deficiency is "well-known" to stimulate lipid accumulation (Chisti 2007). Theoretically, the higher the lipid content of microalgae cells, the higher their calorific value and hence the higher their methane yield. However, a high lipid content does not usually correlate with a high methane yield. Therefore, the content of inert organic matter, rather than the content energy-rich macromolecules, is believed to have a stronger impact on the final methane yield (González-Fernandez et al. 2012).

The high content of proteins observed in several microalgae species results in high concentrations of ammonia nitrogen during anaerobic degradation. Ammonia

Table 2 Cell wall composition of microalgae

Microalgae species	Cell wall (% w/w)	Cell wall composition (%)			References
		Carbohydrates	Proteins	c.n.i.[a]	
Chlorella vulgaris (F)	20.0	30.00	2.46	67.54	Abo-Shady et al. (1993)
Chlorella vulgaris (S)	26.0	35.00	1.73	63.27	Abo-Shady et al. (1993)
Kircheriella lunaris	23.0	75.00	3.96	21.04	Abo-Shady et al. (1993)
Klebsormidium flaccidum	36.7	38.00	22.60	39.40	Domozych et al. (1980)
Ulothrix belkae	25.0	39.00	24.00	37.00	Domozych et al. (1980)
Pleurastrum terrestre	41.0	31.50	37.30	31.20	Domozych et al. (1980)
Pseudendoclonium basiliense	12.8	30.00	20.00	50.00	Domozych et al. (1980)
Chlorella Saccharophila	–	54.00	1.70	44.30	Blumreisinger et al. (1983)
Chlorella fusca	–	68.00	11.00	20.00	Blumreisinger et al. (1983)
Chlorella fusca	–	80.00	7.00	13.00	Loos and Meindl (1982)
Monoraphidium braunii	–	47.00	16.00	37.00	Blumreisinger et al. (1983)
Ankistrodesmus densus	–	32.00	14.00	54.00	Blumreisinger et al. (1983)
Scenedesmus obliquus	–	39.00	15.00	46.00	Blumreisinger et al. (1983)

[a]c.n.i. stands for content not identified

is highly permeable through cell membranes and can affect methane yields due to ammonia inhibition. The acclimation period, substrate composition and operating conditions typically determine the inhibitory concentrations of ammonia, which can vary from 0.05 to 2 g/L (Rajagopal et al. 2013). Thermophilic conditions enhance the inhibition effect (Sialve et al. 2009). In this context, methanogenic communities can acclimate to high concentrations of ammonia, increasing the inhibition threshold level, even if methanogenic productivity remains low.

3 Pretreatments for Increasing the Anaerobic Biodegradability

The conversion of microalgae into biogas is often limited by the hydrolysis step of the AD process. In the 1950s, researchers already noticed that microalgae remained intact after AD in a reactor operating at 30 days of hydraulic retention time (HRT) (Golueke et al. 1957). This phenomenon also occurs when biodegrading other complex organic substrates, such as activated sludge and lignocellulosic biomass, in which organic compounds have low bioavailability and/or low biodegradability. This bottleneck may be overcome by applying a previous pre-treatment step, which is already the case in full-scale WWTPs treating sewage sludge or in the agroindustrial field. Overall, biomass pretreatment methods aim at increasing organic matter solubilisation and, therefore, making those compounds more readily available to the anaerobic bacteria present in the digester, which would ultimately increase the process rate and the methane yield (Passos et al. 2014a).

Particularly, the main reason why microalgae have slow and/or low biodegradability is due to the nature of their cell wall structure and composition. Most species have a complex cell wall composed of recalcitrant components, especially those grown in open ponds treating wastewater. Nonetheless, the characteristics of these cell walls may vary depending on the strain and environmental/operational conditions. Species with a glycoprotein-based, frustule-covered, or a bacterial-like peptidoglycan cell walls, are more sensitive to disruption with pre-treatment techniques than those with silica- or polysaccharide-based cell walls (Bohytskyi et al. 2014). The main constituents of microalgae biomass are carbohydrates, proteins, lipids, carotenoids and lignin. Nonetheless, most of them are polysaccharides, e.g. cellulose, hemicellulose, chitin/chitosan-like molecules, pectin and alginate. A recent study found that, although proteins, lipids and a considerable amount of carbohydrates were present in the cell walls of refractory microalgae species, microalgae resistance was not correlated to the presence of a unique monomer. The authors concluded that the responsible compounds were most likely to be sporopollenin, lignin-like materials and heteropolysaccharides (Montingelli et al. 2015). However, it is hypothesised that the cross-link of these compounds into a complex network building layers around the cell could eventually work as a barrier to anaerobic microbial community (Klassen et al. 2016).

Pretreatment techniques may be classified into four main categories: mechanical, thermal, chemical and biological methods. These methods are based on different mechanisms and, therefore, support different disruption efficiencies. For instance, mechanical techniques, such as microwave, ultrasound and ball-milling, act by reducing the particle size and increasing the superficial contact area; while biological pretreatments act by inducing an enzymatic breakdown of complex molecules. In a study comparing different techniques, physical pretreatments (i.e. thermal and ultrasound) showed the highest effectiveness in protein solubilisation, which was mediated by the release of alogenic organic matter and cell wall breakage, while enzymatic pretreatments increased carbohydrate solubilisation, which was mediated by the biodegradation of cell wall compounds rather than by cell disruption (Ometto et al. 2014). In this experiment, the highest biogas increase in batch tests was obtained for enzymatic pretreated microalgae (270% increase).

Most studies up-to-date were conducted using batch experiments. These tests are mainly used for comparing pretreatments and/or pretreatment conditions. However, continuous experiments with acclimated microorganisms are needed for validating and quantifying the potential methane yield and for estimating the energy balance of the process. Among the studies published so far, most of those dealing with continuous AD of microalgae evaluated the effect of thermal pretreatment. The results reported showed increases from 32 to 108% compared to non-pretreated microalgae (ranging from 0.12 to 0.27 L CH_4/g VS) (Table 3). The best results were obtained during microalgae thermal pretreatment at 75–95 °C for 10 h (70% increase) (Passos and Ferrer 2014) and 120 °C for 2 h (108% increase) (Schwede et al. 2013). Moreover, the energy balance calculations showed that after applying a low-temperature pretreatment at 75 °C, the energy balance shifted from neutral to positive with a 2.7 GJ net energy production per day (Passos and Ferrer 2014). In fact, most recent reviews in microalgae pretreatment concluded that thermal pretreatment is the optimal method, by combining the highest methane improvement and the lowest energy input (Jankowska et al. 2017; Passos et al. 2014a, b; Rodriguez et al. 2015).

Additionally, enzymatic pretreatment has recently been the focus of research on microalgae pretreatment. Studies in continuous mode showed increases of 260% in methane yield compared to non-pretreated microalgae, although biomass was highly recalcitrant in this experiment, i.e. 0.05 L CH_4/ g COD (Mahdy et al. 2015). The enzymatic pretreatment of *Scenedesmus* sp. in a first step anaerobic membrane bioreactor (AnMBR) with rumen microorganisms also showed promising results in terms of methane yield (0.203 L CH_4/g COD) and COD removal (70%) (Giménez et al. 2017).

Although many novel pretreatment methods are being investigated, such as pulse electric field, ozonation or solvent addition, the energy and economic aspects for pilot and full-scale viability must be analysed. The main pros and cons of microalgae pretreatment techniques are summarised in Table 4. Thus, energy demand and scalability are major issues when evaluating pretreatment viability. Although thermal pretreatment seems advantageous, biomass thickening or dewatering is crucial. On the other hand, despite thermochemical pretreatments have

Table 3 Microalgae pretreatment for improved AD in continuous reactors

Microalgae species	Pretreatment conditions	AD conditions	Methane yield increase	References
Scenedesmus sp. and *Chlorella* sp.	Thermal: 100 °C, 8 h	CSTR[a]: 3.7% TS, 28 days HRT[b]	33% (0.270 L CH_4/g VS)	Chen and Oswald (1998)
Scenedesmus sp., *Monorraphidium* sp. and diatoms biomass	Thermal: 75 and 95 °C, 10 h	CSTR: 37 °C, 0.7 g VS/Ld, 20 days HRT	70% (0.180 L CH_4/g VS)	Passos and Ferrer (2014)
Pediastrum sp., *Micractinium* sp. and *Scenedesmus* sp.	Thermal: 60 °C, 2–6 h	AVR[c]: 20 °C, 1.2 g VS/Ld, 91 days SRT[d]	32% (0.136 L CH_4/g VS)	Kinnunen et al. (2014)
Nannochloropsis salina	Thermal: 100–120 °C, 2 h	CSTR: 38 °C, 2.0 g VS/Ld, 120 days HRT	108% (0.130 L CH_4/g VS)	Schwede et al. (2013)
Oocystis biomass	Thermal: 130 °C, 15 min	CSTR: 37 °C, 0.7 g VS/Ld, 20 days HRT	42% (0.120 L CH_4/g VS)	Passos and Ferrer (2015)
Chlorella vulgaris	Thermal: 120 °C, 40 min	CSTR: 35 °C, 1.5 g COD/Ld, 15 days HRT	48% (0.126 L CH4/g COD)	Sanz et al. (2017)
Scenedesmus sp., *Monorraphidium* sp. and diatoms biomass	Microwave: 70 MJ/kg VS, 26 g TS/L	CSTR: 35 °C, 0.8 g VS/Ld, 20 days HRT	60% (0.272 L CH_4/g VS)	Passos et al. (2014b)
Chlorella vulgaris	Enzymatic: protease (0.585 UA), 65 g TS/L	CSTR: 35 °C, 1.5 g COD/Ld, 20 days HRT	260% (0.128 L CH_4/g COD)	Mahdy et al. (2015)
Scenedesmus sp.	Enzymatic: rumen microorganisms fermenter	AnMBR[e]: 38 °C, 0.2 g COD/Ld, 31 days HRT, 100 days SRT	0.203 L CH_4/g COD	Giménez et al. (2017)

Notes [a]CSTR stands for complete stirred tank reactor, [b]HRT stands for hydraulic retention time, [c]AVR stands for accumulating volume reactor, [d]SRT stands for sludge retention time, and [e]AnMBR stands for anaerobic membrane bioreactor supported positive microalgae biodegradability increases, further studies should evaluate the risk of contamination in continuous bench and pilot-scale reactors. An alternative cost-effective microalgae pretreatment method may be the use of environmentally friendly and low-cost chemicals such as lime (CaO). A recent study found that the methane yield increased by 25% in BMP tests after pretreating microalgae at 72 °C with CaO (Solé-Bundó et al. 2017). Biological pretreatments constitute another promising pretreatment technology. Experiments conducted so far have still not elucidated the best pretreatment conditions, resulting in lower biogas production increases compared to thermal and thermochemical methods. In

Table 4 Comparison of pretreatment methods for increasing microalgae anaerobic biodegradability (Passos et al. 2014a)

Pretreatment	Control parameters	Anaerobic biodegradability increase	Pros	Cons
Thermal (<100 °C)	Temperature; exposure time	√√	Lower energy demand; scalability	High exposure time
Hydrothermal (>100 °C)	Temperature; exposure time	√√	Scalability	High heat demand; need for thickened or dewatered biomass; risk of formation of refractory compounds
Thermal with steam explosion	Temperature; exposure time; pressure	√√√	Scalability	High heat demand; Need for thickened or dewatered biomass; risk of formation of refractory compounds Investment cost
Microwave	Power; exposure time	√√	–	High electricity demand; scalability; need for biomass dewatering
Ultrasound	Power; exposure time	√	Scalability	High electricity demand; need for biomass dewatering
Chemical	Chemical dose; exposure time	√	Low energy demand	Chemical contamination; risk of formation of inhibitors; high cost
Thermochemical	Chemical dose; exposure time; temperature	√√	Low energy demand	Chemical contamination; risk of formation of inhibitors; high cost
Enzymatic	Enzyme dose; exposure time; pH, temperature	√	Low energy demand	Cost, sterile conditions

addition, purified enzymes may be expensive and jeopardise the economic viability of the process. However, this limitation may be overcome via enzyme production through other microorganisms, via enzyme expression through the microalgae cells to be digested and via in situ production of hydrolytic enzymes by inoculated living bacteria or fungi (Klassen et al. 2016).

Finally, future research should focus on investigating the mechanisms underlying microalgae cell wall damage and/or disruption with pretreatments, since the analysis of organic matter solubilisation has been shown insufficient to predict the increase in methane yields. The determination of soluble macromolecules, microscopic images and microbiology analyses is important for better understanding how, where and in which scale pretreatments affect microalgae cell structure and which compounds become more readily available. Moreover, it is crucial to conduct experiments in continuous mode and in pilot and full-scale reactors for evaluating the process performance.

4 Anaerobic Co-digestion of Microalgae

AD of raw microalgae or microalgae residues after the generation/extraction of value-added products (i.e. lipids, ethanol and hydrogen) is typically characterised by low methane yields and the occurrence of ammonia inhibition. Despite these limitations, AD is still regarded as a key technology to maximise resource recovery from microalgae and make algae industry economically feasible. AD also aids the mobilisation the nutrients (N and P) needed for algae cultivation (Ward et al. 2014). Anaerobic co-digestion, the simultaneous digestion of two or more substrates, is an established and cost-effective option to overcome the drawbacks of mono-digestion and boost the biogas production of AD plants (Mata-Alvarez et al. 2014). Besides improving the feasibility of AD plants, co-digestion also allows treating several wastes in a single facility and "share/reduce" treatment costs (Neumann et al. 2015).

Algae have been successfully co-digested with a large range of co-substrates such as sewage sludge, animal manures, food waste, energy crops, glycerol, paper waste and fat, oil and grease (FOG). Although the improvement of the methane production is mainly a consequence of the increased organic loading rate (OLR) rather than to the occurrence of synergisms during AD, microalgae have been primarily co-digested with carbon-rich co-substrates, which allows increasing the digester OLR while controlling ammonia concentration. Several studies have optimised the co-substrate dose by balancing the feedstock C/N ratio with optimum values for algae co-digestion ranging between 12 and 27 (Ehimen et al. 2011; Fernández-Rodríguez et al. 2014). However, optimising co-substrate selection and dosage based on the C/N ratio is an oversimplification since this approach does not take into account the characteristics of each co-substrate (Astals et al. 2014; Herrmann et al. 2016). The maximum dose of some co-substrates such as glycerol and FOG is limited by secondary inhibitory mechanisms, while the deficiency of alkalinity or essential nutrients limits the dosage of energy crops and paper waste

(Schwede et al. 2013; Zhong et al. 2013). The maximum dosing rate of self-sufficient co-substrates such as food waste or sewage sludge is typically limited by the anaerobic digestion plant capacity and co-substrate availability. Regardless of the co-substrate, anaerobic co-digestion stands as a suitable option to reach OLR higher than 2 g VS/L/d in algae digesters, since the operation of algae mono-digesters at OLR higher than 2 g VS/L/d has resulted in inhibitory ammonia concentrations and caused the accumulation of volatile fatty acids (VFAs) (i.e. higher risk of process failure) or even process failure (Yen and Brune 2007; Park and Li 2012; Herrmann et al. 2016).

The integration of algae cultivation in WWTP to substitute the conventional activated sludge reactor, to treat the anaerobic digestion supernatant, or to polish the WWTP final effluent followed by their co-digestion with sewage sludge is attracting a lot of attention (Sahu et al. 2013; Beltran et al. 2016; Peng and Colosi 2016). The cultivation of algae on anaerobic digestion supernatant (diluted or pretreated) is of special interest since it (1) reduces the nutrient load to the headworks, which represents about 20% of the WWTP nutrient load; (2) mitigates greenhouse gases emissions by using CO_2 from biogas combustion for algae growth; and (3) produces algae as on-site co-substrate, which lowers the uncertainty about co-substrate availability and seasonality (Rusten and Sahu 2011; Yuan et al. 2012). Even though this scenario appears very promising, it remains uncertain if the amount of algae able to grow on digester supernatant is enough to make a significant difference on the WWTP methane production (Hidaka et al. 2017). Conversely, the addition of large amounts of algae (or any other nitrogen-rich co-substrate) should be carefully evaluated since it will increase the digester and supernatant nitrogen concentration. In this regard, Mahdy et al. (2017), who co-digested algae and cattle manure, showed that inoculum acclimation could provide anaerobic digestion stable performance at nitrogen concentrations as high as 4 gNH_4^+-N/L and 700 mgNH_3-N/L. Likewise, Arnell et al. (2016) plant-wide simulation study warned of the impact of co-digesting nitrogen-rich waste on the WWTP water train, e.g. aeration requirement, methanol consumption, effluent quality. Finally, the cultivation of microalgae on pig and cattle manure effluent supernatant, and its subsequent co-digestion, has also been studied with the aim of increasing the methane production and moving the nutrients from the supernatant to the biosolid (Wang et al. 2016a, b; Mahdy et al. 2017).

5 Design and Operational Considerations

Biogas production using microalgae as substrate has been studied since the 1950s. The first report addressing the anaerobic digestion of microalgal biomass was published by Golueke et al. (1957). This early study reported a biogas production of 0.5 m^3/kg of volatile solids of algal biomass. During the last decade, an intensive research has been conducted in order to develop solar energy fixation processes using microalgae to transform light into chemical energy and anaerobic digestion to transform such biomass into biomethane.

When considering biogas production from microalgae, two scenarios should be considered. The first one relies on coupling biogas production to a microalgae-based biodiesel production process. Microalgae have received great attention as a potential source of oil for biodiesel production due to the ability of certain types of microalgae to accumulate lipids and to the higher biomass productivities achieved when compared with land-based crops (Chisti 2007; Mata et al. 2010; Weyer et al. 2010). When the primary use of microalgae is biodiesel production, the lipids extraction processes employed (usually involving solvents) will generate a "residual" biomass suitable for biogas production. However, recent concerns have been raised by life cycle analyses when considering biodiesel production from microalgae due to potentially low energetic yield when based on traditional technology (Scott et al. 2010; Sialve et al. 2009; Stephens et al. 2010). Indeed, a negative energy balance has been estimated for biodiesel process from microalgae as a result of harvesting and drying steps, which are highly energy intensive (Lardon et al. 2009; Scott et al. 2010). In this context, the production of biogas as a sole fuel using whole microalgae has been proposed. This option would entail a much simpler process, with less and simpler unit operations. However, energy in the form of methane possesses nowadays a low economic value.

Hydrolysis is known to be the rate-limiting step of anaerobic digestion of solid substrates, which is specially the case when using microalgae as a substrate. Thermophilic digestion has been proposed as a way to enhance microalgae biomass hydrolysis and the overall anaerobic digestion performance. The high temperatures applied during thermophilic anaerobic digestion (50–57 °C) accelerate biochemical reactions, increasing both the efficiency of organic matter degradation and the potentially applicable organic loading rates. However, higher degradation and loading rates will increase the concentration of ammonia nitrogen in the digester. Contradictory results have been reported when addressing the thermophilic anaerobic digestion of microalgal biomass (Capson-Tojo et al. 2017; Cea-Barcia et al. 2015; Zamalloa et al. 2012a). Indeed, the benefits of the thermophilic digestion of microalgae still need to be confirmed and most likely, the optimum temperature for anaerobic digestion might be dependent on the microalgae species.

The nitrogen content of microalgae biomass is relevant since ammonia release during anaerobic digestion is expected to be an issue of concern as a result of the above-discussed inhibition of AD. This will be especially critical when oil-extracted microalgae are used as substrate, since lipids extraction increases the proportion of nitrogen per gram of biomass. If anaerobic digestion is performed at solids concentrations over 4–5%, ammonia concentration in digester could reach inhibitory levels for the anaerobic microbial community (Torres et al. 2013). Even though the use of ammonia tolerant inocula may provide conditions for successful operation (Mahdy et al. 2017), measurements need to be taken in order to ensure a stable process performance. In this context, co-digestion of microalgae biomass with carbon-rich substrates or wastes could be an alternative. As previously discussed, indeed, the benefits derived from the co-digestion of microalgae biomass with glycerol, activated sludge and others wastes have consistently showed (Fernandez-Rodriguez et al. 2014; Herrmann et al. 2016; Neumann et al. 2015).

Continuous stirred tank reactors like those used for sewage sludge digestion or other organic substrates are the most popular bioreactor configuration for the conversion of microalgae biomass into biogas. Indeed, most of the reported studies used that configuration with hydraulic retention times ranging from 20 to 40 days (Jankowska et al. 2017). However, other alternative bioreactor configurations such as UASB reactors have been proposed (Tartakovsky et al. 2015). Unfortunately, the low solids retention times of granular-based reactors may not provide an efficient conversion of microalgae. The use of membrane bioreactors has also been proposed, which represents an interesting opportunity to provide the required solids retention for effective microalgae digestion (Zamalloa et al. 2012b) and to tackle the problem of ammonia inhibition. Hence, medium exchange without biomass washout can be implemented in membrane bioreactors to reduce the toxicity mediated by NH_3 built-up, although the operating costs associated to this operational strategy still need to be evaluated under full-scale implementation.

6 Process Modelling

Process modelling is defined as the mathematical representation of a certain process or system, which could be either based on the underlying mechanisms or phenomena (model-based) or on the experimentally generated input/output data (data-based). Mathematical modelling is being increasingly used as a tool for diagnosis, hypothesis formulation, prototyping, scenarios evaluation, process design and optimisation. Thus, the anaerobic degradation of organic biomass, including microalgae, can be modelled using different models. In this context, the Anaerobic Digestion Model 1 (ADM1) has been the most popular, accepted and applied model in research and industrial applications (Batstone and Keller 2002). Nonetheless, there are plenty of other modelling approaches that have been reviewed in the literature (Batstone 2006; Donoso-Bravo et al. 2011; Tomei et al. 2009).

Regardless of the model used to describe the methane production from microalgae, the most important issue is the proper selection of the model parameters. In the specific case of microalgae, as discussed in the above sections, the disruption of microalgae cell wall is considered the limiting reaction step, especially if non-pretreated microalgae are fed to the anaerobic digester. Therefore, both the disintegration and the hydrolysis coefficients, required in ADM1, have to be carefully estimated. However, it is worth to point out that the elimination of the disintegration step has been recommended due to the fact that the use of a two-hydrolysis step possesses some correlation and identification problems, especially for sewage sludge approaches (Batstone et al. 2015). A description of how the modelling of methane production from microalgae has been addressed in different operation modes is given and discussed below.

(Semi) continuous operation. The ADM1 has been used to represent the AD of microalgae by tweaking the original ADM1 with the inclusion of the *Contois* equation instead of the first-order equation to represent the hydrolysis step (Mairet et al. 2011). The *Contois* equation takes into account both the particulate material and the microbial population responsible for this process, while the first-order equation only considers the particulate substrate concentration. The model outperformed the original ADM1 with experimental data from a digester operating for 140 d. An interesting application in this study was the representation of the semi-continuous feeding mode by considering successive batch reactor operations changing the initial conditions for each daily pulses. Moreover, another study tested the same modified ADM1 above-mentioned in an integrated system of wastewater treatment and AD of microalgae (Passos et al. 2015). In this work, the authors found that an appropriate characterisation of the microalgae composition was of paramount importance in order to have a proper model performance due to population changes over time (in particular, the variations in the inert and organic content of the biomass). In addition, a reduced mechanistic 3-reaction model, obtained after principal component analysis, was developed and calibrated with the ADM1 (Mairet et al. 2012). The model was composed of a double hydrolysis reaction to describe the production of volatile fatty acids and a methanogenic reaction. The performance of this model was quite similar to the same simulation results obtained with the ADM1 model, despite its complexity was much lower.

Batch operation. Batch tests, namely biochemical methane potential (BMP) assays, are widely used to assess the kinetics of biodegradation of different substrate. From this test, many parameters such as the hydrolysis coefficient or the inert fraction of the microalgae may be determined when a proper kinetic expression is used (Donoso-Bravo et al. 2010). The first-order model has been a popular option to draw parameters by fitting the accumulated biogas production (Eq. 1).

$$B(t) = B_{\max}(1 - \mathrm{e}^{-k_h * t}) \tag{1}$$

This has been done using as a substrate a residual microalgae (i.e. after lipids extraction) (Neumann et al. 2015) or raw microalgae (Fernández-Rodríguez et al. 2014). Some of the values found up-to-date in literature are shown in Table 5. Moreover, the synergism in the co-digestion of microalgae with waste-activated sludge was assessed by the application of a first-order equation to describe the hydrolysis reaction and the *Monod* equation to model the methanogenesis (Lee et al. 2017). To our knowledge, the *Contois* equation has not been yet used to describe the performance of BMP assays.

Global approaches. Apart from the classic modelling application in continuous or batch mode, other new approaches such as the global WWTP plant-wide model that aims at representing an integrated process have been recently developed. This approach intends to implement a model-based on nonlinear programming to evaluate the best configuration of a microalgae-based biorefinery in which AD is also incorporated (Rizwan et al. 2015).

Table 5 Kinetic parameters from the first-order equation in the AD of microalgae

References	k_H (1/d)	B_{max} ($mLCH_4/gVS$)	Microalgae
Neumann et al. (2015)	0.09	413 (13)	*B. braunii*
Fernández-Rodríguez et al. (2014)	0.49 (0.08)	62	*D. salina*
Lee et al. (2017)	0.07[a]	–	*Chlorella* sp.
Wang et al. (2016a, b)	0.148	180.3	*Chlorella* sp.
Zhen et al. (2016)	0.187	106.9 (3.2)	*Scenedesmus* sp.—*Chlorella* sp.

[a]Modified first-order equation

ADM1 simulation of continuous anaerobic digestion of microalgae: The effect of parameter selection. Figure 1 shows the performance of a virtual anaerobic digester operating in continuous mode with microalgae as a feedstock. The results obtained from experiments investigating the AD of raw and residual microalgae were used to perform the simulations (Fernández-Rodríguez et al. 2014; Neumann et al. 2015). To this aim, the inlet COD was fixed at 50 gCOD/L and the volume of the reactor at 1000 m^3, while the organic loading rate was increased by changing the inlet microalgae flow rate. In addition, the macromolecular composition of the raw microalgae was set at 58, 22 and 20% for proteins, carbohydrates and lipids, respectively (Passos et al. 2015). The simulation considered an inert fraction of the organic matter of 24%, estimated from the BMP results. The macromolecular composition of the residual microalgae was set at 64.5, 31.3 and 4.2% for proteins, carbohydrates and lipids, respectively (adapted from Neumann et al. 2015). In this

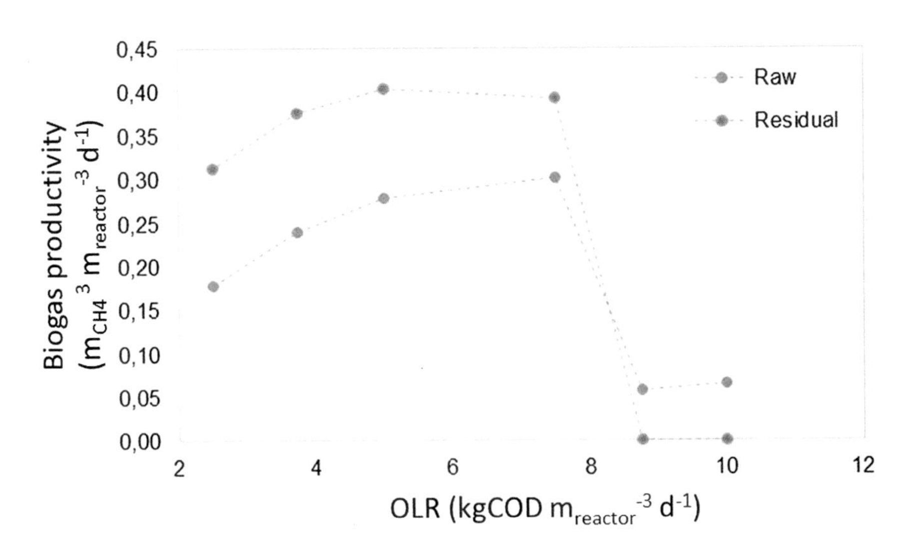

Fig. 1 Simulation of the influence of the organic loading rate on the continuous anaerobic degradation of raw and residual microalgae biomass

case, the simulation considered an inert fraction of the organic matter of 60%, given the amount of methane produced in the BMP test compared to the assay carried out with raw microalgae. The values of the hydrolysis coefficient are shown in Table 5.

The AD of the residual microalgae outperformed the AD of raw microalgae in continuous mode as OLR increased, which may be explained by the low biodegradability of the raw microalgae (Fig. 1). However, methane production in the digester fed with residual microalgae dropped to zero at high OLR values, likely due to the low values of the hydrolytic constant. In contrast, the digester operated with raw microalgae supported a low but stable methane productivity, likely due to the retention of the hydrogenotrophic methane population inside the reactor.

7 Biogas Upgrading to Biomethane

Biogas from the anaerobic digestion of microalgae is typically composed of CH_4 (60–75%), CO_2 (25–30%), H_2S (0–1%), O_2 (0–1%), N_2 (0–4%) and trace levels of NH_3, volatile fatty acids (VFAs) and siloxanes (the latter present in microalgae grown in domestic wastewater) (Alzate et al. 2012). Biogas composition determines the final energy use of this renewable energy feedstock, which ranges from on-site combustion for heat (boilers) or heat/electricity generation (internal combustion engines, turbines, fuel cells), use as a vehicle fuel, and injection into natural grad grids (Bailón and Hinge 2012). In this context, while boilers and internal combustion engines require a removal of H_2S below 0.02–0.1% levels (depending on the manufacturer), micro-turbines and turbines can stand H_2S concentrations in the range of 1–7%. However, the latter require an efficient removal of siloxanes (<0.03–0.1 ppm_v), while internal combustion engines and boilers can cope with concentrations of 5–28 mg Si m^{-3}. Nowadays, the technical requirements for biogas injection into natural gas grids or biogas used as a vehicle fuel are country-specific, although a European draft for biogas quality is currently under approval (Table 6). This entails the need for a biogas-upgrading step prior biogas valorization, which will be stricter when biogas is to be injected into natural gas networks (in the form of biomethane).

Biogas-upgrading technologies can be classified into physical/chemical and biological as a function of the mechanisms governing pollutants removal from biogas. Nowadays, O_2 and N_2 can be only removed by physical/chemical methods (such as membrane separation or low-pressure PSA) (Muñoz et al. 2015), while the removal of CO_2, H_2S, NH_3, VFAs and even siloxanes can be carried using both platform technologies.

Today, the market of CO_2 removal is mainly dominated by water scrubbing (with a 41% of the market share), followed by chemical scrubbing (22%), pressure swing adsorption (21%), membrane separation (10%) and organic solvent scrubbing (6%) (Thrän et al. 2014). Physical/chemical technologies for CO_2 removal from biogas exhibit a high efficiency and robustness at the expenses of high investment and operating costs. Typical CH_4 concentrations in the biomethane

Table 6 Technical specifications for biomethane injection into natural gas grids according to the European draft FprEN 16726

Total S (mg/m^3)	H_2S + COS (mg/m^3)	RSH + Mercaptans (mg/m^3)	O_2 (%)	CO_2 (%)	CO (%)	Volatile Si (mg/m^3)	Amines (mg/m^3)	H_2O (°C) 70 bar	HC (°C) 1–70 bar
20	5	6	0.001–1	2.5–4	0.1	0.1–1	10	−8	−2

produced by the above-mentioned scrubbing, membrane and adsorption technologies range from 95 to 98% (Bauer et al. 2013). However, the CO_2 footprint of these technologies is high as a result of the direct release to the atmosphere of the CO_2 separated and their high energy demand (which represents 3–12% of the energy content present in the raw biogas). Table 7 summarises the fundamentals and the typical design-operating parameters of the main physical/chemical technologies for CO_2 separation from biogas.

Biological CO_2 removal from biogas is still in an early stage of investigation, hydrogenotrophic CO_2 reduction to CH_4 and photosynthetic CO_2 assimilation being the two most promising technologies under scale up. Hydrogenotrophic CO_2 removal, also named power-to-gas, is based on the bioconversion of CO_2 to CH_4 using H_2 as an electron donor and CO_2 as a carbon source and electron acceptor by hydrogenotrophic archaea. Equation 2 describes the stoichiometry of this CO_2 reduction, which can be conducted either directly into the anaerobic digestion (via H_2 supplementation) or in an external bioreactor supplemented with H_2 and biogas:

$$4H_2 + CO_2 \rightarrow CH_4 + 2H_2O \quad \Delta nG^0 = -130.7\,\text{kJ/mol} \tag{2}$$

From an economic and environmental viewpoint, hydrogenotrophic CO_2 removal should be based on H_2 produced from water electrolysis using the excess of renewable electricity (i.e. wind power generated during the night). The main limitation of this technology derives from the limited gas–liquid H_2 mass transfer as a result of the low aqueous solubility of this gas (Diaz et al. 2015). On the other hand, photosynthetic CO_2 removal is based on the intensification of the symbiosis between microalgae and quimioautotrophic bacteria at a high pH (=enhancement in the CO_2 and H_2S biogas–liquid mass transfer) in photobioreactors as a platform technology to simultaneously remove CO_2, H_2S, NH_3 and VFAs from biogas at a low energy cost and with a low environmental impact. In these systems, microalgae use the solar energy to fix the CO_2 from biogas via photosynthesis (Meier et al. 2015). Residual nutrients from the effluents of the anaerobic digesters can be used to support microalgae growth, which will significantly reduce the operation cost of the upgrading process and partially mitigate the eutrophication potential of the digestate. This technology has been successfully implemented in open high rate algal ponds interconnected to external absorption columns at 2–3 times lower operating costs than their physical/chemical counterparts (Toledo-Cervantes et al. 2017).

The other major biogas pollutant, H_2S, can be removed using physical/chemical and biological technologies already available at commercial scale (Abatzoglou and Boivin 2009). Adsorption (with and without chemical reaction) and in situ chemical precipitation still represent the two most widely implemented technologies worldwide despite their high operating cost (3.2 and 2.4 cts €/m^3, respectively). Similarly to their CO_2 removal counterparts, these physical/chemical technologies exhibit high efficiencies and a high robustness. Likewise, biotechnologies such as biotrickling filtration and microaerobic anaerobic digestion support high removal efficiencies (>99%) at significantly lower operating cost (1.5 and 0.28 cts €/m^3,

Table 7 Physical/chemical technologies for CO_2 removal from biogas (Bauer et al. 2013; Muñoz et al. 2015)

Technology	Fundamentals	Design parameters	Operational parameters
Water scrubbing	Pressurised water is used for the absorption of CO_2 from biogas in a packed bed. CO_2 separation is based on the higher aqueous solubility of CO_2 compared to that of CH_4 (24 times more soluble)	1 absorption column + 2 stripping columns Concentrations of CH_4 > 96% and of CO_2 < 2%	Operating pressure = 6–10 bar Recycling water flow rates = 0.18–0.23 m^3water/Nm^3_{biogas} Electricity consumption = 0.24 kWh/Nm^3
Chemical Scrubber	Absorption + reaction in solvents based on amines or basic solutions (NaOH, KOH, CaOH, K_2CO_3, etc.)	1 absorption column + 1 stripping column	Operating pressure = 1–2 bar Electricity consumed = 0.13 kWh/Nm^3 Thermal energy for solvent regeneration 0.55 kWh/Nm^3
Organic solvent scrubbing	CO_2 absorption based on polyethylene glycol solvents (Selexol® o Genosorb®) with a 5 times higher CO_2 solubility than water	1 absorption column + 2 desorption column Concentrations of CH_4 = 96–98.5%	Electricity consumed = 0.22 kWh/Nm^3 Thermal energy for solvent regeneration: 0.4–0.51 kWh/Nm^3
Pressure swing adsorption	Selective separation of CO_2 over CH_4 based on a selective adsorption or size exclusion in the adsorbent bed	Adsorbent materials: Activated carbon, silica gel, Zeolites 4 columns operated sequentially Concentrations of CH_4 = 96–98%	Electricity consumed = 0.26 kWh/Nm^3
Membrane separation	Selective permeation of CO_2 and H_2S through semi-permeable membranes	Gas–gas or liquid–gas configurations Single stage or multiple stage configurations	Electricity consumed = 0.26 kWh/Nm^3

respectively) (Gabriel et al. 2013; Muñoz et al. 2015). Among biological methods, photosynthetic H_2S removal is attracting a significant attention based on its simultaneous occurrence during CO_2 capture in algal–bacterial photobioreactors, which will drastically reduce the operating cost of biogas upgrading (Table 8).

Finally, the removal of volatile fatty acids and siloxanes is mainly conducted in conventional adsorption units due to its compact nature and extensive design

Table 8 Technologies for the removal of H_2S from biogas

Technology	Fundamentals	Design parameters	Operational parameters
Adsorption	Adsorption + reaction in an adsorbent packed bed	1 Adsorption column + 1 desorption column Adsorbent: Fe_2O_3, $Fe(OH)_3$ and ZnO	Empty bed residence time = 1–15 min Adsorption capacity of activated carbon: 0.1–0.2 g H_2S/g carbon
Chemical precipitation	Addition to the digester of $FeCl_2$, $FeCl_3$ and $FeSO_4^2$ salts to promote the in situ precipitation of FeS	Levels of H_2S in the treated biogas > 100–150 ppm_v	Dosing ratio = 0.035 kg $FeCl_3$/kg Total solid
Photosynthetic H_2S removal	Aerobic oxidation of H_2S by chemolitotrophic bacteria using the O_2 produced photosynthetically by microalgae in the photobioreactor	H_2S removals > 99%	Liquid to biogas ratio 0.5–2 between the absorption column and the HRAP
Biotrickling filtration	Aerobic or anoxic oxidation of H_2S in a packed bed column containing a biofilm of chemolitotrophic bacteria supplied with nutrients from a recirculating aqueous solution	H_2S removals > 99%	Empty bed residence time = 2–10 min
Microaerobic anaerobic digestion	O_2 dosing in the headspace of the anaerobic digester to support the partial oxidation of H_2S to elemental sulphur that accumulates in the digester headspace	H_2S removals > 99%	Empty bed residence time = 5 h O2/biogas flow rate ratio = 0.3–3%

experience. However, both VFAs and siloxanes are biodegradable molecules and their removal from biogas could be eventually carried out using biotechnologies, which would a priori support a better environmental and economic performance (Accettola et al. 2008).

Acknowledgements The financial support from MINECO and the FEDER funding programme is gratefully acknowledged (CTM2015-70442-R). The project has received funding from the European Union's Horizon 2020 research and innovation programme under grant agreement No. 689242. David Jeison acknowledges the support provided by CRHIAM Centre (CONICYT/FONDAP/15130015).

References

Abatzoglou, N., & Boivin, S. (2009). A review of biogas purification processes. *Biofuels Bioproducts Biorefining, 3,* 42–71.

Abo-Shady, A. M., Mohamed., Y. A., & Lasheen T. (1993). Chemical composition of the cell wall in some green algae species. *Biologia Plantarum, 35*(4), 629–632.

Accettola, F., Guebitz, G., & Schoeftner, R. (2008). Siloxane removal from biogas by biofiltration: Biodegradation studies. *Clean Technologies and Environmental Policy, 10,* 211–218.

Alzate, M. E., Muñoz, R., Rogalla, F., Fdz-Polanco, F., & Pérez-Elvira, S. I. (2012). Biochemical methane potential of microalgae: Influence of substrate to inoculum ratio, biomass concentration and pretreatment. *Bioresource Technology, 123,* 488–494.

Angelidaki I., & Ahring B. K., 2000. Methods for increasing the biogas potential from the recalcitrant organic matter contained in manure. *Water Science and Technology, 41*(3), 189–194.

Arnell, M., Astals, S., Åmand, L., Batstone, D. J., Jensen, P. D., & Jeppsson, U. (2016). Modelling anaerobic co-digestion in Benchmark Simulation Model No. 2: Parameter estimation, substrate characterisation and plant-wide integration. *Water Research, 98,* 138–146.

Astals, S., Batstone, D. J., Mata-Alvarez, J., & Jensen, P. D. (2014). Identification of synergistic impacts during anaerobic co-digestion of organic wastes. *Bioresource Technology, 169,* 421–427.

Bailón, L., & Hinge, J. (2012). Report: Biogas and bio-syngas upgrading. Danish Technological Institute. http://www.teknologisk.dk/_root/media/52679_ReportBiogas%20and%20syngas%20upgrading.pdf.

Batstone, D. J. (2006). Mathematical modelling of anaerobic reactors treating domestic wastewater: Rational criteria for model use. *Review Environment Science Bio/Technology, 5,* 57–71.

Batstone, D. J., & Keller, J. (2002). Industrial applications of the IWA anaerobic digestion. *Water Science and Technology, 1,* 199–206.

Batstone, D. J., Puyol, D., Flores-Alsina, X., & Rodríguez, J. (2015). *Mathematical modelling of anaerobic digestion processes: Applications and future needs*. Rev: Environment Science Bio/Technology. https://doi.org/10.1007/s11157-015-9376-4.

Bauer, F., Hulteberg, C., Persson, T., & Tamm, D. (2013). Biogas upgrading—Review of commercial technologies. SGC Rapport 2013:270. SGC. http://vav.griffel.net/filer/C_SGC2013-270.pdf.

Beltrán, C., Jeison, D., Fermoso, F. G., & Borja, R. (2016). Batch anaerobic co-digestion of waste activated sludge and microalgae (*Chlorella sorokiniana*) at mesophilic temperature. *Journal of Environmental Science and Health—Part A, 51*(10), 847–850.

Blumreisinger, M., Meindl, D., & Loos, E. (1983). Cell wall composition of chlorococcal algae. *Phytochemistry, 22*(7), 1603–1604.

Bohutskyi, P., Betenbaugh, M. J., & Bouwer, E. J. (2014). The effects of alternative pretreatment strategies on anaerobic digestion and methane production from different algal strains. *Bioresource Technology, 155,* 366–372.

Brown, M. R., Jeffrey, S. W., Volkman, J. K., & Dunstan, G. A. (1997). Nutritional properties of microalgae for mariculture. *Aquaculture, 151,* 315–331.

Capson-Tojo, G., Torres, A., Munoz, R., Bartacek, J., & Jeison, D. (2017). Mesophilic and thermophilic anaerobic digestion of lipid-extracted microalgae *N-gaditana* for methane production. *Renewable Energy, 105,* 539–546.

Cea-Barcia, G., Moreno, G., & Buitron, G. (2015). Anaerobic digestion of mixed microalgae cultivated in secondary effluent under mesophilic and thermophilic conditions. *Water Science and Technology, 72*(8), 1398–1403.

Chen, P. H., & Oswald, W. J. (1998). Thermochemical pretreatment for algal fermentation. *Environment International, 24*(8), 889–897.

Chisti, Y. (2007). Biodiesel from microalgae. *Biotechnology Advances, 25*(3), 294–306.

Díaz, I., Pérez, C., Alfaro, N., & Fdz-Polanco, F. (2015). A feasibility study on the bioconversion of CO_2 and H_2 to biomethane by gas sparging through polymeric membranes. *Bioresource Technology, 185,* 246–253.

Derenne, S., Largeau, C., Berkaloff, C., Rousseau, B., Wilhelm, C., & Hatcher, P. G. (1992). Non-hydrolysable macromolecular constituents from outer walls of *Chlorella fusca* and *Nanochlorum eucaryotum*. *Phytochemistry, 31*(6), 1923–1929.

Domozych, D. S., Stewart, K. D., & Mattox, K. R. (1980). The comparative aspects of cell wall chemistry in the green algae (Chlorophyta). *Journal of Molecular Evolution, 15*(1), 1–12, ISSN: 1432-1432.

Donoso-Bravo, A., Mailier, J., Martin, C., Rodríguez, J., Aceves-Lara, C. A., & Vande Wouwer, A. (2011). Model selection, identification and validation in anaerobic digestion: A review. *Water Research, 45,* 5347–5364.

Donoso-Bravo, A., Pérez-Elvira, S. I., & Fdz-Polanco, F. (2010). Application of simplified models for anaerobic biodegradability tests. Evaluation of pre-treatment processes. *Chemical Engineering Journal, 160,* 607–614.

Ehimen, E. A., Sun, Z. F., Carrington, C. G., Birch, E. J., & Eaton-Rye, J. J. (2011). Anaerobic digestion of microalgae residues resulting from the biodiesel production process. *Applied Energy, 88*(10), 3454–3463.

Fernandez-Rodriguez, M. J., Rincon, B., Fermoso, F. G., Jimenez, A. M., & Borja, R. (2014). Assessment of two-phase olive mill solid waste and microalgae co-digestion to improve methane production and process kinetics. *Bioresource Technology, 157,* 263–269.

Gabriel, D., Deshusses, M. A., & Gamisans, X. (2013). Desulfurization of biogas in biotrickling filter. In: John Wiley & Sons (Ed.), *Air pollution prevention and control: Bioreactors and bioenergy* (1st ed., pp. 513–523). Wiley: Hoboken.

Gelin, F., Boogers, I., Noordeloos, A. A. M., Damsté J. S. S., Riegman, R., & De Leeuw J. W. (1997). Resistant biomacromolecules in marine microalgae of the classes eustigmatophyceae and chlorophyceae: Geochemical implications. *Organic Geochemistry, 26*(11–12), 659–675.

Giménez, J. B., Aguado, D., Bouzas, A., Ferrer, J., & Seco, A. (2017). Use of rumen microorganisms to boost the anaerobic biodegradability of microalgae. *Algal Research, 24,* 309–316.

Golueke, C. G., Oswald, W. J., & Gotaas, H. B. (1957). Anaerobic digestion of Algae. *Applied Microbiology, 5*(1), 47–55.

González-Fernández, C., Sialve, B., Bernet, N., & Steyer, J. P. (2012). Impact of micro- algae characteristics on their conversion to biofuel. Part II: Focus on biomethane production. *Biofuels, Bioproducts and Biorefining, 6*(2), 205–218.

Grobbelaar, J. U. (2004). Algal nutrition. In A. Richmond (Ed.), *Handbook of microalgal culture: Biotechnology and applied phycology*, Hoboken: Wiley-Blackwell.

Herrmann, C., Kalita, N., Wall, D., Xia, A., & Murphy, J. D. (2016). Optimised biogas production from microalgae through co-digestion with carbon-rich co-substrates. *Bioresource Technology, 214,* 328–337.

Hidaka, T., Takabe, Y., Tsumori, J., & Minamiyama, M. (2017). Characterization of microalgae cultivated in continuous operation combined with anaerobic co-digestion of sewage sludge and microalgae. *Biomass and Bioenergy, 99,* 139–146.

IEA, Task 40 and Task 37 Joint Study. http://task40.ieabioenergy.com/wp-content/uploads/2013/09/t40-t37-biomethane-2014.pdf.

Jankowska, E., Sahu, A. K., & Oleskowicz-Popiel, P. (2017). Biogas from microalgae: Review on microalgae's cultivation, harvesting and pretreatment for anaerobic digestion. *Renewable and Sustainable Energy Reviews, 75,* 692–709.

Kadouri, A., Derenne, S., Largeau, C., Casadevall, E., & Berkaloff, C. (1988). Resistant biopolymer in the outer walls of *Botryococcus braunii*, B race. *Phytochemistry, 27*(2), 551–557.

Kinnunen, V., Craggs, R., & Rintala, J. (2014). Influence of temperature and pretreatments on the anaerobic digestion of wastewater grown microalgae in a laboratory-scale accumulating volume reactor. *Water Research, 57,* 247–257.

Klassen, V., Blifernez-Klassen, O., Wobbe, L., Schlüter, A., Kruse, O., & Mussgnug, J. H. (2016). Efficiency and biotechnological aspects of biogas production from microalgal substrates. *Journal of Biotechnology, 234,* 7–26.

Lardon, L., Helias, A., Sialve, B., Steyer, J. P., & Bernard, O. (2009). Life-cycle assessment of biodiesel production from microalgae. *Environmental Science and Technology, 43*(17), 6475–6481.

Lee, E., Cumberbatch, J., Wang, M., & Zhang, Q. (2017). Kinetic parameter estimation model for anaerobic co-digestion of waste activated sludge and microalgae. *Bioresource technology, 228,* 9–17.

Loos, E., & Meindl, D. (1982). Composition of the cell wall of Chlorellafusca. *Planta, 156*(3), 270–273.

Mahdy, A., Mendez, L., Ballesteros, M., & González-Fernández, C. (2015). Protease pretreated *Chlorella vulgaris* biomass conversion to methane via semi-continuous anaerobic digestion. *Fuel, 158,* 35–41.

Mahdy, A., Fotidis, I. A., Mancini, E., Ballesteros, M., González-Fernández, C., & Angelidaki, I. (2017). Ammonia tolerant inocula provide a good base for anaerobic digestion of microalgae in third generation biogas process. *Bioresource Technology, 225,* 272–278.

Mairet, F., Bernard, O., Cameron, E., Ras, M., Lardon, L., Steyer, J.-P., et al. (2012). Three-reaction model for the anaerobic digestion of microalgae. *Biotechnology and Bioengineering, 109,* 415–425.

Mairet, F., Bernard, O., Ras, M., Lardon, L., & Steyer, J.-P. (2011). Modeling anaerobic digestion of microalgae using ADM1. *Bioresource Technology, 102,* 6823–6829.

Mata, T. M., Martins, A. A., & Caetano, N. S. (2010). Microalgae for biodiesel production and other applications: A review. *Renewable and Sustainable Energy Reviews, 14*(1), 217–232.

Mata-Alvarez, J., Dosta, J., Romero-Güiza, M. S., Fonoll, X., Peces, M., & Astals, S. (2014). A critical review on anaerobic co-digestion achievements between 2010 and 2013. *Renewable and Sustainable Energy Reviews, 36,* 412–427.

Meier, L., Pérez, R., Azócar, L., Rivas, M., & Jeison, D. (2015). Photosynthetic CO_2 uptake by microalgae: An attractive tool for biogas upgrading. *Biomass and Bioenergy, 73,* 102–109.

Montingelli, M. E., Tedesco, S., & Olabi, A. G. (2015). Biogas production from algal biomass: A review. *Renewable and Sustainable Energy Reviews, 43,* 961–972.

Muñoz, R., Meier, L., Diaz, I., & Jeison, D. (2015). A critical review on the state-of-the-art of physical/chemical and biological technologies for an integral biogas upgrading. *Reviews in Environmental Science and Biotechnology, 14,* 727–759.

Neumann, P., Torres, A., Fermoso, F. G., Borja, R., & Jeison, D. (2015). Anaerobic co-digestion of lipid-spent microalgae with waste activated sludge and glycerol in batch mode. *International Biodeterioration and Biodegradation, 100,* 85–88.

Okuda, K. (2002). Structure and phylogeny of cell coverings. *Journal of Plant Research, 115,* 283–288.

Ometto, F., Quiroga, G., Psenicka, P., Whitton, R., Jefferson, B., & Villa, R. (2014). Impacts of microalgae pre-treatments for improved anaerobic digestion: Thermal treatment, thermal hydrolysis, ultrasound and enzymatic hydrolysis. *Water Research, 65,* 350–361.

Park, J. B. K., Craggs, R. J., & Shilton, A. N. (2011). Wastewater treatment high rate algal ponds for biofuel production. *Bioresource Technology, 102*(1), 35–42.

Park, S., & Li, Y. (2012). Evaluation of methane production and macronutrient degradation in the anaerobic co-digestion of algae biomass residue and lipid waste. *Bioresource Technology, 111,* 42–48.

Passos, F., Uggetti, E., Carrère, H., & Ferrer, I. (2014a). Pretreatment of microalgae to improve biogas production: A review. *Bioresource Technology, 172,* 403–412.

Passos, F., Hernández-Mariné, M., García, J., & Ferrer, I. (2014b). Long-term anaerobic digestion of microalgae grown in HRAP for wastewater treatment. Effect of microwave pretreatment. *Water Research, 49,* 351–359.

Passos, F., & Ferrer, I. (2014). Microalgae conversion to biogas: Thermal pretreatment contribution on net energy production. *Environmental Science and Technology, 48*(12), 7171–7178.

Passos, F., Gutiérrez, R., Brockmann, D., Steyer, J. P., García, J., & Ferrer, I. (2015). Microalgae production in wastewater treatment systems, anaerobic digestion and modelling using ADM1. *Algal Research, 10,* 55–63.

Passos, F., & Ferrer, I. (2015). Influence of hydrothermal pretreatment on microalgal biomass anaerobic digestion and bioenergy production. *Water Research, 68,* 364–373.

Peng, S., & Colosi, L. M. (2016). Anaerobic digestion of algae biomass to produce energy during wastewater treatment. *Water Environment Research, 88*(1), 29–39.

Rajagopal, R., Massé, D.I., Singh, G. 2013. A critical review on inhibition of anaerobic digestion process by excess ammonia. *Bioresource Technology, 143,* 632–641. https://www.sciencedirect.com/science/article/pii/S0960852413009498

Rizwan, M., Lee, J. H., & Gani, R. (2015). Optimal design of microalgae-based biorefinery: Economics, opportunities and challenges. *Applied Energy, 150,* 69–79.

Rodriguez, C., Alaswad, A., Mooney, J., Prescott, T., & Olabi, A. G. (2015). Pre-treatment techniques used for anaerobic digestion of algae. *Fuel Processing Technology, 138,* 765–779.

Rusten, B., & Sahu, A. K. (2011). Microalgae growth for nutrient recovery from sludge liquor and production of renewable bioenergy. *Water Science and Technology, 64,* 1195–1201.

Sahu, A. K., Siljudalen, J., Trydal, T., & Rusten, B. (2013). Utilisation of wastewater nutrients for microalgae growth for anaerobic co-digestion. *Journal of Environmental Management, 122,* 113–120.

Sanz, J. L., Rojas, P., Morato, A., Mendez, L., Ballesteros, M., & González-Fernández, C. (2017). Microbial communities of biomethanization digesters fed with raw and heat pre-treated microalgae biomasses. *Chemosphere, 168,* 1013–1021.

Schwede, S., Kowalczyk, A., Gerber, M., & Span, R. (2013). Anaerobic co-digestion of the marine microalga *Nannochloropsis salina* with energy crops. *Bioresource Technology, 148,* 428–435.

Scott, S. A., Davey, M. P., Dennis, J. S., Horst, I., Howe, C. J., Lea-Smith, D. J., et al. (2010). Biodiesel from algae: Challenges and prospects. *Current Opinion in Biotechnology, 21*(3), 277–286.

Sialve, B., Bernet, N., & Bernard, O. (2009). Anaerobic digestion of microalgae as a necessary step to make microalgal biodiesel sustainable. *Biotechnology Advances, 27*(4), 409–416.

Simpson, A. J., Zang, X., Kramer, R., & Hatcher, P. G. (2003). New insights on the structure of algaenan from *Botryoccocus braunii* race A and its hexane insoluble botryals based on multidimensional NMR spectroscopy and electrospray-mass spectrometry techniques. *Phytochemistry, 62*(5), 96–783.

Solé-Bundó, M., Carrère, H., Garfí, M., & Ferrer, I. (2017). Enhancement of microalgae anaerobic digestion by thermo-alkaline pretreatment with lime (CaO). *Algal Research, 24,* 199–206.

Stephens, E., Ross, I. L., King, Z., Mussgnug, J. H., Kruse, O., Posten, C., et al. (2010). An economic and technical evaluation of microalgal biofuels. *Nature Biotechnology, 28*(2), 126–128.

Tartakovsky, B., Lebrun, F. M., & Guiot, S. R. (2015). High-rate biomethane production from microalgal biomass in a UASB reactor. *Algal Research-Biomass Biofuels and Bioproducts, 7,* 86–91.

Thrän, D., Persson, T., Daniel-Gromke, J., Ponitka, J., Seiffert, M., Boldwin, D., & et al. (2014). Biomethane—status and factors affecting market development and trade. IEA.

Toledo-Cervantes, A., Estrada, J. M., Lebrero, R., & Muñoz, R. (2017). A comparative analysis of biogas upgrading technologies: Photosynthetic versus physical/chemical processes. *Algal Research, 25,* 237–243.

Tomei, M. C., Braguglia, C. M., Cento, G., & Mininni, G. (2009). Modeling of anaerobic digestion of sludge. *Critical Reviews in Environment Science and Technology, 39,* 1003–1051.

Torres, A., Fermoso, F.G., Rincón, B., Bartacek, J., Borja, R., & Jeison, D. (2013). Challenges for cost-effective microalgae anaerobic digestion. In R. Chamy & F. Rosenkranz (Eds.) *Biodegradation—Engineering and Technology*. Intech: Croatia.

Wang, M., Lee, E., Zhang, Q., & Ergas, S. J. (2016a). Anaerobic co-digestion of swine manure and microalgae chlorella sp.: Experimental studies and energy analysis. *Bioenergy Research, 9* (4), 1204–1215.

Wang, M., Lee, E., Dilbeck, M.P., Liebelt, M., & Zhang, Q., Ergas, S. J. (2016b). Thermal pretreatment of microalgae for biomethane production: Experimental studies, kinetics and energy analysis. *Journal of Chemical Technology and Biotechnology*. https://doi.org/10.1002/jctb.5018.

Ward, A. J., Lewis, D. M., & Green, B. (2014). Anaerobic digestion of algae biomass: A review. *Algal Research-Biomass Biofuels and Bioproducts, 5,* 204–214.

Weyer, K. M., Bush, D. R., Darzins, A., & Willson, B. D. (2010). Theoretical maximum algal oil production. *Bioenergy Research, 3*(2), 204–213.

Yen, H. W., & Brune, D. E. (2007). Anaerobic co-digestion of algal sludge and waste paper to produce methane. *Bioresource Technology, 98*(1), 130–134.

Yuan, X., Wang, M., Park, C., Sahu, A. K., & Ergas, S. J. (2012). Microalgae growth using high-strength wastewater followed by anaerobic co-digestion. *Water Environment Research, 84*(5), 396–404.

Zamalloa, C., Boon, N., & Verstraete, W. (2012a). Anaerobic digestibility of *Scenedesmus obliquus* and *Phaeodactylum tricornutum* under mesophilic and thermophilic conditions. *Applied Energy, 92,* 733–738.

Zamalloa, C., De Vrieze, J., Boon, N., & Verstraete, W. (2012b). Anaerobic digestibility of marine microalgae *Phaeodactylum tricornutum* in a lab-scale anaerobic membrane bioreactor. *Applied Microbiology and Biotechnology, 93*(2), 859–869.

Zhen, G., Lu, X., Kobayashi, T., Kumar, G., & Xu, K. (2016). Anaerobic co-digestion on improving methane production from mixed microalgae (*Scenedesmus* sp., *Chlorella* sp.) and food waste: Kinetic modeling and synergistic impact evaluation. *Chemical Engineering Journal, 299,* 332–341.

Zhong, W., Chi, L., Luo, Y., Zhang, Z., Zhang, Z., & Wu, W. M. (2013). Enhanced methane production from Taihu Lake blue algae by anaerobic co-digestion with corn straw in continuous feed digesters. *Bioresource Technology, 134,* 264–270.

Chapter 13
Biofuels from Microalgae: Photobioreactor Exhaust Gases in Oxycombustion Systems

Ihana Aguiar Severo, Juliano Smanioto Barin, Roger Wagner, Leila Queiroz Zepka and Eduardo Jacob-Lopes

Abstract The aim of this chapter is to present a comprehensive overview of integrated bio-oxycombustion systems with photobioreactors. Divided into seven distinct topics, the chapter discusses issues related to fundamentals of oxycombustion, the operational implications for oxycombustion-enhanced performance, oxygen produced by photosynthesis, volatile organic compounds as energy source, photobioreactors design, the process integration in bio-oxycombustion systems, and the hurdles of bio-oxycombustion technology, summarizing a range of useful strategies directed to the sustainable development of industrial combustion systems.

Keywords Biological carbon capture and utilization · Microalgae
Oxyfuel · Volatile organic compounds · Gaseous fuels · Process integration

1 Introduction

Carbon capture and storage or use (CCS/CCU) is recognized as one of the options to mitigate the increase of atmospheric carbon dioxide (CO_2) concentration (Koytsoumpa et al. 2017). However, through biological carbon capture and utilization (BCCU) as a concept of bioconversion of greenhouse gases (GHG) into value-added metabolic products, oxycombustion has gained considerable attention in recent years (Jajesniak et al. 2014).

Oxycombustion is a promising carbon capture technology due to its ability to reduce emissions by up to 90%, improving the energy efficiency of industrial combustion systems. However, the main barrier to be overcome from

---

I. Aguiar Severo · J. S. Barin · R. Wagner · L. Q. Zepka (✉) · E. Jacob-Lopes
Department of Food Science and Technology, Federal University of Santa Maria (UFSM), Santa Maria, RS 97105-900, Brazil
e-mail: lqz@pq.cnpq.br

E. Jacob-Lopes
e-mail: jacoblopes@pq.cnpq.br

E. Jacob-Lopes et al. (eds.), *Energy from Microalgae*, Green Energy and Technology, https://doi.org/10.1007/978-3-319-69093-3_13

oxycombustion is the obtaining of a high-purity, low-cost oxygen supply, in order to save fuel and energy (Chen et al. 2012b).

In this context, photobioreactors could be the key to getting around this problem. This equipment can provide substantial oxygen (O_2) concentrations through water photolysis reactions during microalgae cultivation. In theory, it is possible to generate on average 0.73 kg of O_2 for every 1 kg of CO_2 bioconverted, demonstrating the production potential of this substance in photobioreactors (Jacob-Lopes et al. 2010, 2017).

In addition, these bioprocesses produce several volatile organic compounds (VOCs), which have considerable energy value, besides releasing, in the photobioreactor exhaust gases, substantial concentrations of unconverted CO_2, which could improve the thermal performance of combustion systems (Jacob-Lopes and Franco 2013).

Therefore, in order to satisfy the oxygen supply required in oxycombustion systems, a promising technological route has been developed through the integrated bio-oxycombustion process. This bioprocess refers to the simultaneous production of two metabolic bioproducts: O_2 and VOCs from the direct conversion of GHG. These compounds are released with photobioreactor exhaust gases, which can be subsequently integrated as oxidizers and gaseous fuels, respectively, in industrial combustion processes. Furthermore, the unconverted CO_2 can be potentially used as nitrogen diluent. With this in mind, the aim of this chapter is to present a comprehensive overview of integrated bio-oxycombustion systems with photobioreactors.

2 Fundamentals of the Oxycombustion

Carbon capture from large point source emitters is a fast-developing technology that can mitigate the impact of anthropogenic CO_2 production. Oxycombustion has proven to be a potential capture technology mainly due to its perceived superiority in relation to efficiency and simplicity (Olajire 2010). Several authors have provided comprehensive information about the different aspects of oxycombustion technology (Buhre et al. 2005; Wall et al. 2009; Toftegaard et al. 2010; Scheffknecht et al. 2011; Chen et al. 2012a, b; Yin and Yan 2016; Khalil et al. 2017; Gładysz et al. 2017).

In a conventional combustion system, air is used as the oxidizer, and the coming CO_2 from the flue gas is diluted by N_2 of air, resulting in a reduced CO_2 concentration per capture (about 15% v/v). In oxycombustion, a combination of practically pure oxygen (usually 95% v/v) and recycled flue gas is used as the oxidizer for burning the fuel. Such flue gas is composed mainly of CO_2 and H_2O, which is used to control the flame temperature in the burner and fill the volume removed N_2, ensuring that there is enough gas to carry heat through the system (Stanger et al. 2015). Carbon dioxide concentration in the flue gases increases by approximately 17–70%, depending on the fuel used, and can then be captured, stored, or used (Buhre et al. 2005).

By determining the physical and chemical processes that the fuel experiences during oxycombustion, characteristics such as heat and mass transfer, temperature, stability and flame velocity, ignition, and pollutant formation are affected globally (Chen et al. 2012a, b). The main impacts are related to the differences in properties of CO_2, the diluent gas in oxycombustion, and N_2, the diluent in the combustion with air (Yin and Yan 2016). Table 1 shows the different physical properties and chemical effects of main gases resulting from oxycombustion (CO_2 and H_2O) and conventional combustion with air (N_2 and O_2), which induces substantial changes in combustion processes.

The total heat and mass transfer in a furnace include radiative and convective heat transfer and depend especially on the flame temperature and gas properties. Radiation is the principal mode of heat transfer in combustion processes, playing a dominant role in the furnace. The entire flame is considered to be a constant source of radiation, and its radiative energy release rate is improved when the emissivity (ε) is higher, that is, when the capacity of a substance to emit heat is greater. Thus, unlike diatomic molecules, such as N_2, triatomic molecules such as CO_2 and H_2O are radiating species and have higher partial pressures, and consequently, the absorptivity and emissivity of the flue gas substantially increase (Chen et al. 2012b).

As for convection, there is a greater contribution to heat exchange, which is influenced by flow velocity of gases, density, viscosity, thermal conductivity, and specific heat capacity, which are also functions of flame temperature. The rate of convective heat transfer coefficient in both oxycombustion and combustion with air can be expressed in terms of dimensionless numbers, such as the Reynolds and the Prandtl numbers, and by fluid thermal conductivity. Thus, the thermal conductivity of CO_2 is slightly higher than that of N_2, not significantly altering the heat transfer. However, the lower kinematic viscosity of CO_2 and its higher density, due to the higher molecular weight (44.09) when compared to N_2 (28.01), results in a larger Reynolds number and, therefore, a higher convective heat transfer coefficient (Yin and Yan 2016). In terms of specific heat capacity, it is observed that at 1000 °C, the N_2 presents a $C\rho$ of 34.18 kJ/k mol, whereas CO_2 has $C\rho$ 57.83 kJ/k mol, further highlighting the high heat transfer of these gases in oxycombustion conditions (Cengel 2003).

In relation to flame temperature, it is necessary to recirculate between 60 and 80% of the oxycombustion of gases into the furnace, aiming to moderate excess temperature due to the increase of oxygen concentration injected, and also to achieve a similar profile of heat transfer in relation to combustion with air. High O_2 concentrations increase the adiabatic flame temperature, which is the largest attained temperature in the combustion products without heat exchanging inside or outside the system, and this occurs due to lack of N_2 dilution. In this case, to moderate excess temperature, the proportion of recycled flue gas and the O_2 concentration to be injected must be adjusted in order to achieve the same flame temperature as in the combustion with air. On the other hand, if the recycled flue gas amount is higher, it will result in a lower average O_2 concentration for furnace

Table 1 Physical properties of the main gases diluents in oxycombustion and conventional combustion with air at 1 atm pressure and at 1000 °C (Griffiths and Barnard 1995; Cengel 2003)

Chemical compound	Chemical formula	Molecular weight [g/mol]	Density (ρ) [kg/m^3]	Specific heat capacity (cρ) [kJ/mol K]	Thermal conductivity (k) [W/m K]	Thermal diffusivity (α) [m^2/s]	Emissivity and absorptivity	Kinematic viscosity (v) [m^2/s]	Mass diffusivity[a] (m^2/s)	Prandtl number
Nitrogen	N≡N	28.01	0.24	34.18	0.082	9.83×10^{-3}	~0	2.00×10^{-4}	1.70×10^{-4}	0.70
Carbon dioxide	O=C=O	44.09	0.38	57.83	0.097	4.37×10^{-3}	>0	1.31×10^{-4}	1.30×10^{-4}	0.75
Water	H–O–H	18.01	0.16	45.67	0.136	1.89×10^{-3}	>0	3.2×10^{-4}	–	4.60
Oxygen	O=O	31.99	0.27	36.08	0.087	8.67×10^{-3}	~0	2.09×10^{-4}	–	0.63

[a] Mass diffusivity refers to the binary diffusion of O_2 in CO_2 and N_2

entry. In this case, flame temperature and gas temperature are lower. This way, low O_2 concentrations may result in lower stability and flame propagation velocity and, consequently, fuel may not burn completely. In parallel, there is a delay in the flame ignition in oxycombustion and this may vary according to the particle size of fuel and its properties, temperature, gas properties, heating rate, and aerodynamic impacts (Wall et al. 2009; Toftegaard et al. 2010).

Finally, the formation and emission of pollutants in oxycombustion should be considered. Due to the atmosphere rich in CO_2 and H_2O, extremely acidic gases such as SO_x and NO_x are formed, causing fouling and corrosion in the exhaust gas output device, which may affect combustion efficiency and damage the equipment. However, the emission is less intense due to pollutant reduction during flue gas recycling, lower formation of thermal NO by N_2 removal, and higher CO concentrations (Stanger and Wall 2011; Normann et al. 2009).

3 Operational Implications for Oxycombustion-Enhanced Performance

3.1 Oxygen Supply

Oxycombustion technology requires highly pure oxygen to function effectively. For this purpose, there are some technologies that separate oxygen from air, such as cryogenic air distillation, adsorption, absorption, and polymeric membranes. However, only the first option, which requires an air separation unit (ASU), presents maturity for large-scale application. The other options are in the early stages of research and development (R&D) and cannot be applied to the full-scale operations (Olajire 2010; Leung et al. 2014).

Conventionally, an ASU for oxycombustion should produce an oxygen stream with purity ranging from 95 to 99%. Energy consumption of separation increases as a function of oxygen purity. The purer the oxygen, the greater the amount of energy consumption involved in the separation process, directly influencing the composition of the gases formed, oxycombustion performance, as well as overall cost of the plant (Banaszkiewicz et al. 2014).

In terms of capacity, ASUs have been designed with design features to meet total oxygen production from 1000 tons (30,000 Nm^3/h) to 5500 tons (165,000 Nm^3/h). Today, the world's largest plant with an ASU for oxygen supply operates at a capacity of 4000 ton/d O_2 (Linde Group 2017). Therefore, assuming that, on a 500 MW oxycombustion power plant operating on an industrial scale, the oxygen supply should be around 10,000 ton/d (Higginbotham et al. 2011), 3 more ASU plants would necessarily have to operate simultaneously, or an ASU with greater capacity than the existing ones should be developed. At the same time, the expected energy consumption to separate 1 ton of oxygen from the air would be 150–200 kWh/t O_2 produced (540 kJ/kg), and the electrical energy necessary for this process would be approximately 80 MW, causing a significant reduction of about 7–11% in the efficiency of net electricity generation (Chorowski and Gizicki 2015).

3.2 Oxygen-Enrichment Methods in Combustion Systems

In addition to purity issues, another important point is the site for oxygen injection production. Oxygen enrichment in combustion processes provides many benefits as mentioned above; however, if the feeding system is not properly designed, problems such as furnace wall damage, non-uniform heating, and increased pollutant emissions can be potentiated (Baukal 2013). According to Daood et al. (2012), techniques for oxygen enrichment in oxycombustion are significantly different from one another, due to the different equipment design requirements, but are similar in regard to the reduced gas flows through the burner, increased residence time in combustion zones, and improvement in fuel burnout.

Thus, there are four main oxygen-enrichment methods in oxycombustion systems, as shown in Fig. 1. One is by adding O_2 in the incoming combustion air stream, also referred to as premix enrichment. Some systems use almost 100% oxygen at the main combustion inlet. However, performance is lower due to the large difference in the oxidizer speed of pure O_2 when compared to air (IHEA 2007). According to Lacava et al. (2006), most burners show enhanced performance and boost productivity with low-level enrichment (about 26% O_2), and only some operate at higher enrichment levels (about 35% O_2). Generally, when O_2 is added to the premix, the flame intensifies, the mixture between fuel/oxidizer is adequate, and the gas stream is dried. However, there is a greater risk of burner damage and explosion, due to the higher temperature, besides higher NO_x emission (Toftegaard et al. 2010).

The second method is the strategic injection of oxygen beside, beneath, or through the air/fuel flame, also referred to as O_2 lancing. This method is generally used for low O_2 levels. Its main advantage is that the flame can be better controlled, and released heat is evenly distributed. Nevertheless, furnace design has to be reconsidered (Baukal 2013).

The third method is to separate the injection of combustion air and O_2 into the burner, referred to as air/oxygen/fuel combustion. O_2 concentration in the burner will possibly be the same, as is the case for operation with air. In addition, it has the flexibility to operate with dual fuels (liquid and gaseous) and the enrichment of higher O_2 levels; however, significant risks are associated with the injection of nearly pure oxygen into a high-temperature stream of fuel and flue gas (Baukal 2013).

The last method consists in the complete replacement of air by high-purity O_2, referred to as oxyfuel combustion, where O_2 and fuel remain, and separation and mixing only occur when they are inserted into the furnace. For safety reasons,

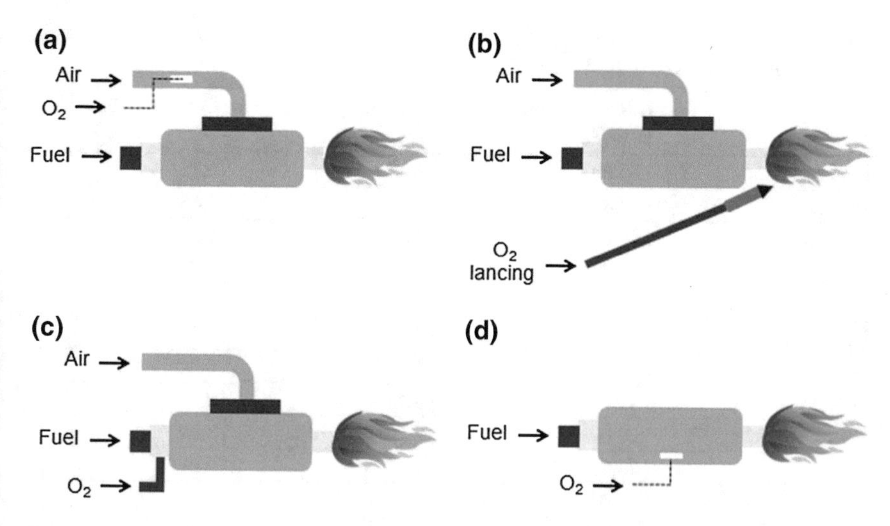

Fig. 1 Scheme of oxygen-enrichment methods in oxycombustion systems. **a** premix enrichment with air; **b** air/fuel flame (O_2 lancing); **c** air/oxygen/fuel combustion; and **d** oxyfuel combustion

there is no premix due to the high-level O_2 (>90%), which is extremely reactive. By this type of enrichment, an overall improvement in the combustion process is achieved, despite the higher operating costs (Baukal 2013).

3.3 *Fuel Supply*

Most industrial combustion processes require large amounts of energy, which is commonly generated by the burning of fossil fuels. These fuels are composed of hydrocarbons and sulfur, which readily combine with oxygen to produce a particular compound, and release a rather large amount of heat (Cengel 2003). Table 2 shows the main fuels (solids, liquids, and gaseous) with different heating value, oxygen supply, and estimated CO_2 emissions. For burning of 1 kg of natural gas, for example, one of the most commonly used gaseous fuels in combustion, it is necessary to provide about 2.11 kg of oxygen, which emits an average of 2.63 kg of CO_2 and presents potential energy of about 47 MJ/kg. The oxygen supply ranges from 2.00 to 3.73 kg; i.e., the required amount of O_2 can be almost 4 times the amount of fuel burned.

On the other hand, the use of oxygen-enrichment systems, besides improving the combustion efficiency, reduces energy loss and also increases fuel economy, depending on the exhaust gas temperature and the percentage of oxygen in the combustion air (ITP 2005). According to the US Department of Energy and the Industrial Heating Equipment Association (IHEA), the conversion to oxygen-enriched combustion is followed by an increase in furnace temperature and

Table 2 Heating value, oxygen supply, and estimated CO_2 emissions for combustion of different fuels (Griffiths and Barnard 1995; Cengel 2003)

Fuel	Heating value (MJ/kg)	O_2 supply (kg_{O_2}/kg_{fuel})	CO_2 emissions (kg_{fuel}/kg_{CO_2})
Methane	50.00	2.00	2.75
Ethane	47.80	3.73	1.46
Propane	46.35	3.63	1.00
Butane	45.75	3.58	0.75
Ethanol	27.70	2.08	0.95
Natural gas	47.00	2.11	2.63
Gasoline	44.40	3.50	0.38
Diesel oil	43.40	3.46	0.19
Petroleum coke	29.00	2.69	3.30
Coal	23.00	2.50	2.89

a simultaneous decrease in furnace gas flow around the product. Considering an oxycombustion furnace operating at a temperature of 1000 °C, and combustion air composed of 95% oxygen, fuel reduction is about 68%. This shows a fuel saving of approximately 35% in relation to a conventional combustion system. Additionally, the control of parameters such as air supply (fuel/air ratio), removal of combustion gases, carrier gas velocity, vapor pressure, and oxygen purity assists in fuel supply to achieve optimum energy efficiency in the furnace.

The remarkable advantages of oxycombustion show the feasibility of its implementation in power generation industries, despite their current operation only on pilot-scale. Meantime, original research and review articles have highlighted many barriers associated with the main operating parameters of the technology, which must be overcome to achieve industrial scale, as shown in Table 3.

4 Oxygen Produced by Photosynthesis

Green plants and photosynthetic microorganisms, such as cyanobacteria and microalgae, perform photosynthesis. Commonly, it is necessary mainly CO_2, which is converted into organic compounds, and light energy to carry out photosynthesis, releasing oxygen molecules and water through a sequence of different chemical reactions in distinct cellular compartments. This mechanism can be subdivided into two stages: light reactions or photochemical step, which occur only when the cells are illuminated, and dark reactions or carbon fixation step, which are not directly influenced by light, also occurring in the dark (Fay 1983).

During photosynthesis, more specifically in light reactions, there is the formation of highly energetic compounds, such as ATP (adenosine triphosphate) and NADPH

Table 3 Critical issues in oxycombustion systems

Parameter	Technical barrier
Oxygen supply	An oxycombustion plant requires large amounts of high-purity oxygen. The only option available on the market is ASU, which requires intense energy demand, operating expenses (OPEX), and capital expenditure (CAPEX)
Cost	The technology is expensive. Demand for electricity can increase plant cost by 70–80%
Scale-up	Although there is an oxygen–air separation process commercially available, it has not been deployed at the scale required for large power plants applications
Energy integration	Steam required for regeneration can only be extracted at conditions defined by the power plants steam cycle. Additionally, mitigation can result in the generation of significant quantities of waste heat. Energy integration can improve plant efficiency
Auxiliary power for CO_2 mitigation	Auxiliary power is also required to operate CO_2 mitigation technologies. This decreases the power plant's net electrical generation and significantly reduces net power plant efficiency
Mechanical integration	Any CO_2 mitigation system must fit within the boundaries of the power plant. This is a significant barrier when dealing with existing plants that have fixed layouts and limited open space
Flue gas pollutants	Constituents of the combustion exhaust gases, mainly sulfur, can damage the equipment and reduce its useful life
Water usage	A significant amount of water is used in current technologies for cooling during CO_2 compression

(nicotinamide adenine dinucleotide phosphate), essential for the assimilation of inorganic carbon and for oxygen production (Williams and Laurens 2010). This process begins in two photosystems (I and II), where pigments such as chlorophyll are responsible for absorbing mainly photons and transferring energy to an electron-accepting substance (located in the thylakoid membranes). From this stage, the excited chlorophyll recovers 6 lost electrons, where the energy is used for the water photolysis, also referred to as Hill reactions (Heldt and Piechulla 2011). By removing the light electrons, water molecules decompose into H^+ ions, releasing oxygen atoms to form the gaseous O_2 molecule, a significant product of microalgae metabolism. Figure 2 shows the schematic representation of water photolysis and oxygen generation during photosynthesis in a microalgae eukaryotic cell. This is an important aspect of photosynthesis, because all the oxygen generated in the process comes from the water photolysis (Barber 2017). The reaction can be described, in chemical terms, as follows (Eq. 1):

$$2H_2O \longrightarrow O_2 + 4H^+ + 4e^- \tag{1}$$

Additionally, the theoretical and realistic conversion efficiencies of water photolysis can be obtained by biological estimates, in terms of quantum efficiency, i.e., through the energy fraction of absorbed photons, or calculated from the solar

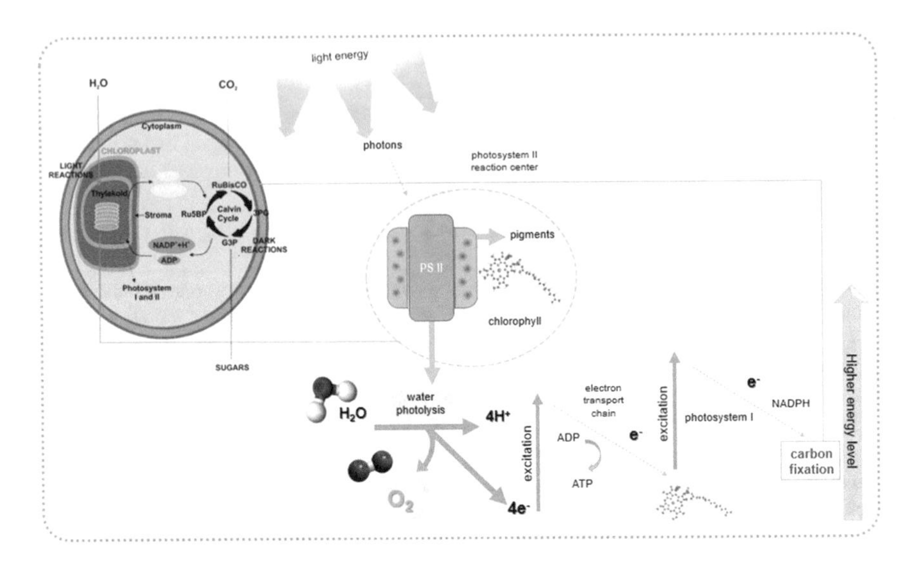

Fig. 2 Schematic representation of the oxygen generation in the photosynthesis

energy conversion point of view, through the solar spectrum. According to Bergene (1996), this ratio provides the value of process efficiency. In this work, the theoretical upper efficiency of water photolysis by microalgae was 0.11. Comparatively, commercial photovoltaic solar cells convert solar energy with efficiency in the range of 0.10–0.15.

Another important measure is the photosynthetic quotient (PQ), which provides more accurate values of the components involved in photosynthesis. The PQ is the molar ratio between released oxygen (gross primary production) in water photolysis during light reactions and CO_2 converted during the Calvin–Benson–Bassham cycle, and it varies as a function of the nitrogen source, carbon/nitrogen ratio assimilated, microalgae species used, type of organic molecule produced, luminous intensity, and photoperiods (Eriksen et al. 2007; Smith et al. 2012).

To accurately measure the photosynthetic activity, the PQ can be calculated according to Eq. 2 (Kliphuis et al. 2010):

$$PQ = \frac{OPR}{CUR} \tag{2}$$

where OPR is the oxygen production rate, and CUR is the carbon dioxide consumption rate.

Generally, the experimental values of the PQ are close to 1.0 (Burris 1981). Table 4 shows the experimental values of the PQ found in different microalgae. Jacob-Lopes et al. (2010) found a PQ of 0.74, which result corroborates the theoretical value estimated through the photosynthetic equation, establishing that each 1 kg of CO_2 consumed corresponds to a release of 0.73 kg O_2.

Table 4 Photosynthetic quotients (PQ) found by different species of microalgae

Microalgal species	Bioreactor type	PQ	References
Arthrospira platensis	Membranes	1.38	Cogne et al. (2005)
Chlamydomonas reinhardtii	Bubble column	1.00	Eriksen et al. (2007)
Chlorella sp.		1.30	
Aphanothece microscopica Nägeli	Bubble column	0.74	Jacob-Lopes et al. (2010)
Chlorella sorokiniana	Bubble column	1.40	Kliphuis et al. (2010)
Tetraselmis striata	Bubble column	1.50	Holdt et al. (2013)
Synechococcus PCC7002	Bubble column	1.30–1.40	Bernal et al. (2014)
Synechocystis sp.			
Anabaena PCC7120			
Chaetoceros wighamii	Bubble column	1.26	Spilling et al. (2015)

Both the efficiency values and PQ show the ability of these photosynthetic microorganisms to convert the solar energy and, consequently, the water photolysis. However, these quantitative relations can only be considered if parameters such as photobioreactor configuration, light incidence, mixing, and ecological aspects are properly determined.

5 Volatile Organic Compounds as Energy Source

Besides the biological oxygen generation, other products are biotransformed by microalgae photosynthetic cultures, being the volatile organic compounds (VOCs) of great relevance. These compounds correspond to the larger fraction (gas phase) of carbon bioconverted in photobioreactors that satisfy the global mass balance in the system, in addition to biomass (solid phase), carbonates, bicarbonates, and extracellular polymers (liquid phase) (Jacob-Lopes and Franco 2013).

The VOCs are organic chemical molecules with high vapor pressure and low boiling point, passing freely through biological membranes, which causes them to easily evaporate into the atmosphere (Dudareva et al. 2013). Additionally, VOCs are among the fastest growing molecules in aquatic ecosystems, and many of these compounds with specific biological activity are generated and released from the metabolism of photosynthetic microorganisms, both in marine and in freshwater phytoplankton (Goldstein and Galbally 2007).

According to Zepka et al. (2015), the VOCs produced by microalgae can be divided into terpenoids, phenylpropanoids/benzenoids, carbohydrate derivates, fatty acids derivates, and amino acid derivates, besides specific compounds not represented in those major classes. Microalgae are able to generate and release substantial amounts of VOCs belonging to different classes of compounds, such as alcohol, aldehydes, ketones, hydrocarbons, esters, terpenes, carboxylic acids, and sulfurized compounds, with chains that can contain up to 10 carbon atoms (Muñoz et al. 2004; Fink 2007; Sun et al. 2012).

Many studies of commercial interest have been conducted to identify VOCs produced by microalgae and cyanobacteria and point out their potential uses. Compounds such as β-cyclocyclal, 2-methyl-1-butanol, and 3-methyl-1-butanol were excreted in the extracellular fraction of *Microcystis aeruginosa* (Hasegawa et al. 2012). A wide variety of compounds, such as β-ionone, hexanol, hexanal, propanol, butanol, among others, were produced by *Phormidium autumnale* (Santos et al. 2016). In a study by Eroglu and Melis (2010), the microalgae *Botryococcus braunii* synthesized long-chain hydrocarbons, which can be commercially exploited for the synthesis of chemicals and biofuels feedstock. Schirmer et al. (2010) found in different cyanobacteria alkanes, such as heptadecane, pentadecane, and methyl heptadecane, besides alkenes, that have desirable properties for combustion. All these compounds have great potential as biofuels.

Most research on microalgae VOCs is focused on their use as industrial chemicals. Meantime, there are few studies demonstrating the feasibility of applying these compounds as fuels. Recently, Jacob-Lopes et al. (2017) developed a bioprocess in an attempt to make feasible the VOCs production in photobioreactors for use as gaseous fuels. A total of 17 compounds of different chemical structures were produced by microalgae *Scenedesmus obliquus* and released from photobioreactor exhaust gases (Fig. 3), which can potentially be used as energy source in combustion systems. Therefore, assuming that the estimated energy potential of these compounds is approximately 86.30 MJ/kg, and comparing them quantitatively with other conventional fuels, VOCs total energy content is superior to the value of natural gas (47.00 MJ/kg) and diesel oil (43.40 MJ/kg), for example. The several VOCs generated in photobioreactors could, therefore, be used for the gaseous fuels production, representing an important step in the consolidation of strategies to reduce dependence on fossil fuels and the expansion of renewable energy sources.

6 Photobioreactors Design

A photobioreactor can be defined as a lighted system designed for the development of photosynthetic reactions. In order for the CO_2 bioconversion in photosynthetic products to occur efficiently, it is necessary to consider some basic requirements, such as adequate light energy and CO_2, dissolved oxygen concentration, efficient mixing system, temperature control, nutrient availability, and scale-up (Wang et al. 2012).

A wide variety of cultivation systems have been reported for microalgae-based processes. Photobioreactors are generally classified into two designs: open or closed systems (Borowitzka 1999). Open systems are most commonly used in large-scale processes and are based on circular ponds and raceway tanks. They are simple to operate, cheap, and easy to expand. However, performance is poor, since the culture

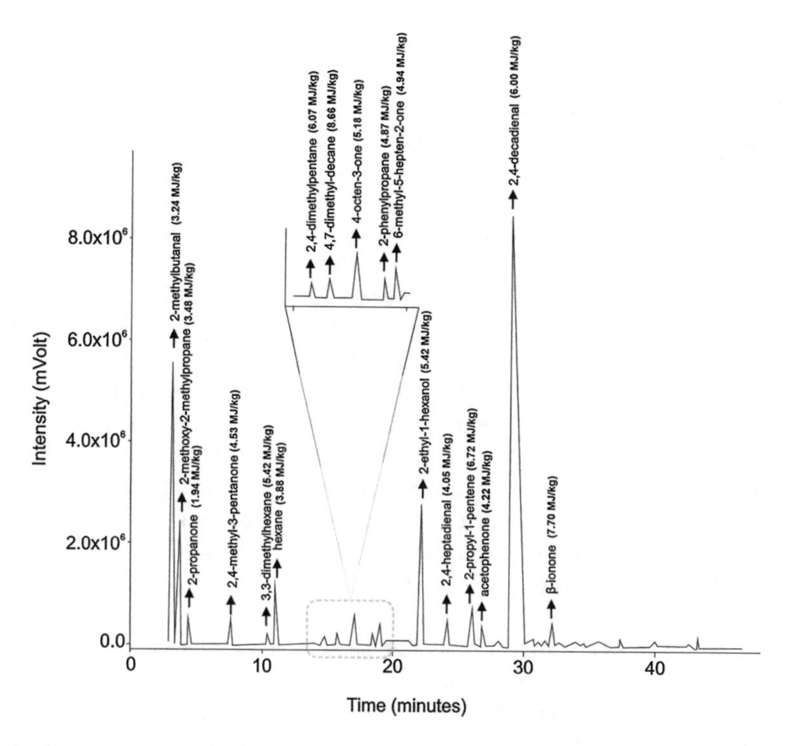

Fig. 3 Identification of VOCs produced by *Scenedesmus obliquus* and released from the photobioreactor exhaust gases. Adapted of Jacob-Lopes et al. (2017)

medium is exposed to variations in weather conditions, affecting the light intensity and temperature, besides low mass transfer, high evaporation rate, and susceptibility to contamination, which makes it unfeasible for an effective CO_2 conversion (Razzak et al. 2017).

On the other hand, closed systems included flat-plate, bubble column, airlift, tubular, hybrid, and biofilm photobioreactors, which enable high rates of CO_2 biotransformation in a wide variety of high-value bioproducts (Medipally et al. 2015; Tao et al. 2017). Moreover, they provide an easily controlled medium, safe against contamination. Despite their greatest potential for commercial application, closed systems are more expensive, due to the requirement of very transparent material, like glass or acrylic (Vasumathi et al. 2012). Another limiting factor is that losses of about 70% of non-bioconverted carbon are predicted when high CO_2 loads are injected (Jacob-Lopes et al. 2009).

Given these varied configurations, currently, one of the most widely accepted configurations for mass culture of microalgae is the closed tubular photobioreactors. This type is basically designed to achieve a maximum surface/volume (S/V) ratio and can be classified based on the horizontal, vertical, inclined, or helical arrangement of the tubes. They are suitable for CO_2 conversion due to their homogeneous mixture, greater gas transfer, smaller hydrodynamic stress, and uniform light distribution, which implies enhanced performance on the microalgae growth. In addition, they can be operated easily, their cell density is 5–6 times higher than that of open ponds, and their capacity can reach up to 25,000 L and occupy a restricted area of about 10 m^2 (Raeesossadati et al. 2014; Jacob-Lopes et al. 2015; Pawar 2016). However, in addition to the drawbacks related to overheating, the main critical issue in these systems is photo-inhibition, energy consumption, high costs, and dissolved oxygen (DO) accumulation (Huang et al. 2017).

As oxygen is a product of photosynthetic metabolism, its formation and solubilization in tubular photobioreactors indicate high inorganic carbon consumption rates, reaching O_2 generation rates of up to 10 mg/L min, even with a very frequent gas exchange (Chisti 2007). To prevent inhibition by O_2 accumulation, the DO concentration in the culture medium should not exceed the maximum tolerable value of 400% of the saturation level achieved in the presence of air. In a study by Raso et al. (2012), O_2 concentration increased from 75 to 250%; air saturation inhibited the growth of microalgae. To improve productivity in tubular photobioreactors, the oxygen level must be controlled or removed. However, optimal control parameters have not yet been well established to improve productivity in these systems.

With regard to operation, in a tubular photobioreactor, the airlift column circulates the broth with the culture medium to ensure light penetration through the solar collector, in which place most of the photosynthesis occurs, with ensuing DO accumulation. This, in turn, cannot be easily removed from the tubes (Molina-Grima et al. 2001). Therefore, in theory, if the oxygen is not removed within about one minute after accumulation, the inhibitory effect on the cells will occur immediately (Huang et al. 2017). In this case, the collecting tubes should be designed with restricted length for continuous DO removal, as well as the insertion of degasser systems.

Despite the fact that tubular photobioreactors are currently the most suitable configurations for application in oxygen generation, the main bottleneck of this type of equipment is its configuration, especially characterized by the geometry of these systems.

Due to the currently limited operational scale, the conventional configurations meet the basic requirements of the photosynthetic process. However, for a potential scale-up of the production process, operational failures must be overcome. Parameters such as the ratio of height/diameter column (H/D) are fundamental to build industrial photobioreactors. For this reason, hybrid photobioreactors compensate the drawbacks caused by limitation of S/V ratio and scale-up, since these systems can be based on a proper H/D ratio, generating configurations of reactors with heavy workloads in contrast to very long tubes or shallow ponds (Jacob-Lopes et al. 2016). If these aspects are considered, photobioreactors could be a fundamental step forward for the consolidation of the industrial biological oxygen generation.

7 Process Integration in Bio-oxycombustion Systems

Process integration has been widely used to further increase production systems efficiency. This concept focuses on the combination of technologies, in which the raw materials used can generate various types of products. Biobased systems may be suitable to minimize environmental impact, use of fossil inputs, and capital expenditures and to maximize the overall efficiency of an energy generation process or industry, provided they are obtained by total chain integration (Budzianowski and Postawa 2016).

Microalgae-mediated processes have recently seen growing demands for research and technological development, due to the versatility of these microorganisms in the CO_2 biotransformation within photobioreactors into valuable metabolic products (Jacob-Lopes et al. 2010).

Therefore, process integration using microalgae is a sustainable and economically viable route for improved sustainability, and it can be achieved by two types of integration basically: (i) mass integration, through effluents reuse and water recycling, and (ii) energy integration by heat recovery (Moncada et al. 2016).

By way of example, Fig. 4 describes a bioprocess, which represents the gain in thermal performance of a bio-oxycombustion furnace integrated into a photobioreactor. The thermal images show the superiority of use of the photobioreactor exhaust gases when compared to the injection of different oxidizers and at different cell residence times, during petroleum coke burning (Jacob-Lopes et al. 2017).

In this context, for bio-oxycombustion system proposed, mass integration occurs by direct conversion of GHG, especially CO_2 (gaseous effluent integration) in photobioreactors. Subsequently, part of the CO_2 is converted into photosynthetic metabolism by-products, such as biomass, inorganic salts, exopolymers, O_2, and VOCs. In parallel, energy integration is made by recovering of the photobioreactor gaseous phase, which contains the compounds of interest: VOCs (heat integration), O_2, and unconverted CO_2 released from the exhaust gases. These are integrated into

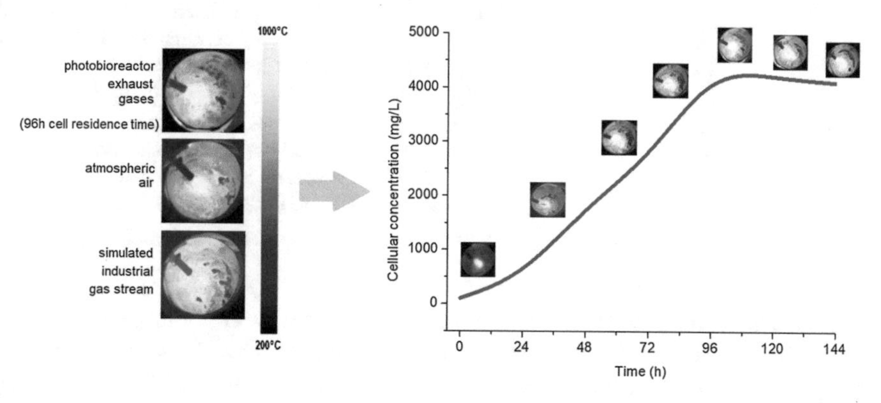

Fig. 4 Thermal performance of the integrated bio-oxycombustion system. Adapted of Jacob-Lopes et al. (2017)

Table 5 Challenges facing bio-oxycombustion technology

R&D challenges to integrated bio-oxycombustion systems scale-up	Comments
Photobioreactor design	Aspects associated with engineering, maintenance, economics, and microalgae species are the key to the construction of industrial photobioreactors for oxygen production
Collection of photobioreactor exhaust gases	Closed photobioreactors would be potentially suitable equipment for the oxygen supply and VOCs generation. For the removal mainly of the accumulated O_2, it would be necessary to design a degassing zone equipped with valves to control the flow and pressure of the gaseous fluid
Humidity of the photobioreactor exhaust gases	The gaseous phase of photobioreactor contains water vapor. When recovering exhaust gases, water should be removed in a separate unit to not interfere in the combustion
Pre-heating of the gases for injection	After removal of humidity, gases can be cooled; it would be necessary to do their pre-heating for injection into the burner system so as to avoid system thermal efficiency reduction
Injection site in the furnace	The injection zone must be defined so as to optimize energy utilization potential of O_2 and VOCs
Concentration of O_2 and VOCs	The bio-oxycombustion system requires high loads of the substances released from the photobioreactor exhaust gases. The photobioreactors currently available are not able to meet this demand, due to the lack of an ideal configuration
Process integration	Due to technical barriers of oxycombustion, process integration should be taken into account in order to balance the cost of CO_2 capture, oxygen, and fuel supply and to improve energy performance
Process life cycle analysis	Although bio-oxycombustion eliminates N_2 from flue gas and presents a potential increase in thermal efficiency, issues related to GHG emissions, more specifically CO_2, and energy consumption must be properly addressed, in order to reduce the environmental impact over its entire life cycle
Economic impacts	Microalgae-based processes are currently economically viable only on the fine chemicals production. It is necessary to develop new technological routes to the potential bulk chemical production

a combustion furnace such as oxidizer, gaseous fuels, and nitrogen diluent, respectively. After oxidation of the fuel, the resulting combustion CO_2 returns to the photobioreactor (mass integration) partially or totally, integrating the process globally.

8 Challenges Facing Bio-oxycombustion Technology

The implementation of bio-oxycombustion technology is a cost-effective means of BCCU, which could significantly reduce emissions from various industrial manufacturing sectors. R&D needs regarding fundamentals and performance of the oxycombustion system, scale-up of photobioreactors, and the integration and optimization of processes are identified in Table 5, for that the integrated bio-oxycombustion system can be fully scalable in the future.

9 Final Considerations

The growing development of oxycombustion systems has proven to be a viable strategy to mitigate CO_2 and increase the thermal efficiency of industrial processes. The integration of this technology with microalgae-based processes is considered an important engineering approach to promote sustainable development. Therefore, the full use of the photobioreactors exhaust gases could provide overall improvements in the thermal performance of integrated bio-oxycombustion systems. However, the CO_2 industrial biotransformation into O_2 and VOCs is very limited due to lack of an ideal photobioreactor design. Conversely, considering that combustion systems have extensive infrastructure, it would be necessary to design a photobioreactor that would operate at large volumes for the production of these substances in a mature industrial process. In this sense, for that bio-oxycombustion technology to present viability, efficiency, and productivity, operational problems must be solved in order to meet industrial demand for photobioreactors with applicability in full scale at field conditions.

References

Banaszkiewicz, T., et al. (2014). Comparative analysis of oxygen production for oxy-combustion application. *Energy Procedia, 51,* 127–134.

Barber, J. (2017). A mechanism for water splitting and oxygen production in photosynthesis. *Nature Plants, 3,* 17041.

Baukal, C. E. (2013). *Oxygen-enhanced combustion* (2nd ed.). Boca Raton, FL: CRC Press.

Bergene, T. (1996). The efficiency and physical principles of photolysis of water by microalgae. *International Journal of Hydrogen Energy, 21,* 189–194.

Bernal, O. I., et al. (2014). Specific photosynthetic rate enhancement by cyanobacteria coated onto paper enables engineering of highly reactive cellular biocomposite “leaves”. *Biotechnology and Bioengineering, 111,* 1993–2008.

Borowitzka, M. A. (1999). Commercial production of microalgae: Ponds, tanks, tubes and fermenters. *Journal of Biotechnology, 70,* 313–321.

Budzianowski, W. M., & Postawa, K. (2016). Total chain integration of sustainable biorefinery systems. *Applied Energy, 184,* 1432–1446.

Buhre, B. J. P., et al. (2005). Oxy-fuel combustion technology for coal-fired power generation. *Progress in Energy and Combustion Science, 31,* 283–307.
Burris, J. E. (1981). Effects of oxygen and inorganic carbon concentrations on the photosynthetic quotients of marine algae. *Marine Biology, 65,* 215–219.
Cengel, Y. A. (2003). *Heat transfer: A practical approach* (2nd ed.). New York: McGraw-Hill.
Chen, C., et al. (2012a). Oxy-fuel combustion characteristics and kinetics of microalgae *Chlorella vulgaris* by thermogravimetric analysis. *Bioresource Technology, 144,* 563–571.
Chen, L., et al. (2012b). Oxy-fuel combustion of pulverized coal: Characterization, fundamentals, stabilization and CFD modeling. *Progress in Energy and Combustion Science, 38,* 156–214.
Chisti, Y. (2007). Biodiesel from microalgae. *Biotechnology Advances, 25,* 294–306.
Chorowski, M., & Gizicki, W. (2015). Technical and economic aspects of oxygen separation for oxy-fuel purposes. *Archives of Thermodynamics, 36,* 157–10.
Cogne, G., et al. (2005). Design, operation, and modeling of a membrane photobioreactor to study the growth of the cyanobacterium *Arthrospira platensis* in space conditions. *Biotechnology Progress, 21,* 741–750.
Daood, S. S., et al. (2012). Deep-staged, oxygen enriched combustion of coal. *Fuel, 101,* 187–196.
DOE. U.S. Department of Energy. Available at: https://energy.gov/.
Dudareva, N. et al. (2013). Biosynthesis, function and metabolic engineering of plant volatile organic compounds. *New Phytologist, 198,* 16–32.
Eriksen, N. T., et al. (2007). On-line estimation of O_2 production, CO_2 uptake, and growth kinetics of microalgal cultures in a gastight photobioreactor. *Journal Applied Phycology, 19,* 161–174.
Eroglu, E., & Melis, A. (2010). Extracellular terpenoid hydrocarbon extraction and quantitation from the green microalgae *Botryococcus braunii* var. Showa. *Bioresource Technology, 101,* 2359–2366.
Fay, P. (1983). *The blue-greens (cyanophyta-cyanobacteria)* (5ª ed., p. 88). London: Edward Arnold Publishers, Studies in Biology 160.
Fink, P. (2007). Ecological functions of volatile organic compounds in aquatic systems. *Marine and Freshwater Behaviour and Physiology, 40,* 155–168.
Gładysz, P., et al. (2017). Thermodynamic assessment of an integrated MILD oxyfuel combustion power plant. *Energy* (in press).
Goldstein, A. H., & Galbally, I. E. (2007). Known and unexplored organic constituents in the Earth's atmosphere. *Environmental Science and Technology, 41,* 1415–1421.
Griffiths, J. F., & Barnard, J. A. (1995). *Flame and combustion* (3rd ed.). London, UK: Chapman and Hall.
Hasegawa, M., et al. (2012). Volatile organic compounds derived from 2-keto-acid decarboxylase in *Microcystis aeruginosa. Microbes and Environments, 27,* 525–528.
Heldt, H.-W., & Piechulla, B. (2011). *Plant biochemistry* (4ª ed., p. 618). German edition: Academic Press in an imprint of Elsevier.
Higginbotham, P., et al. (2011). Oxygen supply for oxyfuel CO_2 capture. *International Journal of Greenhouse Gas Control, 55,* S194–S203.
Holdt, S. L., et al. (2013). A novel closed system bubble column photobioreactor for detailed characterisation of micro- and macroalgal growth. *Journal of Applied Phycology, 26,* 825–835.
Huang, Q., et al. (2017). Design of photobioreactors for mass cultivation of photosynthetic organisms. *Engineering, 3,* 318–329.
IHEA. Industrial Heating Equipment Association. (2007). *Improving process heating system performance: A sourcebook for industry.* Prepared for the United States Department of Energy Office of Energy Efficiency and Renewable Energy, Industrial Technologies Program.
Jacob-Lopes, E., et al. (2015). Microalgal biorefineries, biomass production and uses (Chap. 5). In E. Atazadeh (Ed.), InTech.
Jacob-Lopes, E., et al. (2016). Bioprocesso de conversão de dióxido de carbono de emissões industriais, bioprodutos, seus usos e fotobiorreator híbrido. Patent WO 2016041028 A1.
Jacob-Lopes, E., et al. (2017). Process and system for re-using carbon dioxide transformed by photosynthesis into oxygen and hydrocarbons used in an integrated manner to increase the thermal efficiency of combustion systems. Patent WO 2017/112984 A1.

Jacob-Lopes, E., & Franco, T. T. (2013). From oil refinery to microalgal biorefinery. *Journal of CO_2 Utilization, 2,* 1–7.
Jacob-Lopes, E., et al. (2009). Development of operational strategies to remove carbon dioxide in photobioreactors. *Chemical Engineering Journal, 153,* 120–126.
Jacob-Lopes, E., et al. (2010). Biotransformations of carbon dioxide in photobioreactors. *Energy Conversion and Management, 51,* 894–900.
Jajesniak, P., et al. (2014). Carbon dioxide capture and utilization using biological systems: Opportunities and challenges. *Bioprocessing & Biotechniques, 4,* 3.
Khalil, A. E. E., & Gupta, A. K. (2017). The role of CO_2 on oxy-colorless distributed combustion. *Applied Energy, 188,* 466–474.
Kliphuis, A. M. J., et al. (2010). Photosynthetic efficiency of *Chlorella sorokiniana* in a turbulently mixed short light-path photobioreactor. *Biotechnology Progress, 26,* 687–696.
Koytsoumpa, E. I., et al. (2017). The CO_2 economy: Review of CO_2 capture and reuse technologies. *The Journal of Supercritical Fluids* (in press).
Lacava, P. T., et al. (2006). Thermal analysis of an enriched flame incinerator for aqueous residues. *Energy, 31,* 528–545.
Linde Group. (2017). Available at: http://www.linde-engineering.com/en/index.html.
Leung, D. Y. C., et al. (2014). An overview of current status of carbon dioxide capture and storage technologies. *Renewable and Sustainable Energy Reviews, 39,* 426–444.
Medipally, S. R., et al. (2015). Microalgae as sustainable renewable energy feedstock for biofuel production. *BioMed Research International, 2015,* 519513.
Molina-Grima, E., et al. (2001). Tubular photobioreactor design for algal cultures. *Journal of Biotechnology, 92,* 113–131.
Moncada, J., et al. (2016). Design strategies for sustainable biorefineries. *Biochemical Engineering Journal, 116,* 122–134.
Muñoz, J., et al. (2004). Effects of ionic strength on the production of short chain volatile hydrocarbons by *Dunaliella salina* (Teodoresco). *Chemosphere, 54,* 1267–1271.
Normann, F., et al. (2009). Emission control of nitrogen oxides in the oxy-fuel process. *Progress in Energy and Combustion Science, 35,* 385–397.
Olajire, A. A. (2010). CO_2 capture and separation technologies for end-of-pipe applications—a review. *Energy, 35,* 2610–2628.
Pawar, S. (2016). Effectiveness mapping of open raceway pond and tubular photobioreactors for sustainable production of microalgae biofuel. *Renewable and Sustainable Energy Reviews, 62,* 640–653.
Raeesossadati, M. J., et al. (2014). CO_2 bioremediation by microalgae in photobioreactors: Impacts of biomass and CO_2 concentrations, light, and temperature. *Algal Research, 6,* 8–85.
Raso, S., et al. (2012). Effect of oxygen concentration on the growth of *Nannochloropsis* sp. at low light intensity. *Journal of Applied Phycology, 24,* 863–871.
Razzak, S. A., et al. (2017). Biological CO_2 fixation with production of microalgae in wastewater—a review. *Renewable and Sustainable Energy Reviews, 76,* 379–390.
Santos, A. B., et al. (2016). Biogeneration of volatile organic compounds produced by *Phormidium autumnale* in heterotrophic bioreactor. *Journal of Applied Phycology, 28,* 1561–1570.
Scheffknecht, G., et al. (2011). Oxy-fuel coal combustion—a review of the current state-of-the-art. *International Journal of Greenhouse Gas Control, 5,* 16–35.
Schirmer, A. et al. (2010). Microbial Biosynthesis of Alkanes. *Science, 329*, 559–562.
Smith, L. M., et al. (2012). Quantifying variation in water column photosynthetic quotient with changing field conditions in Narragansett Bay, RI, USA. *Journal of Plankton Research, 34,* 437–442.
Spilling, K., et al. (2015). Interaction effects of light, temperature and nutrient limitations (N, P and Si) on growth, stoichiometry and photosynthetic parameters of the cold-water diatom *Chaetoceros wighamii*. *PLoS One, 10,* 1–18.

Stanger, R., & Wall, T. (2011). Sulphur impacts during pulverised coal combustion in oxy-fuel technology for carbon capture and storage. *Progress in Energy and Combustion Science, 37,* 69–88.

Stanger, R., et al. (2015). Oxyfuel combustion for CO_2 capture in power plants. *International Journal of Greenhouse Gas Control, 40,* 55–125.

Sun, S.-M., et al. (2012). Volatile compounds of the green alga, *Capsosiphon fulvescens*. *Journal of Applied Phycology, 24,* 1003–1013.

Tao, Q., et al. (2017). Enhanced biomass/biofuel production and nutrient removal in an algal biofilm airlift photobioreactor. *Algal Research, 21,* 9–15.

Toftegaard, M. B., et al. (2010). Oxy-fuel combustion of solid fuels. *Progress in Energy and Combustion Science, 36,* 581–625.

Vasumathi, K. K., et al. (2012). Parameters influencing the design of photobioreactor for the growth of microalgae. *Renewable and Sustainable Energy Reviews, 16,* 5443–5450.

Wall, T., et al. (2009). An overview on oxyfuel coal combustion—state of the art research and technology development. *Chemical Engineering Research and Design, 87,* 1003–1016.

Wang, B., et al. (2012). Closed photobioreactors for production of microalgal biomasses. *Biotechnology Advances, 30,* 904–912.

Williams, P. J. B., & Laurens, L. M. L. (2010). Microalgae as biodiesel & biomass feedstocks: Review and analysis of the biochemistry, energetics & economics. *Energy & Environmental Science, 3,* 554–590.

Yin, C., & Yan, J. (2016). Oxy-fuel combustion of pulverized fuels: Combustion fundamentals and modeling. *Applied Energy, 162,* 742–762.

Zepka, L. Q., et al. (2015). Biogeneration of volatile compounds from microalgae. Chapter: Flavour Generation, 257–260.

Chapter 14
Recent Patents on Biofuels from Microalgae

Ahmad Farhad Talebi, Meisam Tabatabaei and Mortaza Aghbashlo

Abstract To reduce greenhouse gas emissions and to prevent their devastative impacts of human health and the environment, bioenergy carriers have been at center of attention to supply global energy demand. Microalgae as solar energy-driven factories could efficiently convert carbon dioxide to a variety of hydrocarbons that can be used as biofuels. With the aim of realizing the current status of algal biofuels, respective patents were surveyed in this chapter using various databases, i.e., World Intellectual Property Organization, United States Patent and Trademark Office, and European Patent Office database. Information derived from the aforementioned databases was categorized into three: upstream, mainstream, and downstream strategies. The upstream strategies included patents on selection of algal strain and genetic engineering approaches while the mainstream strategies reviewed and discussed innovations pertaining to improving algal cultivation systems, production media and nutrients supply, and CO_2 supply. Finally, in the downstream strategies section, the inventions aimed at enhancing harvesting and dewatering of microalgae cells and lipid extraction were presented.

Keywords Patent mining · Innovation · Microalgae · Biofuels
Biodiesel

A. F. Talebi
Faculty of Microbial Biotechnology, Semnan University, Semnan, Iran

M. Tabatabaei (✉)
Agricultural Biotechnology Research Institute of Iran, Karaj, Iran
e-mail: meisam_tabatabaei@abrii.ac.ir

M. Tabatabaei
Biofuel Research Team, Karaj, Iran

M. Aghbashlo
College of Agriculture and Natural Resources, University of Tehran, Karaj, Iran
e-mail: maghbashlo@ut.ac.ir

E. Jacob-Lopes et al. (eds.), *Energy from Microalgae*, Green Energy and Technology, https://doi.org/10.1007/978-3-319-69093-3_14

1 Introduction

It is advisable that innovations presenting promising commercial applications be patent-protected. Going through the patent filing process, it is important to ensure that the set of claims drafted cover all the patentable concepts included in an application. Since this might not be technically feasible in some cases, more than a single patent application might be required to cover all commercial aspects of a given invention. A great deal of patent applications in the area of biofuels production is filed on a daily basis, but only a few are accompanied with a concert commercialization agenda, resulting in considerable investment problems. In line with that, reviews of the recent patent publications in the field could be of great assistance to the inventors and patent examiners in biofuels-related domain. It should be noted that business interests might undergo substantial changes over the course of time, affecting the commercial value of a certain patent as well.

Having reviewed the patent publications filed since 70s, a large number of patents could be found focusing on various aspects of biofuels production from microalgae, e.g., strain selection, process efficiency, and process management. This is mainly ascribed to the unique features of microalgae, i.e., being a non-agricultural crop (triggering no food vs. fuel conflict), possibility of cultivation in various aquatic environments (freshwater, saline water, and wastewater), as well as huge biomass production and high lipid content.

Most recent patent applications concerning microalgal fuels are focused on improving the cellular metabolic flux of the organisms or on improving/optimizing the algal biofuel production processes (Thompson 2013). As an example, the distribution of patents for two candidate microalgae for biodiesel production is presented in Table 1 (de la Jara et al. 2016).

It should be noted that this chapter is not advocating or determining the patentability of any technologies related to microalgal fuel but rather striving to shed light on the latest development in the field in the form of patent applications. More specifically, patent applications are reviewed and discussed under upstream, mainstream, and downstream strategies.

Table 1 Distribution of patents for two candidate microalgae for biodiesel production (adopted from de la Jara et al. 2016)

Patent application details	Microalgae	
	Chlorella sp.	*Dunaliella* sp.
Year of first publication	1964	1978
Year of first publication in energy area	1977	2008
Evolution of published patents	2011	Steady state
Application category (in order of development *vs.* time)	Production methodology, food, energy	Culture media, food, energy

2 Microalgae as Biofuel Feedstock

The introduction of commercially viable microalgal feedstock for biofuels production is among the most important bottlenecks of the whole process and therefore, numerous published patents exist in this regards. Algal biomass could undergo any of the following main pathways to be converted into biofuels: (1) extraction and transesterification of triglycerides to produce biodiesel; (2) fermentation of carbohydrates to produce bioalcohols (e.g., bioethanol and biobutanol); (3) anaerobic digestion to produce biogas (biomethane); and (4) gasification or other thermochemical conversions of the biomass (Craggs et al. 2011). The details of processes and systems to convert algal biomass for production of biofuels are the subject of the patents such as US7977076B2, US2012/0329099, US2012/0288917, US2012/0283496, WO2009158028A2, CN102120938A, US8308949, US8211308. In all these inventions, it has been stressed that microalgae cultivation in an aquatic environment could harness sunlight energy in the form of carbohydrates by photosynthesis. This is environmentally important as the combustion of petroleum-derived energy carriers interferes with the carbon pool, i.e., the energy of sunlight stored in the past, leading to an increased level of atmospheric carbon while algal biofuels capture these released carbon atoms on a current basis.

A patent analysis by using a major generic keyword, i.e., "microalgae biodiesel" in the title, abstract or full text, led to 49 hits in the Worldwide EN database (https://www.epo.org). The related patents were also mined in the USPTO Patent Full-Text and Image Database (http://www.uspto.gov/) from 1976 to 2017, and 107 hits were found. While using the free trial version of Matheo Patent software, the number of patents found using the same keyword was only one since the year 2011. However, by using "microalgae" as keyword, more than 2048 US patents describing microalgae potential applications were found since the year 1979. These patents are mostly focused on limited applications such as introduction of novel cultivation conditions (e.g., SU678065, SU663723, and SU686686), or for water treatment (e.g., SU701570). After a lag phase between 1960s and 1990s, the publication of patents related to the general term "microalgae" experienced a slight increase. Interestingly, the introduction of the industrial applications of microalgal biotechnology led to an exponential growth in the number of patents following the year 2007.

Apart from the liquid biofuels produced from microalgae such as bioethanol and biodiesel, production of other bioenergy carriers such as biohydrogen and biomethane from this feedstock is also of substantial interest. Researchers have also developed genetically modified algae to produce specific biofuel precursors. In the subsequent sections, some examples of algal potentials in bioenergy production are presented. Contrary to the algae-related applications published in last decades (usually disclosing methods that were interesting, but not necessarily commercial), in the past 6 years, both aspects of novelty and commercial viability have been considered while some patents have also taken environmental considerations (e.g., life cycle assessments) into account.

2.1 Biodiesel

"How do scientists make biodiesel from microalgae?" The answer lies in the ability of microalgae to synthesize huge amounts of free fatty acids and accumulate them in storage form of lipids, i.e., triglycerides (TAGs), in their cells. These lipids are then used as feedstocks for biodiesel production. The lipid bodies are extracted and the TAGs would be converted to the corresponding methyl/ethyl esters through a reaction called transesterification. The term "biodiesel" has been considerably more frequently cited in patent databases than bioalcohol or hydrogen (de la Jara et al. 2016). Further details on converting algal oil to biodiesel, from extraction to catalytic conversion, have been provided in the following patents: US7977076B2, US4341038, US8475543B2, US20110136189, and US20160053191A1.

2.2 Bioalcohols

Some microalgae contain carbohydrates (generally not cellulose) that can be used as feedstock for alcoholic fermentation. It should be highlighted that bioalcohols production process from microalgae is very simplified owing to the fact that microalgae cells do not require lignin and hemicelluloses for their structural support and therefore, no chemical and enzymatic pretreatment step would be required to extract sugars. Nevertheless, a simple and economically justified physical pretreatment process such as extrusion and mechanical shear is still required to break down the cell wall to release the fermentable sugars for subsequent conversion into bioalcohols such as bioethanol (John et al. 2011). In spite of all the aforementioned favorable features, fermentative fuels produced from algae are still faced by some challenges such as low fermentable carbohydrate content of algae biomass compared with other starchy crops such as maize. Some nutritional limitations such as nitrogen starvation and genetic manipulation have been considered in the published patents as promising solutions to overcome this shortcoming. Suitable starting materials, methods of fermentation, involved strains and enzymes as well as general processes and systems to produce and isolate alcohols from microalgal feedstock have been the subject of numerous patents such as US patent applications 5578472A, 7135308B1, 7507554, and 9260730.

2.3 Biohydrogen

Hydrogen is promoted by many because of possessing the potential to be a clean sustainable energy carrier. Photoautotrophic H_2 producing green algae, including *Chlamydomonas reinhardtii*, has been shown to metabolize H_2 under anoxic conditions. In the photosynthetic system, H_2 can be produced through hydrogenase-catalyzed reduction of protons by the electrons generated from photosynthetic oxidation of water. Sunlight acts as generator to supply the required energy to continue this stepwise reaction (Lee 2010). In spite of the numerous research efforts put into improving algal H_2 production, the limited rate of this reaction has so far made H_2 production commercially impractical.

2.4 Biogas (Biomethane)

Following lipid (oil) extraction or ethanol fermentation, the remaining algal biomass could also be anaerobically digested to produce biogas. If upgraded, i.e., by increasing the methane content of the gas stream, biogas can be used directly for heating or for electricity generation. Conversion of wet algal biomass to biomethane along with a liquid fertilizer (anaerobic digestate) is an advisable strategy to harness the majority of the remained energy and nitrogen contained in the algal biomass after oil and/or carbohydrates extraction (Scott et al. 2010). Biogas obtained from microalgae was found 7–13% more enriched in terms of its methane content in comparison with the biogas obtained from maize (Sialve et al. 2009), but the production was still limited by the relatively high N content of biomass or in another word, the ammonia inhibition effect. An average yield of approximately 0.30 m^3 (0.20 kg) CH_4/kg algal biomass (energy value equivalent to 1 L of petrol (34 MJ) for each cubic meter of biogas) was reported by Craggs et al. (2011).

2.5 Biomass to Liquid (BTL) Fuel

Biomass to liquid (BTL) is an integrated process aimed at fuel by thermal conversion of biomass. More specifically, the biomass gasified through biosyngas production systems is converted into liquid fuels through different processes such as Fischer–Tropsch (Medina et al. 2010). Carbon monoxide and hydrogen can be produced from gasification of any biomass such as algal feedstock. Methanol for instance can then be produced through direct reaction between these gases. US Patents 20090321349, 8163041B1 as well as WO2012109720A1 and WO2014057102A1 present further details on the catalysts and methods involved in the conversion of syngas to liquid fuels.

3 Microalgal Biofuels-Related Patents

Based on a logical order, the microalgal biofuels-related patents could be categorized into three major groups, i.e., patents concerning upstream, mainstream, and downstream strategies. The patents presented were selected from the Matheo patent database software which contains more than 80 million patents.

3.1 Upstream Strategies

It is usually easier to optimize existing biofuel technologies rather than exploring newly emerged capacities. Accordingly, to overcome the existing shortcomings of algal biofuels production, selection of more productive strains to boost productivity is considered as the first approach. As the subsequent approaches, reprogramming of algal metabolism either by modifying growth conditions or by genetic engineering tools could be considered.

3.1.1 Selection of Algal Strain

Selection of appropriate algal strains is a prerequisite to successful and economically viable algae-based biofuel industry. In fact, proper strain selection results in better breeding, engineering, and adaptation of strains to reach the most desirable phenotypes. This is because each microalgal genus may need specific requirements for growth and cultivation conditions, harvesting equipment, downstream processes as well as extraction protocols due to cell different physiology or morphology. Therefore, development of non-species specific and common devices can be helpful in achieving a more successful algal-based bioenergy market. Numerous patents are focused on specific genera of *Chlorella*, *Spirulina*, *Dunaliella*, *Haematococcus, Synechocystis, Microcystis, Desmodesmus,* and *Chlamydomonas*. US20140302569A1, US20090211150A1, and WO2008083352A1 are just a few examples. Among algal species, *Chlorella* sp. is of largest interest in this field, probably due to its high growth rate and comparatively lower production costs. For instance, the more economical the produced lipid, the more competitive the produced biodiesel would be. The first patent related to this species in the EPO database dates back the year 1977, but the number of records increased to 36 in the year 2016 (de la Jara et al. 2016).

Further search in Espacenet showed that considering only quantitative growth parameters would not be sufficient for an efficient biodiesel production and high intracellular lipid content would also act as a key criterion for selection of candidate microalgae strains for biodiesel production. On such a basis, *Isochrysis galbana* was introduced as a productive strain by the patent ES2088366A1. Furthermore, strains of *Characium polymorphum* and *Ankistrodesmus braunii* were also studied as oil-rich microalgae which can be used as feedstock for biofuel production by the US patent 20130157344A.

It should be mentioned that the selection of species for scaling-up an algal biofuel production system would also depend on fatty acid (FA) composition and lipid productivity. In line with that, Weiss in a US patent 20080220486A1 claimed that the strains of *Skeletonema costatum* and *Nannochloropsis* sp. could be considered as prone oleaginous microalgal strains; according to their FA profile and growth rate (20 g/m^2/d) under certain conditions.

The quality parameters of biodiesel (i.e., Cetane number, heat of combustion, cold flow properties, oxidative stability, and viscosity) depend on the characteristics of individual FA alkyl esters and are determined by the structural features of the FAs such as chain length, number and the situation of double bonds, and chain branching) (Ramos et al. 2009; Talebi et al. 2013a, b). The type of the produced FAs by algal cells is greatly influenced by genetic characteristics and also by the environmental conditions during cultivation. In general, intrinsic tolerance to higher temperatures and higher CO_2 concentrations could lead to a high biomass growth rate with a huge quantity of lipids. Overall, super microalgal strains could be either isolated or mutated. There are patents which try to introduce methods to mutate and maintain "old" strains to obtain "prone and powerful" ones, e.g., JP10248553A, US20130236951A1, US7935515B2, and CN101412965A. Other publications such as JP8257356A, JP10248553A, and TW291493B highlight the use of thermophilic microalgae when hot flue gases are employed.

3.1.2 Genetic Engineering Approaches

To obtain superior microalgae strains capable of swift cell growth, efficient photosynthesis, enhanced inorganic carbon fixation, as well as producing improved type and quality of fuel genetic engineering approaches have been exploited. Thanks to the developments made in sequencing tools since early 2000s, substantial advances in genetic manipulation of single-celled photosynthetic microalgal model organisms such as *C. reinhardtii* and *Chlorella vulgaris* have been achieved (Talebi et al. 2013a, b).

In general, genetically modified microalgae could be efficiently used for biofuel production, CO_2 sequestration, as well as other bioremediation goals. For instance, the patent US20170191094A1 claimed that the recombinant algae strains harboring at least one of the following exogenous genes were able to produce greater amounts of lipids under nitrogen starvation conditions [acyl-CoA synthetase, acyl-CoA reductase, acetyl-CoA carboxylase, acyl-ACP thioesterase, phosphatidic acid phosphatase, or diacylglycerol 0-acyltransferase (DGAT)]. Moreover, the quality of biodiesel can be also improved by engineering the cells toward the accumulation of lipids with a more desired FAs profiles. Among the strategies considered to engineer FA biosynthesis toward more compatible lipid profiles are the overexpression of FAs enzymes and their up-regulation by transcription factors as well as increasing the availability of precursor molecules (acetyl-CoA) and reducing power (NADPH). Moreover, down-regulation of FA catabolism by inhibiting β-oxidation, or lipase hydrolysis is also among the other available strategies (see patents US8951777B2, WO2011026008A1, WO2013034648A1, and US9593351B2). Microalgae can also be modified to express different enzymes that influence the production of long-chain FAs (e.g., patent WO2010019813A2). Also related to fatty acid synthesis is the polyketide synthase enzyme (PKS), whose impacts on the production of poly unsaturated FAs are discussed in the US patent 20070244192A1. Altering the saturation degree through the introduction or regulation of desaturases, and optimization of FA chain length with thioesterases are among various biomimetic approaches proposed to enhance the quality and yield of the produced biodiesel from microalgae (Talebi et al. 2013a, b).

Apart from enhancing metabolic lipid synthesis, other targets of genetical modification of microalgae concern improving light utilization, enhancing photosynthetic efficiency, as well as modifying carbon assimilation and trophic conversion pathways. For example, methods for enhancing cell growth of microalgae by transgenic expression of a bicarbonate transporter, carbonic anhydrase, and light-driven proton pump were introduced by the US patents 2014/0120623 and US20170211086A1.

Light penetration characteristic (also known as chlorophyll engineering) is a new interesting area being explored by related patents such as the US patent 20090023180A1 presenting methods to increase the efficiency of light utilization of photosynthetic microorganisms.

In conclusion, it worth quoting that although the aim of the upstream strategies is not to cover all topics involved in algal biofuels production, it is important to highlights that potential achievements made in this category could play key roles in overcoming the economic challenges faced by industries dealing with large-scale microalgal biofuels production. Microalgae can be selected or modified to express different valuable byproducts. On the other hand, the nutritional requirements, compatibility to available facilities, tolerance to biotic and abiotic stress could also be engineered through proper upstream processes, since the possibilities of both selection (genetic diversity) and genetic engineering seem infinite.

3.2 *Mainstream Strategies*

Different patents exist with an aim to improve the production of biomass and/or biocompounds. It is worth highlighting that economically produced algal biomass is an essential part of successful large-scale algal fuels industries. This clearly explains major efforts put into this category. Photobioreactor (PBR) designs for cultivating microalgae might be the most important challenge, but there are other mainstream bottlenecks as well, such as controlling systems. While harvesting systems, extraction and conversion of microalgal biomass to various biofuel are the subjects of downstream strategies considered when an ideal large-scale model plant producing algal biofuels is discussed.

3.2.1 Improving Algal Cultivation Systems

As mentioned earlier, sustainable large-scale cultivation of microalgae is a prerequisite for successful production of algal biofuels such as algal biodiesel. To identify the most relevant activities carried out in the development of microalgal cultivation, patents issued pertaining to this topic are summarized in this subcategory. There are three main alternatives for cultivating photoautotrophic algae: (1) open systems such as the routinely used raceway pond systems, (2) closed systems involving PBRs, and (3) hybrid production systems which are combinations of the other two systems. Among these systems, open ponds like raceway pond systems are the most common design employed for large-scale applications (Pulz 2001). A typical raceway pond comprises a closed oval channel, open to the air, and mixed with a paddle wheel to circulate the water and prevent sedimentation. These ponds are usually shallow; i.e., ~0.25–0.4 m deep, to facilitate light penetration and prevent self-shading by algal cells. Limited light penetration through the algal broth would decrease the photosynthesis and consequently the biomass production. Some genetic manipulations aimed at remodeling photosynthesis apparatus to enhance this trait were discussed in Sect. 3.1.2. High rate algal ponds (HRAPs) are shallow, open raceway ponds. HRAPs have been originally used for the treatment of municipal, industrial, and agricultural wastewaters; however, the algal biomass produced from these systems could be converted through various pathways to biofuels.

A semi-closed ocean system enriched by iron was introduced by the US patent 2014/0113331 leading to atmospheric CO_2 sequestration, reduced ocean acidity, as well as efficient cultivation and harvesting of a high deal of algal biomass. Although open systems in general look promising for commercial applications, their several shortcomings led to a widespread search for alternative algal cultivation systems. In light of that, tubular PBRs were introduced through which problems like susceptibility to contamination (as seen in open pond systems), large water consumption, low CO_2 absorption efficiency, the presence of dark zones (or in another word, low light penetration efficiency), and the resultant low photosynthesis efficiency were overcome. The issue of surface–volume ratio, light, CO_2, and nutrients supply as well as the development of tools to control temperature and pH have been extensively studied and have been the subject of numerous patents such as US20090211150A1, US5104803A, US20090291485A1, US2010000571A1, US20090029445A1, and US9605238B2.

Different configurations of tubular PBRs (i.e., horizontal, helical, and flat panels) have also been introduced in a number of patents (e.g., US20100248333A1 and US20080311649A1). Moreover, airlift reactors in which bubbles are used as bubble-columns to mix the media were later introduced (US20110113682A1) to overcome the problems associated with the large surface–volume ratio observed in tubular PBRs. However, bubble-columns and airlift reactors require high gas flows to ensure an efficient circulation would be taking place between the light and dark zones. This could further impose a shearing force on the growing algal cells. A novel PBRs design elaborated in the patent US20150275161A1 eliminated the need for sparging and compressors for suspending cells and mixing carbon dioxide through the introduction of attendant mixing by subtending wave motion. The novel system resulted in reduced initial investment required as well as the elimination of the above-mentioned sharing force. Another example of novel PBRs is discussed in WO2015056267A1.

In conclusion, PBR systems allow for better control of the algal cell growth but are also accompanied with higher energy demands and, therefore, are more costly than open systems to operate (US20090011492A1). On the other hand, areal productivity of airlift PBRs is higher than that of the tubular PBRs, but their volumetric productivity is around half of what achieved using tubular PBRs (Tabernero et al. 2013). More detailed information on cultivating microalgae using PBRs could be found in patents such as US20090130704A, US20140356931A1, WO2015050775A1, US9045724B2, and US8003379B2.

3.2.2 Production Media and Nutrients Supply

As mentioned earlier, by using PBRs, growth-limiting factors such as light, CO_2, nutrients supply, and temperature can be easily controlled. Numerous patents like US20110092726A, US20110107664A1, US20130023044A1, and US20110294196A1 are concerned about efficient nutrient supply. More specifically, their aim is the development of nutrient media to increase biomass production and boost accumulation of valuable compounds. Introduction of novel sources of essential minerals and CO_2 to enhance the economic aspects of the systems has also been among the objectives of such patents.

Historically, biochemical engineering, e.g., nutrients management (such as nitrogen and phosphor starvation), precursor addition as well as design of growth and/or environmental conditions (like salinity, acidity, and photon flux) in microalgae have been used as primary forward tools to enhance desired metabolic productivity (Courchesne et al. 2009). Exploring the respective regulatory mechanisms was the subject of the following patents: GB2501101A, US9295206B2, and WO2015088127A1.

As it was previously mentioned, the environmental conditions as well as provided nutrients could directly affect the FA profile of oleaginous microalgae. Heterotrophic culture system is an example of ways to increase FA concentration in microalgae. Within the heterotrophic culture, the microalgal cells consume an organic source (e.g., glucose, glycerol) instead of CO_2. In spite of increased cost and reduced environmental benefits, several advantages such as an increase in growth speed and lipid concentration are expected. This has been the subject of numerous investigations such as the patents WO9107498A1, US20090209014A1, US5130242A, and US20060094089A1.

Although the mechanisms of wastewater tolerance in microalgal community are yet to be discovered, strains which are naturally adapted and are capable of efficiently growing in wastewaters/effluents are regarded as successful strains to achieve economical biofuel production. This is ascribed to the fact that nutrient-rich municipal, agricultural, and industrial wastewaters could provide an economically sustainable means of cultivation for different strains of microalgae. In addition, such systems offer the advantage of combining wastewater treatment (i.e., heavy metal and nutrients removal) with biofuels production systems (Pittman et al. 2011). Such combination can potentially reduce unit cost energy by 20–25% in addition to eliminating the cost of nutrients and freshwater supply (Craggs et al. 2011). It has been reported that algal cells can efficiently remove contaminating nitrogen (N) and phosphorus (P) nutrients as well as toxic metal pollutants from wastewaters (de-Bashan and Bashan 2010; Ruiz-Marin et al. 2010). The claims of multiple patents, e.g., US20110247977A1, FR3023548A1, WO2014076327A1, GB2484530A, and GB2509710A, are focused biofuel production systems coupled with wastewater remediation. Nevertheless, to achieve desired levels of wastewater treatment with algal systems, maximizing autotrophic production and discovering physiological characteristics of algal cell are of primary importance and has been the subject of the following patent: US2010267122, CN101368193A, US20090294354A1, and US8101080B2.

To benefit all the advantageous features of coupled systems, some sustainable systems such as HRAPs have already been developed (see ES2563852T3). HRAP system in comparison with the conventional wastewater treatment methods requires lower capital and operating costs and needs no intensive advanced technology to operate compared with the conventional mechanical treatment technologies (Craggs et al. 2012). Eight comprehensive patents could be found by searching the term "HRAPs" in the PGPUB production database since the year 2001; for example, the patents US20160122705A1 and US20100252498A1 discuss simultaneous methods for HRAP-based wastewater treatment and algal production in detail.

3.2.3 CO_2 Supply

Most microalgae strains have been evolved to mitigate the environmental impacts. This capability of microalgae strains to remove pollutants was first introduced in the 1970s and twenty years later patents such as JP4075537A, JP314777A, and US5011604A strived to describe the details of removing pollutants from power stations with microalgae. To facilitate the utilization of ambient CO_2, microalgae possess a CO_2-concentrating mechanism (CCM) (Ndimba et al. 2013). Nevertheless, direct CO_2 addition to medium has still been shown to significantly enhance algal productivity. This can be simply achieved through control of pH inhibition, reducing phosphate precipitation and nitrogen loss (mainly by reduced ammonia volatilization), as well as by increasing nutrient assimilation into algal biomass (Park and Craggs 2011).

Since exhaust emission gases usually include NOx and SOx, the attractive idea of biorefining urban air harboring NOx and SOx originated from automobile emissions was first discussed in the year 1998 by the patent US6083740A based on the fact that gaseous pollutants such as CO_2, NOx, and SOx can be used as nutrients for microalgae. In the case of CO_2 supply from power stations, the pretreatment steps such as desulfurization and setting CO_2 concentration as well as transportation to an algal cultivation system should be taken into account (see US20080220486A1).

Many researchers such as Chisti (2008) and Lardon et al. (2009) have emphasized that to minimize operational costs in full-scale applications, carbon dioxide should be supplied by waste gaseous emissions such as flue gas from fossil fuel burning power plants. Providing CO_2 in form of micronized bubbles could improve mass transfer and consequently the CO_2 diffusion in the system (see CN101555455A). Moreover, the US patent 5659977A elaborated on a closed system for carbon sequestration in which CO_2 from the exhaust gas would be introduced as nutrient to the microalgae production plant. It should be mentioned that the electrical energy obtained from the algal biomass could be used to produce artificial illumination and/or drive pumps, motors, and control unit in the microalgae production plant. Having achieved this, such a system could be operated like a sustainable biorefinery platform. Another cyclic system consisting of several integrated processes is discussed in the patent US8510985B2 advocating a method for simultaneous production of energy and some byproducts coupled with pollutant sequestration.

The issue of CO_2 supplementation into algal growth chamber involves different aspects which should be taken into serious consideration, from susceptibility of microalgae strains to certain CO_2 concentrations to the impacts of different sources of CO_2 and nutrients pollution which could be potentially caused by emissions of power stations. These have been the subject of patents like US20130217082A1, WO2010010554A1, US8262776B2, and US20080220486A1.

In brief, sustainable production of biofuels from microalgae still requires technological innovations and highly optimized cultivation systems, and without a positive energy and carbon balance, microalgae cultivation presents a mixed picture.

3.3 *Downstream Strategies*

The issue of transformation of biomass into biofuels is a very wide topic and more than 28,000 scientific papers and 3000 international patents have been published to describe this issue during the last decade only (Faba et al. 2015). Therefore, given the diversity of the subjects, this section is only focused on different methods used to obtain liquid biofuels (more specifically biodiesel) from algal biomass. While solid fuels or other types of energy are generally obtained from the entire algal biomass as feedstock, biodiesel production requires specific processes to transform a certain fraction of algal biomass, i.e., FAs.

Extensive downstream processing, like biomass harvesting and drying, lipid extraction and fuel processing are regarded as major hurdles in algal biofuel commercialization. These steps (especially harvesting and extraction) usually take up more than half of the input energy required for algal biodiesel production. Since the final cost of the marketable produced lipid (regardless of their use in the health or energy market) is determined during these steps, downstream category has a strong impact on the other categories as well as final productivity. In addition to that, downstream processes (especially extraction and transesterification) determine the quality of the obtained biodiesel. In better words, the higher the biochemical quality of the lipid is, the more competitive the final product will be. Therefore, optimization of the steps involved is critical and has been the focus of many patent applications.

3.3.1 Harvesting and Dewatering of Microalgae Cells

The harvesting process involves two sets of operations, i.e., bulk harvesting and concentrating the resultant slurry (Brennan and Owende 2010). Recently published patents such as US6000551 and US7022509B2 are focused on a promising and sustainable way as bulk harvesting alternative, i.e., gas flotation or adsorptive bubble separation process. This method not only removes the need for flocculants application, but it is also capable of lysing algal cells concurrent to gathering. In addition, inventors have also developed a number of hydrophobic chemical treatments for harvesting algal cells from broth. More specifically, by adding a hydrophobic liquid/flocculent with lower salinity, the microalgae suspension forms a top phase comprising the hydrophobic liquid and at least a portion of the microalgal cells and a bottom phase comprising the aqueous solution (see the patents US20110165662A1 and US4958460A).

Based on the microalgae size or density, several methods for dewatering of concentrated algal suspension are available. Conventional processes like filtration, gravity, mesh lining centrifugal sedimentation, chemical coagulation and flocculation, use of adsorbents, magnetic separators as well as ultrasonic aggregation have been explained in the patents US4554390A, US20090317886A1, US20090134091A1, US8399239B2, US8399239B2, US8772004B2, and EP2747890A1.

3.3.2 Lipid Extraction

Once an algal biomass is dewatered and dried, only then high-value products could be extracted. Various methods for algal lipid extraction have been developed among which oil press machine, organic solvent extraction, supercritical fluid, subcritical water, and electrochemical extraction methods have seen significant technological advances over the course of recent years. In general, to achieve a desired extraction productivity, some important consideration should be taken, e.g., optimization of pretreatment steps, type and amounts of selected solvents, etc., which have been the aim of several patent applications (see patents US20150252285A1, US20120083617, US20120238732A1, and US20140243540). It should be noted that the fuel properties of biodiesel are significantly influenced by the extraction method used.

Overall, extracting algal oil from the microalgae cells could be achieved through biological or non-biological cell wall rupturing methods. More specifically, the lysis can be performed with vapor (see the patent US2009081742A1), solvents (such as methylene chloride as elaborated upon in the patent US4554390A), mechanical means (such as flotation processes as elaborated upon in the patents US7022509B2 and US6000551), pulsed electric field (US20110107655A1 and US20040224397), electromagnetic radiation (US2009087900A1 and WO2009142765A2), pulsing ultrasonic waves (US20120125763, US8043496B1, and US20100151540A1), or even through genetic manipulation (autolysis as elaborated upon in the patent US20170022436A1 and US20160130627). In addition, the procedure may use at least one enzymatic (such as a cellulose, protease or glycoproteinase) or physico-chemical treatments to disrupt the cells (WO2010039030A1 and US20120238732A1), followed by repetitive steps to separate different cell compounds which finally leads to extract the oil. No-solvent cell lysis or single step extraction protocol is always promising since the extracted oil can be released from wet microalgae, directly entering the subsequent steps. More information on this methodology can be found in the patents US20120040428A1, US20160265011, and US20110107655A1.

Compared with conventional and mechanical means of oil extraction, in which cells into should be transformed into dried granules and different solvents are used, alternative methods have been developed with an aim to further save time and financial resources (see the patents US2012065416 and US2011225878). For example, FAs can be simultaneously extracted and saponified from dry biomass using ethanol and hydroxide sodium in an argon atmosphere at 75–90 °C (ES2289898A1). Even wet microalgae biomass can be extracted directly (US2009227678A1 and US5928696A). A US patent publication 2014/0113363 entitled "process of producing oil from algae using biological rupturing" introduced an oil extraction bioreactor operatively connected to algae growth reservoirs. In this system, for biologically rupturing algal oil vesicles, a structured enzyme system such as a cellulosome was claimed. Different patents, i.e., US8450111B2 and US20110076748A1, claimed that utilizing a single hydrophilic ionic liquid was effective for a one-step process for the lysis of microalgae cell walls and separation of the lipids contained for use in biodiesel production.

Overall, diverse strategies have been put forth claiming success in oil extraction from diverse set of dried and/or wet algal materials. To draw a distinction, patents US20130210093 and WO2010039030A1 made a comparison among the methods described in the other patents, from routine solvent extraction to supercritical fluid extraction and a modified Bligh and Dyer method. The findings reported showed that more long-chain omega-3 FAs could be obtained using a solvent miscible with water in comparison with the supercritical fluids extraction and hexane extractions.

4 Conclusions

In this chapter, patents published since 70s on biofuels production from microalgae were reviewed and discussed in order to shed light on the state of the art of algal fuel production. Overall, a significant number of patents could be found on various aspects of biofuels production from microalgae with a sensible increase in their number since 70s probably due to the general awareness regarding the depletion of conventional energy resources and the environmental problems associated with the widespread utilization of fossil-oriented energy carriers. The published patents presented herein were classified into three main categories, i.e., upstream, mainstream, and downstream strategies. The upstream strategies were focused on algal strain selection and genetic modification for enhancing the quality and quantity of biomass produced. The mainstream strategies concerned improving cultivation systems and conditions, while downstream strategies dealt with improving harvesting, dewatering, and lipid extraction.

References

Brennan, L., & Owende, P. (2010). Biofuels from microalgae—A review of technologies for production, processing, and extractions of biofuels and co-products. *Renewable and Sustainable Energy Reviews, 14*(2), 557–577.

Chisti, Y. (2008). Biodiesel from microalgae beats bioethanol. *Trends in Biotechnology Advances, 26,* 126–131.

Courchesne, N. M. D., Parisien, Wang, B., & Lan, C. Q. (2009). Enhancement of lipid production using biochemical, genetic and transcription factor engineering approaches. *Journal of Biotechnology, 141,* 31–41.

Craggs, R., Heubeck, S., Lundquist, T., & Benemann, J. (2011). Algal biofuels from wastewater treatment high rate algal ponds. *Water Science and Technology, 63*(4), 660–665.

Craggs, R., Sutherland, D., & Campbell, H. (2012). Hectare-scale demonstration of high rate algal ponds for enhanced wastewater treatment and biofuel production. *Journal of Applied Phycology, 24*(3), 329–337.

de la Jara, A., Assunção, P., Portillo, E., Freijanes, K., & Mendoza, H. (2016). Evolution of microalgal biotechnology: A survey of the European Patent Office database. *Journal of Applied Phycology, 28*(5), 2727–2740.

De-Bashan, L. E., & Bashan, Y. (2010). Immobilized microalgae for removing pollutants: Review of practical aspects. *Bioresource Technology, 101*(6), 1611–1627.

Faba, L., Díaz, E., & Ordóñez, S. (2015). Recent developments on the catalytic technologies for the transformation of biomass into biofuels: A patent survey. *Renewable and Sustainable Energy Reviews, 51,* 273–287.

John, R. P., Anisha, G. S., Nampoothiri, K. M., & Pandey, A. (2011). Micro and macroalgal biomass: A renewable source for bioethanol. *Bioresource Technology, 102,* 186–193.

Lardon, L., Hlias, A., Sialve, B., Steyer, J. P., & Bernard, O. (2009). Life-cycle assessment of biodiesel production from microalgae. *Environmental Science and Technology, 43,* 6475–6481.

Lee, J. W. (2010). *Switchable photosystem-II designer algae for photobiological hydrogen production.* US Patent 7642405B2.

Medina, C., García, R., Reyes, P., Fierro, J. L. G., & Escalona, N. (2010). Fischer-Tropsch synthesis from a simulated biosyngas feed over Co(x)/SiO_2 catalysts: Effect of co-loading. *Applied Catalysis A: General, 373,* 71–75.

Ndimba, B. K., Ndimba, R. J., Johnson, T. S., Waditee-Sirisattha, R., Baba, M., Sirisattha, S., et al. (2013). Biofuels as a sustainable energy source: An update of the applications of proteomics in bioenergy crops and algae. *Journal of proteomics, 93,* 234–244.

Park, J. B. K., & Craggs, R. J. (2011). Nutrient removal and nitrogen balances in high rate algal ponds with carbon dioxide addition. *Water Science and Technology, 63*(8), 1758–1764.

Pittman, J. K., Dean, A. P., & Osundeko, O. (2011). The potential of sustainable algal biofuel production using wastewater resources. *Bioresource Technology, 102*(1), 17–25.

Pulz, O. (2001). Photobioreactors: Production systems for phototrophic microorganisms. *Applied Microbiology and Biotechnology, 57*(3), 287–293.

Ramos, M. J., Fernandez, C. M., Casas, A., Rodriguez, L., & Perez, A. (2009). Influence of fatty acid composition of raw materials on biodiesel properties. *Bioresource Technology, 100*(1), 261–268.

Ruiz-Marin, A., Mendoza-Espinosa, L. G., & Stephenson, T. (2010). Growth and nutrient removal in free and immobilized green algae in batch and semi-continuous cultures treating real wastewater. *Bioresource Technology, 101*(1), 58–64.

Scott, S. A., Davey, M. P., Dennis, J. S., Horst, I., Howe, C. J., Lea-Smith, D. J., et al. (2010). Biodiesel from algae: Challenges and prospects. *Current Opinion in Biotechnology, 21*(3), 277–286.

Sialve, B., Bernet, N., & Bernard, O. (2009). Anaerobic digestion of microalgae as a necessary step to make microalgal biodiesel sustainable. *Biotechnology Advances, 27*(4), 409–416.

Tabernero, A., Martín del Valle, E. M., & Galan, M. A. (2013). Microalgae technology: A patent survey. *International Journal of Chemical Reactor Engineering, 11*(2), 733–763.

Talebi, A. F., Mohtashami, S. K., Tabatabaei, M., Tohidfar, M., Zeinalabedini, M., Hadavand, H., et al. (2013a). Fatty acids profiling: A selective criterion for screening microalgae strains for biodiesel production. *Algal Research, 2,* 258–267.

Talebi, A. F., Tohidfar, M., Tabatabaei, M., Bagheri, A., Mohsenpor, M., & Mohtashami, S. K. (2013b). Genetic manipulation, a feasible tool to enhance unique characteristic of *Chlorella vulgaris* as a feedstock for biodiesel production. *Molecular Biology Reports, 40*(7), 4421–4428.

Thompson, S. P. (2013). Biofuel patent trends: 2012 in review. *Industrial Biotechnology, 9*(1), 13–16.

Printed in the United States
By Bookmasters